Heavy Metals - Their Environmental Impacts and Mitigation

Edited by Mazen Khaled Nazal and Hongbo Zhao

Published in London, United Kingdom

IntechOpen

Supporting open minds since 2005

Heavy Metals - Their Environmental Impacts and Mitigation
http://dx.doi.org/10.5772/intechopen.91574
Edited by Mazen Khaled Nazal and Hongbo Zhao

Contributors
Keiko Sasaki, Kojo T. Konadu, Ettiyagounder Parameswari, Veeraswamy Davamani, Periasami Kalaiselvi, Selvaraj Paul Sebastian, Tamilselvan Ilakiya, Ayushi Verma, Meenu Aggarwal, Olufemi Akinnifesi, Femi Adesina, Germaine Ogunwole, Sylvanus Abiya, Mulugeta Sisay, Narjala Rama Jyothi, Richa Madan, Sangeeta Madan, Athar Hussain, Mir Mohammad Mohammad Ali, Delower Hossain, Al- Imran, Md. Suzan Khan, Maksuda Begum, Mahadi Hasan Osman, Claudia Alejandra Alejandra Aguilar, Yunuen Canedo, Carlos Montalvo, Alejandro Ruiz, Rocio Barreto, Laura Castro, M. Luisa Blázquez, Felisa Gónzalez, Jesús A. Muñoz, Miebaka Moslen, Samrat Devaramani, Bhanuprakash G., Doreswamy B. H., Jayadev, Vincent Salvatore Gallicchio, Juley Harper, Madri Pramod Kumar, Tella Lakshmi Prasad, Kothapalli Nagalakshmi, Nadimikeri Jayaraju, Ballari Lakshmanna, Ayşe Handan Dökmeci, Chioma Hope Adiela

First published in London, United Kingdom, 2021 by IntechOpen
IntechOpen is the global imprint of INTECHOPEN LIMITED, registered in England and Wales, registration number: 11086078, 5 Princes Gate Court, London, SW7 2QJ, United Kingdom
Printed in Croatia

British Library Cataloguing-in-Publication Data
A catalogue record for this book is available from the British Library

Additional hard and PDF copies can be obtained from orders@intechopen.com

Heavy Metals - Their Environmental Impacts and Mitigation
Edited by Mazen Khaled Nazal and Hongbo Zhao
p. cm.
Print ISBN 978-1-83968-121-9
Online ISBN 978-1-83968-122-6
eBook (PDF) ISBN 978-1-83968-402-9

Meet the editors

Dr. Mazen Khaled Nazal received an MS and BS in Chemistry from the University of Jordan in 2003 and 2006, respectively. He obtained a Ph.D. in Chemistry specializing in Analytical and Environmental Chemistry from King Fahd University of Petroleum and Minerals (KFUPM), Saudi Arabia, in 2016. Later, he joined the Center for Environment and Marine Studies (CEMS) Research Institute at KFUPM as a research scientist leading the organic contaminants analysis section. Dr. Nazal has participated in many funded research projects and published numerous research papers in refereed journals. He also has several patents to his name. Dr. Nazal's research interests include developing efficient and selective materials and methods for removing emerging pollutants from different matrices and employing these materials for extraction and enrichment of analytes of interest in environmental, food, and biological samples prior to their instrumental analysis.

Prof. Dr. Hongbo Zhao obtained a BSc, MSc, and Ph.D. from Central South University, Changsha, China. He commenced his Ph.D. studies, instructed by Professor Guanzhou Qiu, in 2012, and obtained his doctoral degree in Minerals Processing Engineering in 2016. Since then, he has been working at Central South University as an associate professor. His research focuses on mineral processing and extractive metallurgy (resource recovery), biohydrometallurgy, solid waste treatment and resource recycling, bioremediation, and the mining environment.

Contents

Preface

Urbanization and industrialization are producing large amounts of heavy metals, which are toxic to humans and the environment. These metals are commonly used for applications such as protecting surfaces, building, painting, in the food industry, and more. In addition to natural sources, these anthropogenic activities introduce heavy metals such as titanium (Ti), cadmium (Cd), chromium (Cr), arsenic (As), mercury (Hg), lead (Pd), nickel (Ni), copper (Cu), zinc (Zn), and others into the environment. In addition, rare earth elements such as neodymium (Nd), europium (Eu), terbium (Tb), dysprosium (Dy), and ytterbium (Yb) are being introduced into our ecosystem including the air, water, soil and sediment, and biological species. These elements are hazardous, toxic, and carcinogenic and they have a negative impact on the ecosystem and human health. Although minute quantities of some metals are required for the physiological processes in the body, excessive amounts can lead to a number of disorders. This book includes fifteen chapters and presents a comprehensive overview of heavy metals including their physiochemical properties, toxicity, transfer in the environment, legislation, environmental impacts, and mitigation measures. Chapters include scientific research and studies. This book provides insight into the available techniques for removing heavy metals, such as adsorption, their qualitative and quantitative analysis, and future prospects for these removal methods.

Mazen Khaled Nazal
Center for Environment and Marine Studies (CEMS),
Research Institute (RI),
King Fahd University for Petroleum and Minerals (KFUPM),
Dhahran, Saudi Arabia

Dr. Hongbo Zhao
Professor,
Central South University,
Key Lab of Bio-hydrometallurgy of Ministry of Education,
Changsha, Hunan, China

Chapter 1

Environmental Pollution with Heavy Metals: A Public Health Concern

Mir Mohammad Ali, Delower Hossain, Al-Imran, Md. Suzan Khan, Maksuda Begum and Mahadi Hasan Osman

Abstract

Heavy metals (HMs) are natural environmental constituents, but their geochemical processes and biochemical equilibrium have been altered by indiscriminate use for human purposes. Due to their toxicity, persistence in the environment and bioaccumulative nature; HMs are well-known environmental contaminants. As result, there is excess release into natural resources such as soil and marine habitats of heavy metals such as cadmium, chromium, arsenic, mercury, lead, nickel, copper, zinc, etc. Their natural sources include the weathering of metal-bearing rocks and volcanic eruptions, while mining and other industrial and agricultural practices include anthropogenic sources. Prolonged exposure and increased accumulation of such heavy metals may have detrimental effects on human life and aquatic biota in terms of health. Finally, the environmental issue of public health concern is the pollution of marine and terrestrial environments with toxic heavy metals. Therefore, because of the rising degree of waste disposal from factories day by day, it is a great concern. Pollution of HMs is therefore a problem and the danger of this environment needs to be recognized.

Keywords: Bioaccumulation, Contamination, Heavy metal, Health hazard, Soil, Toxicity, Water

1. Introduction

Heavy metals (HMs) are an environmental threat and are of grave concern worldwide [1]. Rapid industrialization and urbanization have caused heavy metals to contaminate the atmosphere and it is a problem for human health [2–4]. HMs poses a major environmental threat to living organisms and habitats due to their non-biodegradability, bioaccumulation, environmental stability, persistence and biotoxicity characteristics [5–7]. In order to prevent microbial activities, they can directly influence the physical and chemical properties of sediment, soils and water [8]. They can also disrupt the natural ecosystem and impact the human body acutely and permanently through the food chain [9–11]. The non-degradable HMs can also accumulate in the surface sediment for a

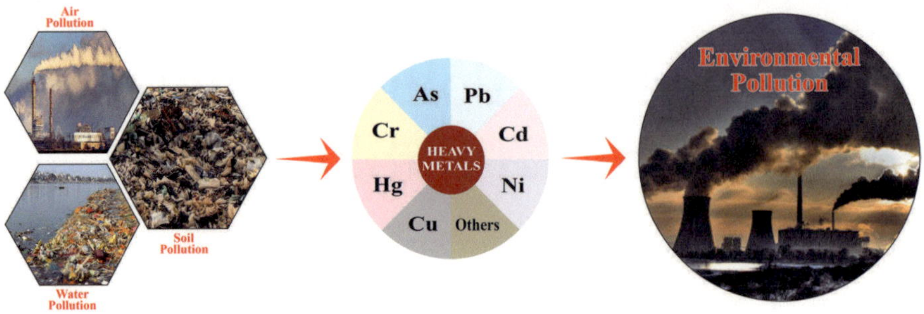

Figure 1.
Sources, metals and the environmental degradation.

long time via the food chain's amplification effect, causing various diseases and complications in the human body [1, 12, 13].

Natural activities (e.g. geological weathering, atmospheric precipitation, wave erosion, wind and bioturbation) and anthropogenic activities (e.g. rapid industrialization, urbanization, agricultural runoff and transport) play a key role in the spread of HMs to the marine habitats of aquatic ecosystems such as rivers and estuaries [2, 4, 14]. In addition, human activities that can produce industrial pollution, the deposition of urban waste and the offensive use of chemical fertilizers and pesticides result in the accumulation and sinking of HMs in aquatic habitat surface sediments [15–17]. The HMs released into the water column have a negative effect on water quality [11, 18] and on surface sediments that alter environmental parameters such as pH, temperature, bioturbation etc. [19, 20]. Sediment quality can therefore play a critical role in identifying the effects of natural and anthropogenic activities [21–23]; sediment quality can also provide information on the anthropogenic impact on the ecosystem and guide environmental policy and management [24].

Farm waste, agricultural runoff, pesticides, solid waste, waste management, effluents from fish processing plants, jute processing, cement manufacturing, oil refining, fertilizer manufacturing, building materials, soup and detergent factories and brickyard waste are the major sources of pollution (**Figure 1**). Due to the potential risk of HMs in water, soil and sediment through the disposal of the effluents mentioned, this riverine water, sediment and environment may be important.

2. Heavy metals

2.1 Definition

Any metal or metalloid of environmental significance is a heavy metal [25]; the term originated in reference to the adverse effects, all denser than iron, cadmium, mercury and lead (**Figure 2**). It has since been extended to some other similarly toxic metal or metalloid, irrespective of density, such as arsenic, chromium, cobalt, nickel, copper, zinc, arsenic, selenium, silver, cadmium, antimony, mercury, thallium and lead are commonly found to be heavy metals [26, 27].

In a general, collective term that refers to a group of metals and metalloids with an atomic density greater than 4 g/cm^3 or 5 times or more than water, the term heavy metal is often referred to as trace elements as they exist in minute concentrations in biological systems [28].

Heavy metals are classified as "metals that occur naturally and have an atomic number greater than 20" [29, 30]. They are a significant class of contaminants that

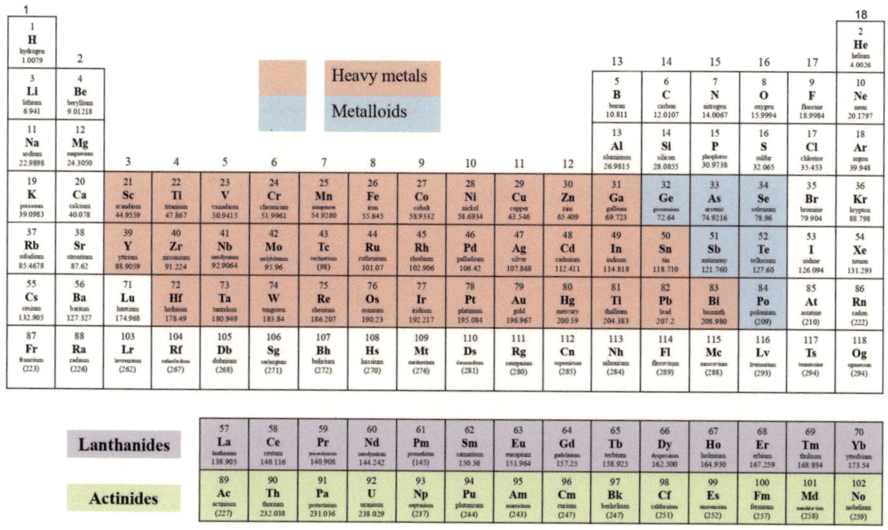

Figure 2.
Position of heavy metals in periodic table.

affect the environment. A serious problem with cultural, ecological and economic consequences is heavy metal contamination in the environment. Because of the toxicity, persistence, and bioaccumulative nature of these materials, research on heavy metals in the atmosphere is an important part of environmental research.

2.2 Sources of heavy metals

Heavy metals may come from natural and anthropogenic processes and end up in various environmental compartments (soil, water, air and their interface). **Figure 3** gives information on natural and anthropogenic sources of heavy metals.

2.2.1 Natural sources

Various natural sources of HMs have been recorded in several studies. Natural emissions of HMs occur under numerous and certain environmental conditions. Volcanic eruptions, sea-salt sprays, forest fires, rock weathering, biogenic sources and particles of wind-borne soil are included in these pollutants. The release of metals from their endemic spheres to different environmental compartments will lead to natural weathering processes. In the form of hydroxides, oxides, sulfides, sulfates, phosphates, silicates and organic compounds, heavy metals can be found.

Lead (Pb), nickel (Ni), chromium (Cr), cadmium (Cd), arsenic (As), mercury (Hg), selenium (Se), zinc (Zn) and copper (Cu) are the most popular heavy metals. While the above-mentioned heavy metals can be present in traces, humans and other mammals still cause significant health problems.

2.2.2 Anthropogenic sources

Industries, irrigation, drainage, mining and metallurgical processes, as well as runoff, also contribute to the release of pollutants into various compartments of the ecosystem. For certain metals, anthropogenic heavy metal processes have been noted to go beyond natural fluxes. In wind-blown dust, metals naturally released are mainly from industrial areas. Car exhaust that releases lead; smelting that releases arsenic, copper and zinc; insecticides that release arsenic and the burning

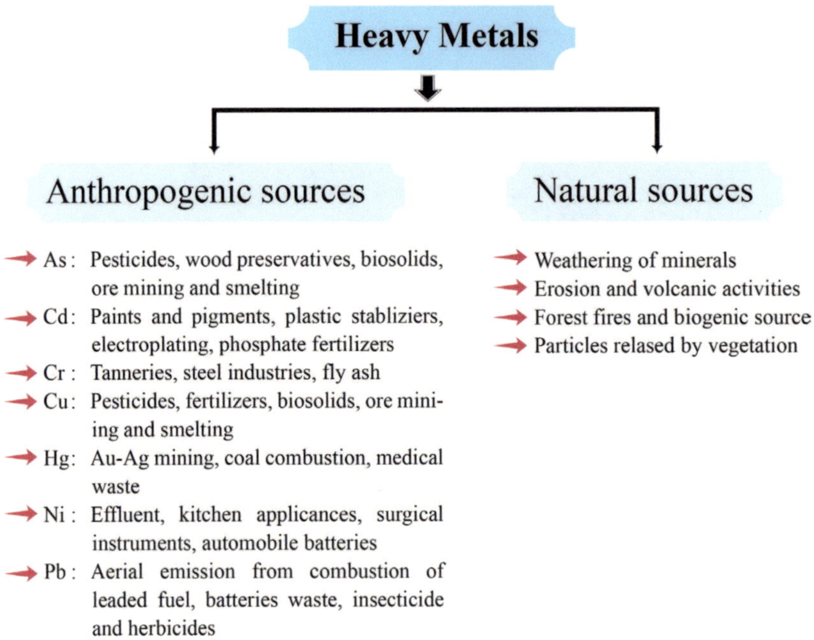

Figure 3.
Sources of heavy metals in the environment.

of fossil fuels that release nickel, vanadium, mercury, selenium and tin are some essential anthropogenic causes that contribute significantly to heavy metal pollution in the environment. Because of the everyday manufacture of products to meet the demands of the large population, human activities have been found to contribute more to environmental pollution.

3. Contamination of heavy metals

Harmful trace metals are a significant threat to both aquatic and terrestrial ecosystems [31]. Upon release from both natural and anthropogenic sources, HMs contaminates natural aquatic bodies, sediments and soils. Ultimately, during volcanic eruptions and complex industrial emissions, heavy metals released into the atmosphere often return to the soil and cause water and soil contamination. They either collect in biota or leach down into ground water because heavy metals in the atmosphere are irreversible. There are significant public health effects of the contamination of biota and groundwater with potentially harmful heavy metals. In riverine settings, the degree of heavy metal pollution can be measured by observing the concentrations and distribution of these elements [32]. **Figure 4** provides a conceptual schematic of the contamination of the marine (riverine) environment with heavy metals. Various physicochemical and climatic effects affect the overall dynamics and biogeochemical cycling of heavy metals in the atmosphere.

3.1 Water

Water is regarded as the life-blood of the biosphere as well. It can dissolve distinct organic and inorganic chemicals and environmental contaminants, as water is a common solvent. They are vulnerable to pollution in both freshwater and marine aquatic environments. Heavy metals are important contaminants in the pollution of

Figure 4.
A conceptual schematic of contamination of heavy metals in aquatic ecosystem.

the marine environment. Water contamination by HMs is a critical environmental issue that adversely affects plants, animals and human health [33]. Also at very low levels, heavy metals are highly harmful to marine species [34]. These elements may cause significant physiological changes in the body and histopathological changes in the tissues of aquatic organisms, such as fish [35]. There are several sources of heavy metal pollution in water. Two main culprits of this industrialization and urbanization are the increased degree of concentration of heavy metal in water. Heavy metals from factories, municipalities and urban areas are carried by runoff [36]. This release of untreated industrial waste into marine bodies is a significant cause of surface and groundwater contamination [37]. Due to the environmental persistence, bioaccumulation, and biomagnification of food chains and the toxicity of these elements, contamination of water bodies with heavy metals is a global issue [38].

3.2 Sediments

The contamination of sediments with HMs is a very important environmental problem with implications for marine life and human health. Sediments act as the main source of HMs in the aquatic environment. Their quality can indicate the contamination status of water [39]. Sediments act as a sink and heavy metal source, releasing them into the column of water [40]. Continued heavy metal accumulation in sediments can also contribute to groundwater pollution of these contaminants [41]. Many physicochemical variables such as temperature, hydrodynamic conditions, redox status, organic matter and microbe content, salinity, and particle size influence the adsorption, desorption, and subsequent concentrations of heavy metals in sediments [42]. The distribution of heavy metals in sediments is influenced by the sediment's chemical composition; grain size and total organic matter content [43]. The pH is an important determinant of the bioavailability of metals in sediments. A decrease in pH increases the competition between metal ions and H^+ for sediment binding locations and may contribute to the dissolution of metal complexes, thereby releasing free metal ions into the water column [44]. Higher toxic heavy metals concentrations in riverine sediments can pose an ecological risk to benthos (bottom-dwelling organisms) [45].

3.3 Soils

Heavy metals and metalloids are released into soils from activities and sources such as manufacturing activities, mine tailings, high metal waste disposal, leaded gasoline and paints, fertilizer land application, animal manures, sewage sludge, pesticides, irrigation of waste water, residues of coal combustion and petrochemical spillage, resulting in soil contamination by heavy metals [46]. Most HMs does not typically experience microbial or chemical degradation and thus, after being released to the atmosphere, their total concentrations last in the sail for a long time. The composition of the parent rock, the degree of weathering and physical, chemical and biological characteristics of soil and environment conditions are factors influencing the presence and distribution of heavy metals in soils [47]. Compared to virgin soils and soils with low inputs, substantial heavy metal enrichment has been recorded in soils receiving more fertilizer input and Cu fungicide [48]. Soils may be polluted with heavy metals from heavy vehicular activity on roads in urban areas. In urban areas, soil samples have elevated levels of Pb, of which 45–85% is bioaccessible [49]. The bioavailability of heavy metals in soils is of great importance for their environmental fate and for their plant uptake. Different HMs has different soil bioavailability and this bioavailability depends on the speciation of metals and the different soil physicochemical characteristics.

3.4 Fish

Aquatic biota is exposed to heavy metals by water, sediments and food on various routes [50]. Different toxic HMs released to freshwater bodies from various natural and anthropogenic sources are introduced to freshwater fish. Heavy metal pollution of fish has become a major global concern because it poses a danger to fish and poses health hazards to buyers of fish [51]. Assessment of the bioaccumulation of HMs in fish species from various aquatic ecosystems is very significant [52]. Assessing the amount of heavy metals in fish tissues is important for the conservation of marine environments and the human consumption of fish [53]. There are high levels of unsaturated fatty acids and low cholesterol levels in fish. They are a major protein source [54].

It is advantageous to use edible fish in human diets and is also recommended in healthy diets. Contamination of fish by toxic heavy metals is considered a risk to human health and has raised concerns about their consumption, especially among more vulnerable groups of people, such as women, children and people at risk of other diseases.

Heavy metal bioaccumulation in freshwater fish depends on different factors, including the characteristics of the fish and the external environmental factors. Fish-related factors include fish age, size (weight and length), feeding habits and physiology of the body, while external environmental factors include water column metal concentration and bioavailability, water physicochemical properties and other climatic factors. Depending on the structure and function of the tissues, the degree of accumulation of heavy metals in the various tissues of fish is usually different. Metabolically active tissues such as the gills, liver and kidneys typically have higher heavy metal accumulations than other tissues such as the skin and muscles. The comparatively higher accumulation of heavy metals in metabolically active fish tissues is normally explained by the induction/occurrence in these tissues of metal-binding proteins called metallothioneins (MTs) upon exposure to heavy metals. Fish gills have been observed as the target tissue for heavy metals such as Ni to accumulate and eliminate [38]. While fish muscles are the tissue of poor heavy metal accumulation [55], from a human consumption point of view, they are essential. Trace metal bioaccumulation in fish muscles is typically species-specific [56].

The bioaccumulation of toxic HMs in freshwater fish has significant environmental, ecological and social consequences; it affects the carnivorous species and human by eating fish [57, 58]. Waterborne HMs are absorbed into fish and penetrate the human body through the food chain, thereby impacting human health [59]. In addition, poisonous HMs affects the health and well-being of fish as well.

4. Trophic transfer of heavy metals

As HMs in the atmosphere is permanent, they accumulate in living organisms and are passed in the food chains from one trophic stage to another. The degree to which heavy metals are deposited in biota depends on their accumulation rate and their removal rate from the body.

Heavy metals, i.e. water, sediments and soil, may enter the body of an organism directly from the abiotic system, or may enter the body of the organism from its food/prey. For example, heavy metals may reach the body of the fish directly from water or sediments through the gills/skin of the fish or from the food/prey of the fish through its food canal. Over successive trophic levels in a food chain, the concentration of a heavy metal can increase or decrease. The retention of heavy metals in an organism's body depends on a variety of factors, such as the speciation of the metal concerned and the physiological mechanisms established by the organism for heavy metal control, homeostasis and detoxification. Because of their lipophilicity, methylated forms of heavy metals like Hg are accumulated to a greater extent in biota and also biomagnified in food chains. In metal-rich environments, such plants have the potential to survive and are called metallophytes. Special mechanisms for

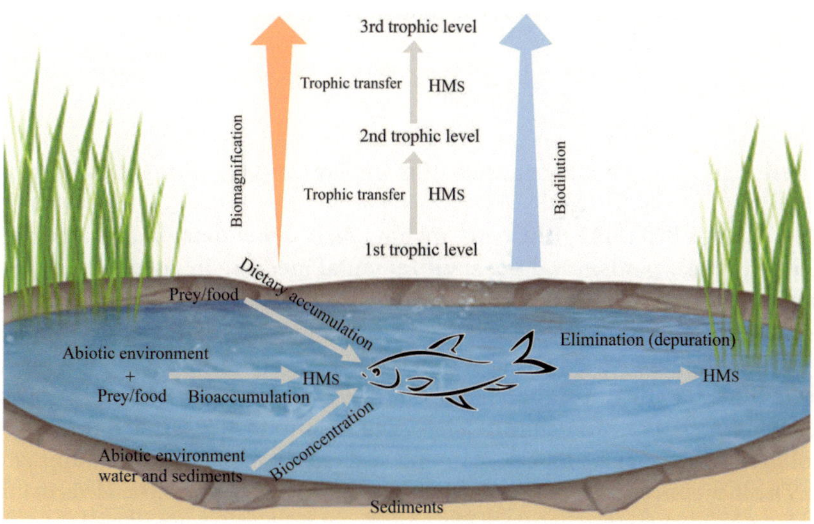

Figure 5.
Trophic transfer of heavy metals in the environment.

dealing with higher heavy metal concentrations in soil have been established by these special plants and are classified into three groups, i.e. excluders, markers, and hyper accumulators [57]. To define the trophic transfer of heavy metals, certain words are used (**Figure** 5).

5. Quantification of trophic transfer

The degree or extent of accumulation of HMs in biota has been quantified using some terminology. Bioconcentration factor (BCF), bioaccumulation factor (BAF), bioaccumulation coefficient (BAC) and so on are some of these quantitative terms. Below, some of these terms are discussed.

5.1 Bioconcentration factor (BCF)

Bioconcentration Factor is representative of the degree of heavy metal enrichment in an organism as opposed to that in its habitat. The ratio of the concentration of a heavy metal in the tissue of an organism to its concentration in the abiotic medium is known as' (water and sediments).

It is calculated by the following equation [60]:

$$BCF = C_{\text{organism tissue}} / C_{\text{abioitic medium}} \tag{1}$$

Where, $C_{organism}$ is the metal concentration in the organism tissue and $C_{\text{abiotic medium}}$ is the metal concentration in the abiotic medium.

The alternative words transfer factor (TF), metal transfer factor (MTF), accumulation factor (AF), bioaccumulation factor (BAF), and biota sediment accumulation factor (BSAF) are used by some writers and measured accordingly. All these indexes, however, indicate the extent of a heavy metal's accumulation in the organism compared to that in the world where it grows/lives.

5.2 Bioaccumulation coefficient (BAC)

Bioaccumulation Coefficient is calculated by the following equation [61]:

$$BAC = C_{\text{plant}} / C_{\text{soil}} \tag{2}$$

Where, C_{plant} is the metal concentration in plant and C_{soil} is the metal concentration in soil.

The values of BCF, BAF, BAC, etc. are obviously dependent on the concentration of the HM in the organism and the environmental medium in question. In view of the fact that the values of these indices are inversely related to the concentration of metals in the environmental medium (water, sediments and soil), the values of these indices should be used with caution to measure the contamination of heavy metals in biota. For example, because of the lower metal content in the environment of the former fish, the BCF value of a heavy metal in the muscles of a fish living in less polluted water may be higher than that of a fish living in more contaminated water. The bio-concentration factor (BCF) values of seven common HMs in crop grains have been shown to decrease exponentially with average soil metal concentrations [62].

6. Effects of heavy metals in human health

A big source of exposure for the human population to heavy metals is food and drinking water. The intensification of HMs in industrial and agricultural activities has led to much industrialization, modern urbanization and rapid economic development around the world [63]. These activities can cause pollution of toxic HMs in water, air, and soils. Heavy metal-contaminated media contribute to the bioaccumulation of these elements from the environment in the human food chains, eventually reaching the human body. There are some health impacts of HMs on the human body depending on the amount and duration of exposure (**Figure 6**).

Among different HMs, Cd is considered as seventh most hazardous and noxious metal that causes an indirect oxidative stress and have carcinogenic and mutagenic effect resulting severe health problem in human body such as kidney damage, prostate dysfunction, bone diseases and cancer [64, 65]. It can also responsible for kidney dysfunction and proteinuria if Cd has prolonged exposure to kidneys.

Arsenic is also causes skin damage, cancer and marked problems with circulatory system [66]. When As cross its permissible limit in drinking water is also causes developmental abnormalities, neurobehavioral sicknesses, cardiovascular diseases and hearing sickness, together with anemia, leukopenia, eosinophilia and carcinoma [67, 68].

Chromium is known as highly toxic HM as it can cross the cell membrane via sulfate transport system and causes denaturation and mutation of nucleic acids and proteins. It also creates critical health issues like skin problems, nasal irritation, hear impairment and lung carcinoma [65, 69, 70].

Lead can also create different health problems like decrease in intelligent quotient, memory loss, infertility, mood swing, weakness of joints, nausea, insomnia, anorexia, or even death [65]. Young and infants are more sensitive to Pb poisoning than adults. Details about effects of different HMs on human health were described in **Table 1**.

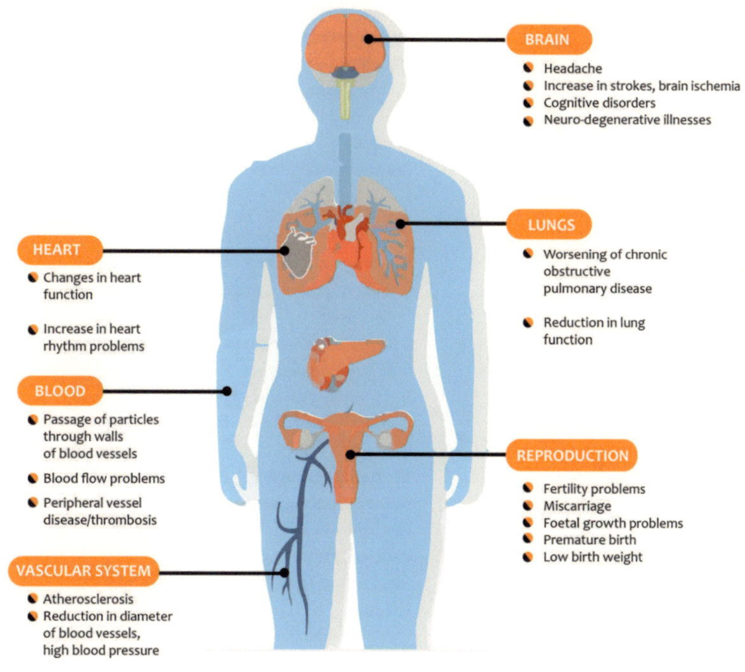

Figure 6.
Effects of heavy metals in different vital organs of human health.

Heavy metals	EPA regulatory limit for drinking water (ppm)	OSHA limit for workplace air (μg/m^3)	Toxic Effects	References
Arsenic (As)	0.01	10.0	**Lower level exposure** Nausea and vomiting Decreased production of red and white blood cells Abnormal heart rhythm Damage to blood vessels A sensation of "pins and needles" in hands and feet also cause muscle damage **Long-term low level exposure** Darkening of the skin and the appearance of small "corns" or "warts" on the palms, soles and torso Diffuse or spotted hyper-pigmentation of the skin Benign skin lesions (hyperkeratosis) and cancer of the skin Liver disease reflected by abnormal porphyry metabolism Chronic inhalation can cause lung cancer Chronic exposure via water can cause cancer of internal organs, particularly the urinary bladder, lung, liver, and kidney **Others** Affects essential cellular processes such asoxidative phosphorylation and ATP synthesis	[28, 63, 71, 72]
Barium (Ba)	2.0	500	**Short term exposure** Gastrointestinal dysfunction (vomiting, abdominal cramps, diarrhea) Difficulties in breathing Increased or decreased blood pressure Numbness around the face and Muscle weakness **High level exposure** High blood pressure Changes in heart rhythm and cardiac arrhythmias Muscle twitching Paralysis Respiratory failure and Possibly death	[63, 73, 74]
Cadmium (Cd)	5.0	5.0	**High level exposure to cadmium fumes** Cause acute bronchitis or even chronic disease, such as emphysema or pulmonary fibrosis and lung cancer Severe irritation in the stomach, leading to vomiting and diarrhea **Long-term low level exposure** Buildup kidney disease, lung damage, and fragile bones **Chronic exposure** Kidney tubular dysfunction Osteoporosis (elderly women with iron deficiency) Lung cancer (only in chronic inhalation) **Others** Carcinogenic Mutagenic Endocrine disruptor Fragile bones Affects calcium regulation in biological systems	[75–81]

Heavy metals	EPA regulatory limit for drinking water (ppm)	OSHA limit for workplace air (μg/m^3)	Toxic Effects	References
Chromium (Cr)	0.1	0.5 to 1000	**Breathing high levels** Irritation to the lining of the nose Nose ulcers and runny nose Breathing problems, such as asthma, cough, shortness of breath, or wheezing **Skin contact can cause** Skin ulcers Allergic reactions consisting of severe redness and swelling of the skin Hair loss **Long term exposure** Damage to liver, kidney circulatory and nerve tissues Skin irritation	[73, 75]
Silver (Ag)	0.10	10.0	**Long-term high level exposure** Arygria, a blue-gray discoloration of the skin and other body tissues Mild allergic reactions such as rash, swelling, and inflammation in some people Breathing problems Lung and throat irritation and Stomach pains	[73, 82]
Mercury (Hg)	2.0	100 for organic Hg 50 for metallic Hg	**Effect on brain** Brain damage Tremor Temper outbursts Loss of memory Disturbance of vision Psychiatric disturbances Restlessness Depression Altered behavior **Renal toxicity of mercury** Glomerulnephritis Kidney failure **Effect on unborn fetus** Paresthesia Blindness Deafness Fetal death and abortion **High level Exposure** Damage the brain resulting irritability, shyness, tremors, changes in vision or hearing, and memory problems Damage kidneys and Developing fetuses **Short-term high levels exposure** Lung damage Gastrointestinal dysfunction (nausea, vomiting, diarrhea) Increases in blood pressure or heart rate Skin rashes and Eye irritation **Others** Autoimmune diseases Drowsiness Fatigue Hair loss Insomnia Lung failure	[28, 83–87]

Heavy metals	EPA regulatory limit for drinking water (ppm)	OSHA limit for workplace air (μg/m³)	Toxic Effects	References
Lead (Pb)	15	50	**High level exposure to Pb** Severely damage the brain and kidneys Miscarriage in pregnant women Damage the organs responsible for sperm production In children causes impaired development, reduced intelligence, short-term memory loss, disabilities in learning and coordination problems, risk of cardiovascular disease Finally death **Chronic high level exposure** Produce anemia Effects in central nervous system, including impaired motor function and cognitive function and even seizures, coma, and death with markedly elevated blood lead levels Impaired heme synthesis Chronic kidney disease Lethargy Impairment of cognitive function Lead produces tumors in experimental animals, but there is not enough evidence to regard lead as a human carcinogen	[75, 88–92]
Selenium (Se)	15	200	**Short-term high level exposure** Nausea Vomiting and Diarrhea **Chronic high level exposure** Produce selenosis; major signs of selenosis are hair loss, nail brittleness, and neurological abnormalities Respiratory irritation Bronchitis Difficulty breathing Coughing Stomach pains **Others:** Dietary exposure of around 300 μg/day affects endocrine function Impairment of natural killer cells activity Hepatotoxicity	[63, 93]
Copper (Cu)	1.3		Brain and kidney damage Liver cirrhosis Chronic anemia Stomach and intestine irritation	[63, 75, 88]
Nickel (Ni)	0.2		Allergic skin diseases such as itching Cancer of the lungs, nose, sinuses, throat through continuous inhalation Immunotoxic, neurotoxic, genotoxic effect Affects fertility Hair loss	[75, 94, 95]
Zinc (Zn)	0.5		Dizziness Fatigue	[63, 96]

Environmental Protection Agency (EPA), Parts Per Million (ppm), Occupational Safety and Health Administration (OSHA).

Table 1.
Toxic effects of heavy metals on human health.

DOI: http://dx.doi.org/10.5772/intechopen.96805

7. Conclusions

Heavy metals and metalloids are critical environmental contaminants for both marine and terrestrial environments and ecosystem. The danger of an environmental chemical is a function of its persistence, toxicity and bioaccumulative ability in the environment. Due to these three characteristics: persistence, bioaccumulation, and toxicity; HMs are considered as harmful. The most dangerous heavy metals and metalloids that are environmentally important include Cr, As, Cd, Pb, Hg, Ni, Cu and Zn. In aquatic and terrestrial food chains, the trophic transition of these elements has major consequences for different lives of ecosystem and human health. So contamination of environment of heavy metals is of great concern, and its treatment from the soil and water around industrial areas is needed urgently in every place. The assessment and monitoring of the concentrations of potentially toxic heavy metals and metalloids in the various environmental segments and in the resident biota is very significant. A detailed environmental chemistry and ecotoxicology analysis of dangerous heavy metals and metalloids demonstrates that steps should be taken to mitigate the effects on human health and the environment of these elements.

Author details

Mir Mohammad Ali[1], Delower Hossain[2]*, Al-Imran[3], Md. Suzan Khan[4], Maksuda Begum[5] and Mahadi Hasan Osman[4]

1 Department of Aquaculture, Sher-e-Bangla Agricultural University, Dhaka, Bangladesh

2 Department of Medicine and Public Health, Sher-e-Bangla Agricultural University, Dhaka, Bangladesh

3 Rural Microenterprise Transformation Project, Palli Karma-Sahayak Foundation, Dhaka, Bangladesh

4 Fisheries and Livestock Unit, Palli Karma-Sahayak Foundation, Dhaka, Bangladesh

5 Department of Poultry Science, Sher-e-Bangla Agricultural University, Dhaka, Bangladesh

*Address all correspondence to: delowervet@gmail.com; delowervet@sau.edu.bd

References

[1] Zhu G, Noman MA, Narale DD, Feng W, Pujari L, Sun J. Evaluation of ecosystem health and potential human health hazards in the Hangzhou Bay and Qiantang Estuary region through multiple assessment approaches. Environmental Pollution 2020;114791.

[2] Nour HE, El-Sorogy AS, Abd El-Wahab M, Mohamaden M, Al-Kahtany K. Contamination and ecological risk assessment of heavy metals pollution from the Shalateen coastal sediments, Red Sea, Egypt. Marine pollution bulletin 2019;144:167-72.

[3] Liu P, Hu W, Tian K, Huang B, Zhao Y, Wang X, et al. Accumulation and ecological risk of heavy metals in soils along the coastal areas of the Bohai Sea and the Yellow Sea: A comparative study of China and South Korea. Environment International 2020;137:105519.

[4] Kahal A, El-Sorogy AS, Qaysi S, Almadani S, Kassem OM, Al-Dossari A. Contamination and ecological risk assessment of the Red Sea coastal sediments, southwest Saudi Arabia. Marine Pollution Bulletin 2020;154:111125.

[5] Khan MB, Dai X, Ni Q, Zhang C, Cui X, Lu M, et al. Toxic Metal Pollution and Ecological Risk Assessment in Sediments of Water Reservoirs in Southeast China. Soil and Sediment Contamination: An International Journal 2019;28:695-715.

[6] Ustaoğlu F, Islam MS. Potential toxic elements in sediment of some rivers at Giresun, Northeast Turkey: A preliminary assessment for ecotoxicological status and health risk. Ecological Indicators 2020;113:106237.

[7] Zhang C, Shan B, Tang W, Wang C, Zhang L. Identifying sediment-associated toxicity in rivers affected by multiple pollutants from the contaminant bioavailability. Ecotoxicology and Environmental Safety 2019;171:84-91.

[8] Omwene PI, Öncel MS, Çelen M, Kobya M. Heavy metal pollution and spatial distribution in surface sediments of Mustafakemalpaşa stream located in the world's largest borate basin (Turkey). Chemosphere 2018;208:782-92.

[9] Lian M, Wang J, Sun L, Xu Z, Tang J, Yan J, et al. Profiles and potential health risks of heavy metals in soil and crops from the watershed of Xi River in Northeast China. Ecotoxicology and environmental safety 2019;169:442-8.

[10] Weber AA, Sales CF, de Souza Faria F, Melo RMC, Bazzoli N, Rizzo E. Effects of metal contamination on liver in two fish species from a highly impacted neotropical river: A case study of the Fundão dam, Brazil. Ecotoxicology and Environmental Safety 2020;190:110165.

[11] Ali MM, Ali ML, Proshad R, Islam S, Rahman Z, Tusher TR, et al. Heavy metal concentrations in commercially valuable fishes with health hazard inference from Karnaphuli river, Bangladesh. Human and ecological risk assessment: an international journal 2020;26:2646-62.

[12] Wang Z, Zhou J, Zhang C, Qu L, Mei K, Dahlgren RA, et al. A comprehensive risk assessment of metals in riverine surface sediments across the rural-urban interface of a rapidly developing watershed. Environmental Pollution 2019;245:1022-30.

[13] Yang QQ, Wang SL, Liu WJ, Yang YW, Jiang SQ. Spatial distribution of perfluoroalkyl acids (PFAAs) and

their precursors and conversion of precursors in seawater deeply affected by a city in China. Ecotoxicology and Environmental Safety 2020;194: 110404.

[14] Ding L, Zhao K, Zhang L, Liang P, Wu S, Wong MH, et al. Distribution and speciation of mercury affected by humic acid in mariculture sites at the Pearl River estuary. Environmental Pollution 2018;240:623-9.

[15] Kinimo KC, Yao KM, Marcotte S, Trokourey A. Distribution trends and ecological risks of arsenic and trace metals in wetland sediments around gold mining activities in central-southern and southeastern Côte d'Ivoire. Journal of Geochemical Exploration 2018;190:265-80.

[16] Lee P-K, Yu S, Jeong Y-J, Seo J, Choi S, Yoon B-Y. Source identification of arsenic contamination in agricultural soils surrounding a closed Cu smelter, South Korea. Chemosphere 2019;217:183-94.

[17] Varol M, Canpolat Ö, Eriş KK, Çağlar M. Trace metals in core sediments from a deep lake in eastern Turkey: Vertical concentration profiles, eco-environmental risks and possible sources. Ecotoxicology and Environmental Safety 2020;189: 110060.

[18] Timofeev I, Kosheleva N, Kasimov N. Contamination of soils by potentially toxic elements in the impact zone of tungsten-molybdenum ore mine in the Baikal region: A survey and risk assessment. Science of the total environment 2018;642:63-76.

[19] Simpson SL, Spadaro DA. Bioavailability and chronic toxicity of metal sulfide minerals to benthic marine invertebrates: implications for deep sea exploration, mining and tailings disposal. Environmental science & technology 2016;50:4061-70.

[20] Singh UK, Kumar B. Pathways of heavy metals contamination and associated human health risk in Ajay River basin, India. Chemosphere 2017;174:183-99.

[21] Nour HE, El-Sorogy AS. Distribution and enrichment of heavy metals in Sabratha coastal sediments, Mediterranean Sea, Libya. Journal of African Earth Sciences 2017;134: 222-9.

[22] Bhuyan MS, Bakar MA, Akhtar A, Hossain MB, Ali MM, Islam MS. Heavy metal contamination in surface water and sediment of the Meghna River, Bangladesh. Environmental nanotechnology, monitoring & management 2017;8:273-9.

[23] El-Sorogy A, Al-Kahtany K, Youssef M, Al-Kahtany F, Al-Malky M. Distribution and metal contamination in the coastal sediments of Dammam Al-Jubail area, Arabian Gulf, Saudi Arabia. Marine pollution bulletin 2018;128:8-16.

[24] Tian K, Huang B, Xing Z, Hu W. Geochemical baseline establishment and ecological risk evaluation of heavy metals in greenhouse soils from Dongtai, China. Ecological Indicators 2017;72:510-20.

[25] Lee P-K, Kang M-J, Yu S, Ko K-S, Ha K, Shin S-C, et al. Enrichment and geochemical mobility of heavy metals in bottom sediment of the Hoedong reservoir, Korea and their source apportionment. Chemosphere 2017;184:74-85.

[26] Duffus JH. " Heavy metals" a meaningless term?(IUPAC Technical Report). Pure and applied chemistry 2002;74:793-807.

[27] Liu R, Bao K, Yao S, Yang F, Wang X. Ecological risk assessment and distribution of potentially harmful trace elements in lake sediments of Songnen

Plain, NE China. Ecotoxicology and environmental safety 2018;163:117-24.

[28] Duruibe JO, Ogwuegbu MOC, Egwurugwu JN. Heavy metal pollution and human biotoxic effects. International Journal of physical sciences 2007;2:112-8.

[29] Hoodaji M, Ataabadi M, Najafi P. Biomonitoring of airborne heavy metal contamination. Air Pollut.–Monit., Model., Health Control 2012;21.

[30] Ali H, Khan E. What are heavy metals? Long-standing controversy over the scientific use of the term 'heavy metals'–proposal of a comprehensive definition. Toxicological & Environmental Chemistry 2018;100:6-19.

[31] Slaveykova VI, Cheloni G. Preface: special issue on environmental toxicology of trace metals. 2018;

[32] Islam MS, Proshad R, Ahmed S. Ecological risk of heavy metals in sediment of an urban river in Bangladesh. Human and ecological risk assessment: an international journal 2018;24:699-720.

[33] Rezania S, Taib SM, Din MFM, Dahalan FA, Kamyab H. Comprehensive review on phytotechnology: heavy metals removal by diverse aquatic plants species from wastewater. Journal of hazardous materials 2016;318:587-99.

[34] ABRAR M, HUSSAIN Z, AKIF M, SOK K, Muhammad A, KHAN AR, et al. Textile effluents and their contribution towards aquatic pollution in the Kabul River (Pakistan). Journal of the Chemical Society of Pakistan 2011;24:106.

[35] Ahmed MK, Parvin E, Islam MM, Akter MS, Khan S, Al-Mamun MH. Lead-and cadmium-induced histopathological changes in gill, kidney and liver tissue of freshwater climbing perch *Anabas testudineus* (Bloch, 1792). Chemistry and Ecology 2014;30:532-40.

[36] Zhuang P, Li Z, McBride MB, Zou B, Wang G. Health risk assessment for consumption of fish originating from ponds near Dabaoshan mine, South China. Environmental Science and Pollution Research 2013;20:5844-54.

[37] Afzal MS, Ashraf A, Nabeel M. Characterization of industrial effluents and groundwater of Hattar industrial estate, Haripur. Advances in Agriculture and Environmental Science: Open Access (AAEOA) 2018;1:70-7.

[38] Rajaei G, Mansouri B, Jahantigh H, Hamidian AH. Metal concentrations in the water of Chah nimeh reservoirs in Zabol, Iran. Bulletin of environmental contamination and toxicology 2012;89:495-500.

[39] Zahra A, Hashmi MZ, Malik RN, Ahmed Z. Enrichment and geo-accumulation of heavy metals and risk assessment of sediments of the Kurang Nallah—feeding tributary of the Rawal Lake Reservoir, Pakistan. Science of the Total Environment 2014;470:925-33.

[40] Fernandes LL, Nayak GN. Heavy metals contamination in mudflat and mangrove sediments (Mumbai, India). Chemistry and Ecology 2012;28:435-55.

[41] Sanyal T, Kaviraj A, Saha S. Deposition of chromium in aquatic ecosystem from effluents of handloom textile industries in Ranaghat–Fulia region of West Bengal, India. Journal of advanced research 2015;6:995-1002.

[42] Zhao S, Shi X, Li C, Zhang H, Wu Y. Seasonal variation of heavy metals in sediment of Lake Ulansuhai, China. Chemistry and Ecology 2014;30:1-14.

[43] Ali Azadi N, Mansouri B, Spada L, Sinkakarimi MH, Hamesadeghi Y, Mansouri A. Contamination of lead (Pb) in the coastal sediments of north

and south of Iran: a review study. Chemistry and Ecology 2018;34:884-900.

[44] Nowrouzi M, Mansouri B, Nabizadeh S, Pourkhabbaz A. Analysis of heavy metals concentration in water and sediment in the Hara biosphere reserve, southern Iran. Toxicology and industrial health 2014;30:64-72.

[45] Pacle Decena SC, Sanita Arguelles M, Liporada Robel L. Assessing Heavy Metal Contamination in Surface Sediments in an Urban River in the Philippines. Polish Journal of Environmental Studies 2018;27.

[46] Alloway BJ. Sources of heavy metals and metalloids in soils. In: Heavy metals in soils. Springer; 2013. page 11-50.

[47] Arunakumara K, Walpola BC, Yoon M-H. Current status of heavy metal contamination in Asia's rice lands. Reviews in Environmental Science and Bio/Technology 2013;12:355-77.

[48] Semu E, Singh BR. Accumulation of heavy metals in soils and plants after long-term use of fertilizers and fungicides in Tanzania. Fertilizer research 1995;44:241-8.

[49] Mackay AK, Taylor MP, Munksgaard NC, Hudson-Edwards KA, Burn-Nunes L. Identification of environmental lead sources and pathways in a mining and smelting town: Mount Isa, Australia. Environmental Pollution 2013;180:304-11.

[50] Youssef DH, Tayel FT. Metal accumulation by three Tilapia spp. from some Egyptian inland waters. Chemistry and Ecology 2004; 20:61-71.

[51] Rahman MS, Molla AH, Saha N, Rahman A. Study on heavy metals levels and its risk assessment in some edible fishes from Bangshi River, Savar, Dhaka, Bangladesh. Food chemistry 2012;134:1847-54.

[52] Yousafzai AM, Ullah F, Bari F, Raziq S, Riaz M, Khan K, et al. Bioaccumulation of some heavy metals: analysis and comparison of *Cyprinus carpio* and *Labeo rohita* from Sardaryab, Khyber Pakhtunkhwa. BioMed Research International 2017;2017.

[53] Ariyaee M, Azadi NA, Majnoni F, Mansouri B. Comparison of metal concentrations in the organs of two fish species from the Zabol Chahnimeh Reservoirs, Iran. Bulletin of environmental contamination and toxicology 2015;94:715-21.

[54] Malakootian M, Mortazavi MS, Ahmadi A. Heavy metals bioaccumulation in fish of southern Iran and risk assessment of fish consumption. 2016;

[55] Khaled A. Trace metals in fish of economic interest from the west of Alexandria, Egypt. Chemistry and Ecology 2009;25:229-46.

[56] Kumar B, Senthil Kumar K, Priya M, Mukhopadhyay D, Shah R. Distribution, partitioning, bioaccumulation of trace elements in water, sediment and fish from sewage fed fish ponds in eastern Kolkata, India. Toxicological & Environ Chemistry 2010;92:243-60.

[57] Ali H, Khan E. Assessment of potentially toxic heavy metals and health risk in water, sediments, and different fish species of River Kabul, Pakistan. Human and Ecological Risk Assessment: An International Journal 2018;24:2101-18.

[58] Ali H, Khan E. Trophic transfer, bioaccumulation, and biomagnification of non-essential hazardous heavy metals and metalloids in food chains/webs—Concepts and implications for wildlife and human health. Human and

Ecological Risk Assessment: An International Journal 2019;25:1353-76.

[59] Dwivedi AC, Tiwari A, Mayank P. Seasonal determination of heavy metals in muscle, gill and liver tissues of Nile tilapia, *Oreochromis niloticus* (Linnaeus, 1758) from the tributary of the Ganga River, India. Zoology and Ecology 2015;25:166-71.

[60] Ali H, Khan E. Environmental chemistry in the twenty-first century. Environmental Chemistry Letters 2017;15:329-46.

[61] Liang S, Gao N, Li Z, Shen S, Li J. Investigation of correlativity between heavy metals concentration in indigenous plants and combined pollution soils. Chemistry and Ecology 2016;32:872-83.

[62] Wang S, Wu W, Liu F, Liao R, Hu Y. Accumulation of heavy metals in soil-crop systems: a review for wheat and corn. Environmental Science and Pollution Research 2017;24: 15209-25.

[63] Dixit R, Malaviya D, Pandiyan K, Singh UB, Sahu A, Shukla R, et al. Bioremediation of heavy metals from soil and aquatic environment: an overview of principles and criteria of fundamental processes. Sustainability 2015;7:2189-212.

[64] Adamis PDB, Panek AD, Leite SGF, Eleutherio ECA. Factors involved with cadmium absorption by a wild-type strain of *Saccharomyces cerevisiae*. Brazilian Journal of Microbiology 2003;34:55-60.

[65] Mishra S, Bharagava RN, More N, Yadav A, Zainith S, Mani S, et al. Heavy metal contamination: an alarming threat to environment and human health. In: Environmental biotechnology: For sustainable future. Springer; 2019. page 103-25.

[66] Scragg A. Environmental biotechnology. 2nd editio. Oxford: Oxford University Press; 2006.

[67] Tchounwou PB, Centeno JA, Patlolla AK. Arsenic toxicity, mutagenesis, and carcinogenesis–a health risk assessment and management approach. Molecular and cellular biochemistry 2004;255:47-55.

[68] Centeno JA, Tchounwou PB, Patlolla AK, Mullick FG, Murakata L, Meza E, et al. Environmental pathology and health effects of arsenic poisoning. Managing Arsenic in the Environment: From Soil to Human Health 2006;311-27.

[69] Stern BR. Essentiality and toxicity in copper health risk assessment: overview, update and regulatory considerations. Journal of Toxicology and Environmental Health, Part A 2010;73:114-27.

[70] Mishra S, Bharagava RN. Toxic and genotoxic effects of hexavalent chromium in environment and its bioremediation strategies. Journal of Environmental Science and Health, Part C 2016;34:1-32.

[71] Tripathi RD, Srivastava S, Mishra S, Singh N, Tuli R, Gupta DK, et al. Arsenic hazards: strategies for tolerance and remediation by plants. Trends in biotechnology 2007;25:158-65.

[72] NAS (National Academy of Sciences/National Research Council). Arsenic in drinking water. National Research Council, Washington, DC 1999;

[73] Martin S, Griswold W. Human health effects of heavy metals. Environmental Science and Technology briefs for citizens 2009;15:1-6.

[74] Jacobs IA, Taddeo J, Kelly K, Valenziano C. Poisoning as a result of barium styphnate explosion. American

journal of industrial medicine 2002;41:285-8.

[75] Salem HM, Eweida EA, Farag A. Heavy metals in drinking water and their environmental impact on human health. ICEHM2000, Cairo University, Egypt 2000;542-56.

[76] Sobha K, Poornima A, Harini P, Veeraiah K. A study on biochemical changes in the fresh water fish, *Catla catla* (Hamilton) exposed to the heavy metal toxicant cadmium chloride. Kathmandu University Journal of Science, Engineering and Technology 2007;3:1-11.

[77] Davison AG, Taylor AJN, Darbyshire J, Chettle DR, Guthrie CJG, O'Malley D, et al. Cadmium fume inhalation and emphysema. The Lancet 1988;331:663-7.

[78] Järup L, Berglund M, Elinder CG, Nordberg G, Vanter M. Health effects of cadmium exposure–a review of the literature and a risk estimate. Scandinavian journal of work, environment & health 1998;1-51.

[79] Goyer R. Issue paper on the human health effects of metals. US Environmental Protection Agency; 2004.

[80] Lalor GC. Review of cadmium transfers from soil to humans and its health effects in the Jamaican environment. Science of the total environment 2008;400:162-72.

[81] Liu J, Goyer RA, Waalkes MP. Toxic effects of metals. Casarett and Doull's Toxicology: The Basic Science of Poisons, seventh edition (CD Klaasen, Editor). McGraw-Hill Medical, New York, NY, USA 2008;931-79.

[82] ATSDR (Agency for Toxic Substances and Disease Registry). Toxicological Profile for Silver. U.S. Department of Health and Human Services, Public Health Service: Atlanta, GA, USA 1990;

[83] Neustadt J, Pieczenik S. Patient Handout: Toxic Metal Contamination: Mercury. Integrative Medicine-Innovision Communications 2007;6:36.

[84] Ainza C, Trevors J, Saier M. Environmental mercury rising. Water, Air, and Soil Pollution 2010;205:47-8.

[85] Gulati K, Banerjee B, Lall SB, Ray A. Effects of diesel exhaust, heavy metals and pesticides on various organ systems: Possible mechanisms and strategies for prevention and treatment. 2010;

[86] NAS (National Academy of Sciences/National Research Council). Toxicological Effects of methyl mercury. National Research Council, Washington, DC 2000;

[87] ATSDR (Agency for Toxic Substances and Disease Registry). Toxicological profile for Mercury. U.S. Department of Health and Human Services, Public Health Service: Atlanta, GA, USA 1999;

[88] Wuana RA, Okieimen FE. Heavy metals in contaminated soils: a review of sources, chemistry, risks and best available strategies for remediation. Isrn Ecology 2011;2011.

[89] Padmavathiamma PK, Li LY. Phytoremediation technology: hyper-accumulation metals in plants. Water, Air, and Soil Pollution 2007; 184:105-26.

[90] Odum HT. Background of Published Studies on Lead and Wetlands. In: Heavy Metals in the Environment. CRC Press; 2016. page 34-53.

[91] ATSDR (Agency for Toxic Substances and Disease Registry). Multiple lead and cadmium exposure study with biological markers incorporated. U.S. Department of

Health and Human Services, Public Health Service: Atlanta, GA, USA 1995;

[92] IPCS, (International Programme on Chemical Safety WHO. Inorganic lead. Environmental Health Criteria Document No. 165. World Health Organization 1995;152-92.

[93] Vinceti M, Wei ET, Malagoli C, Bergomi M, Vivoli G. Adverse health effects of selenium in humans. Reviews on environmental health 2001;16:233-52.

[94] Duda-Chodak A, Blaszczyk U. The impact of nickel on human health. Journal of Elementology 2008;13: 685-93.

[95] Khan MA, Ahmad I, Rahman IU. Effect of environmental pollution on heavy metals content of *Withania somnifera*. Journal of the Chinese Chemical Society 2007;54:339-43.

[96] Hess R, Schmid B. Zinc supplement overdose can have toxic effects. J. Paediatr. Haematol./Oncol 2002;24:582-4.

Chapter 2

Heavy Metal Sources and Their Effects on Human Health

Narjala Rama Jyothi

Abstract

Heavy metals are defined in many ways, based on various factors such as density and atomic weight. Some of the heavy metals are essential as nutrients for humans such as iron, cobalt and, zinc in small quantities but are toxic in higher quantities. But few metals, such as lead, cadmium and, mercury are poisonous even in small quantities. The toxicity of heavy metals is depending on concentration,period of exposure and route of exposure. Heavy metal exposure takes place on human beings through inhalation from the atmosphere, intake through drinking water and, ingestion through the skin by dermal contact. The present chapter describes the definition of heavy metals, sources of these heavy metals, toxicity and, their impact on various environmental segments, such as air, water and, soil.

Keywords: heavy metals, heavy metal toxicity, sources, exposure, environmental impacts

1. Introduction

Heavy metals, the name has so many definitions based on various parameters. Based on density the metals which are having a density values greater than 5 g/cm^3 are considered as heavy metals [1]. According to this study, the heavy metals which would consider as most threat to human beings are lead, cadmium, mercury, and arsenic. Duffs [2] reviewed the usage of the term heavy metals from the history and finally, he concluded that using the term "heavy metals" is meaningless. He established that there is no relation between the density of the metal and to the usage of the term. In the case of heavy metals, metalloid arsenic also included, from this the term heaviness means may be toxicity.

Some of the heavy metals are having so much of biological importance in trace amounts [3] particularly the elements that are present in the 4th period in the modern periodic table. The biological importance of these metals is enzyme functioning (vanadium and manganese), hormone functioning, production (selenium), cellular growth (nickel), and metabolic growth (arsenic). But these metals are required for the human in trace amounts only if their amount in the body increases they cause adverse effects on human health. Overall the heavy metal should be considered as having high density and also biological importance in trace amounts.

There is a lot of importance for the determination of heavy metals in the various environmental segments, such as air, water, and soil due to their carcinogenic and toxic nature. The IARC (International Agency for Research on Cancer) declared arsenic, hexavalent chromium, cadmium, and nickel and their compounds as group 1 carcinogens (proven carcinogens). Arsenic and their compounds cause urinary bladder, liver, and lung cancers. Hexavalent chromium causes lung cancer and nickel and its compounds cause nasal cavity and lung cancers. All these elements cause cancers to human beings by the route of exposure is through inhalation and ingestion [4]. Regarding the availability of various heavy metals in the earth's crust is about 5%, among which iron occupies nearly 95% [5].

Due to their toxicity and carcinogenic nature, most of the researchers all over the world are reported about the determination and health implications of heavy metals in the environment. Some of them are discussed hereunder.

Jyothi and Mohamed Farook [6] reported the sources, exposure, and toxicity of mercury. Suvarapu et al. [7] reviewed the heavy metal concentrations in ambient air. This study is limited to the estimation of toxicity of heavy metals in the Indian atmosphere and another study [8] they reviewed the heavy metal determination in ambient air all over the world. Kim et al. [9] reviewed heavy metal toxicity and chelating therapeutic strategies. Giller et al. [10] reviewed the toxicity of heavy metals in microorganisms in agricultural soils. This study found that the microorganisms in soil are much sensitive to heavy metal toxicity than animals and plants. Yabe et al. [11] summarized heavy metal pollution and its impact on the environment and the human population in Africa. Das et al. [12] reviewed the toxicity of cadmium in plants. Proshad et al. [13] reviewed the toxicity of heavy metals in soils of Bangladesh. In this study, they concentrated on the impact of industrialization on the concentration of heavy metals in soil. Su et al. [14] reported the heavy metal contamination in soil worldwide. In this study, they mentioned the current situation of contamination and remediation methods.

The present chapter describes the heavy metal sources, exposure, and the impact of their toxicity on various environmental segments. Based on the toxicity and non-biodegradable nature of lead, cadmium, mercury, and arsenic, the present study mainly focused on these metals.

2. Types of heavy metals

Based on the survey of literature the metals that are considered as heavy metals are chromium, lead, cadmium, iron arsenic, cobalt, mercury, copper and zinc are the Heavy metal. According to Kim et al. [9] studies heavy metals have been classified in to two types as essential and non- essential (**Table 1**). Essential Heavy metals are less toxic at low concentrations and they act as coenzyme in biological process. For example Hemoglobin and Myoglobin consist of Iron, Vitamin B12 consist of cobalt. Non-essential heavy metals are highly toxic even at very low concentrations, they are non -biodegradable and cause severe toxic effects to living organisms.

ESSENTIAL (Harmless)	NON ESSENTIAL (Toxic)
Zinc (Zn)	Chromium (Cr)
Copper (Cu)	Lead (Pb)
Iron (Fe)	Arsenic (As)
Cobalt (Co)	Mercury(Hg)
	Cadmium (Cd)

Table 1.
Classification of heavy metals.

2.1 Heavy metal toxicity

Some heavy metals are essential to the human biological process, but depending upon their dosage intake leads some unexpected hazardous effects on health and the physiological system. According to [9] studies shows that despite of its beneficiary health effects, heavy metals are acting as carcinogenic agents. Dissolved forms of these metals through different forms as soil pollutants, water pollutant and air pollutants entering into food chain and finally ending in humans, these are leading to severe damage to the cellular system and leading to expose towards cancer. According to the reports of the International agency for research on cancer non-essential heavy metals (As, Cd, Cr) are major cancer- causing agents [9].

2.2 Sources, exposure, and environmental impacts of lead

The sources of lead varies with different countries based on old and new usage of lead products. It is not limited to the processing of gold ore and recycling of used lead products. It is found that the decrease in blood lead levels in the population of the countries in which unleaded gasoline is in usage [15].

A recent study [16] has reported elevated blood levels in pregnant women in a rural village in Bangladesh. In this study, they found more than 30% of women they sampled were had lead levels in blood in the range of above 5 μg/dL. They found the major source of lead exposure to these women were identified as food storage cans. Nearly 18% of food storage cans (out of the tested) were having lead soldering insides and are responsible for lead contamination in these women. Another study in China [17] determined the blood lead levels (BLL's) in children who are taking treatment in lead specialty clinics. In this study, they found the BLL's ranging from 5 to 126 μg/dL. The major reasons they found for the higher lead levels in their blood as industrial sources and folk medicine which is popular in China. Another important thing was determined as it is difficult to find lead poisoning in children due to non-specific symptoms. A very recent study from Australia [18] determined the higher lead levels in children due to the high concentration of lead in soil and pretty dish dust at their premises. This study found that the population who are living in old houses built before 1940 are diagnosed with higher lead levels due to pretty dish dust.

In Nigeria, lead poisoning in the population was observed in the area of Zamfara state which contains gold mining activities. Mahuta [19] reported that in Nigeria, the sources of the lead include mining of gold, lead pipes used for drinking water, and cultural usage of lead.

Based on the studies in all parts of the world it is assumed that the sources of lead are historical usage of lead, industrial activities, and leaded gasoline. Major studies reported that children are the most common victims of lead poisoning. The way of exposure includes the inhalation through the nose and ingestion through drinking water and soil.

There are several ways to minimize the lead levels in the environment such as remediation techniques (in soil), using adsorbents (in water), and using unleaded gasoline (the air). After the identification of leaded gasoline as a source of lead poisoning by US EPA, a major decline in their levels was found by replacing it with unleaded gasoline. Dongre [20] reported the toxicological profile, remedial solutions for lead levels in water by using polymeric materials, such as chitin and chitosan.

Zaltauskaite and Kniuipyte [21] reported the impact of lead concentration in soil on *Eisenia fetida* (earthworm). They found that lead in soil inhibits the growth of earthworms. Lead in soil can enter into human food by the vegetation in the contaminated soil [22]. The Types of carcinogenic effects of lead toxicity was explained in **Figure 1**. The lead toxicity in humans causes intestinal cancer, lung cancer, and central nervous system.

Figure 1.
Carcinogenic effect of Lead.

2.3 Sources, exposure and environmental impacts of cadmium

The major exposure of cadmium by human beings is through contaminated food and water. Cigarette smoke is the way of exposure through the inhalation route. The toxicity of cadmium is due to accumulation in plants and animals for nearly 25–30 years. Microbial fermentation is one of the effective methods to remove cadmium from food [23]. Another major source of cadmium in the environment is phosphate fertilizers and the waste incineration process. Blood cadmium levels are having a huge difference between smokers and non-smokers of cigarettes [1].

The presence of lead and cadmium in the human body can reach the brain and can cause Alzheimer's disease [24]. After the exposure to the cadmium, in human, it can accumulate at the kidney, due to this reason urinary cadmium levels has been considered as biomarkers for cadmium levels in humans [25].

In case of occupational exposure to cadmium includes the workers at battery production, pigment industries, and electroplating. Because of long time accumulation in the human body even in small quantities is toxic and carcinogenic [26]. Another important source of cadmium in the soil is sewage sludge which can make the cadmium almost the same amount as fertilizers consumption (https://www.osti.gov/biblio/5478006, accessed on 1st October/2020). The types of carcinogenic effects of cadmium toxicity were explained in **Figure 2**. Cadmium shows its toxic effects on the gastric system and leads to gastric cancer, breast cancer, lung cancer, and it also affects the excretory system and leads to renal cancer.

2.4 Sources, exposure, and environmental impacts of mercury

Mercury is the metal widely studied all over the world due to its toxic nature and easily entering into the food chain. An extensive review report was published by Jyothi and Mohamed Farook [6] regarding the sources, exposure, and toxicity

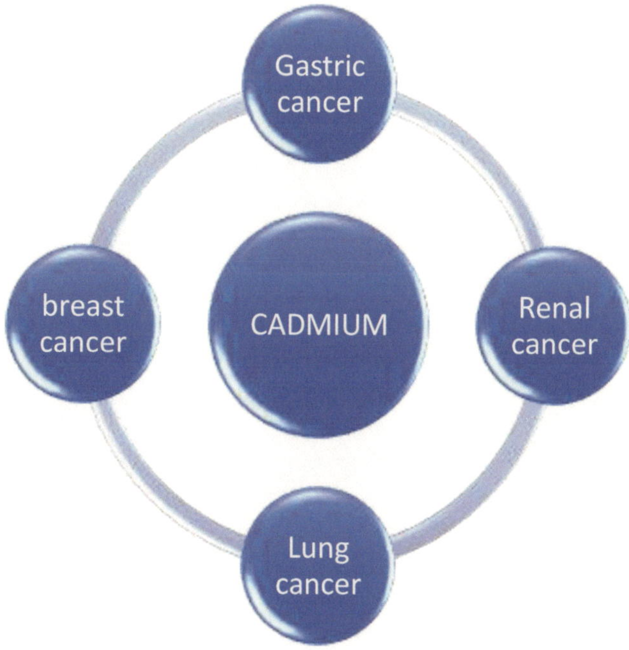

Figure 2.
Carcinogenic effect of cadmium.

of mercury. The toxicity of mercury depends on its chemical composition. Methyl mercury is more toxic than inorganic mercury. Due to its toxic nature and historical incidents like Minamata so many authors were published various facts regarding the sources, transport, and fate of mercury in the environment. Both volcanoes and forest fires are natural sources of mercury in the atmosphere. The burning of fossil fuels in power plants is one of the major anthropogenic sources of mercury (https://www.epa.gov/mercury/basic-information-about-mercury; accessed on 15th October/2020). Because of easy transportation, it is considered a global pollutant [27].

Even exposure to small quantities, shows toxic effects on various physiological systems, such as nervous and digestive systems and organs like lungs and kidneys. Due to this reason WHO declares mercury as one of the top most priority toxic metals (https://www.who.int/news-room/fact-sheets/detail/mercury-and-health; accessed on 20th October/2020). When it enters into water it largely affects the aquatic animal's life and through them, it can enter into the food chain to reach human beings.

Yokoyama [28] reported the methylmercury poisoning control measures and the current situation of its effects on fetuses and infants particularly. In this study, they addressed the global cycle of methyl mercury also. Strode et al. [29] studied the emission of mercury into the North American atmosphere due to gold and silver mining in the 19th century. The types of carcinogenic effects of mercury toxicity was explained in **Figure 3**. Mercury toxicity effects on the rectal system and leads to colorectal cancer. It shows vast effects on the central nervous system leads to brain cancer and lung cancer.

2.5 Sources, exposure, and environmental impacts of arsenic

Arsenic is a metalloid but due to its toxic and carcinogenic nature, it is discussed under the heading of heavy metal toxicity. Abdul et al. [30] reviewed the health effects of arsenic exposure to human beings. According to this study, the majority of

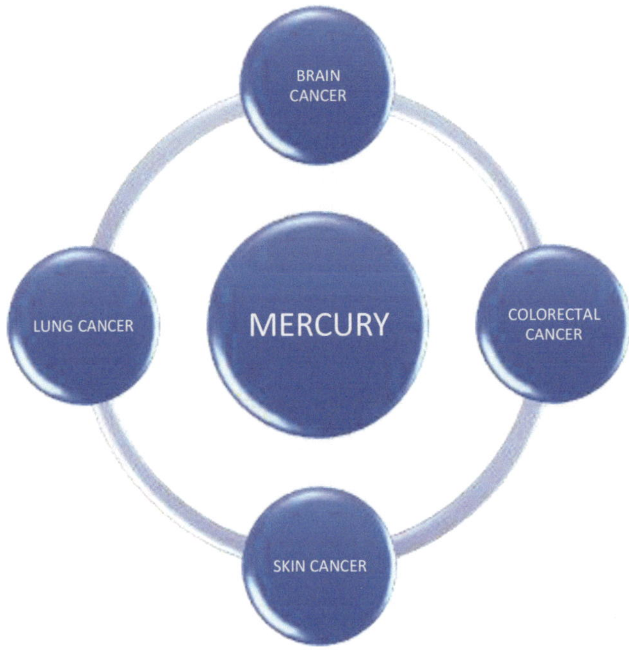

Figure 3.
Carcinogenic effect of mercury.

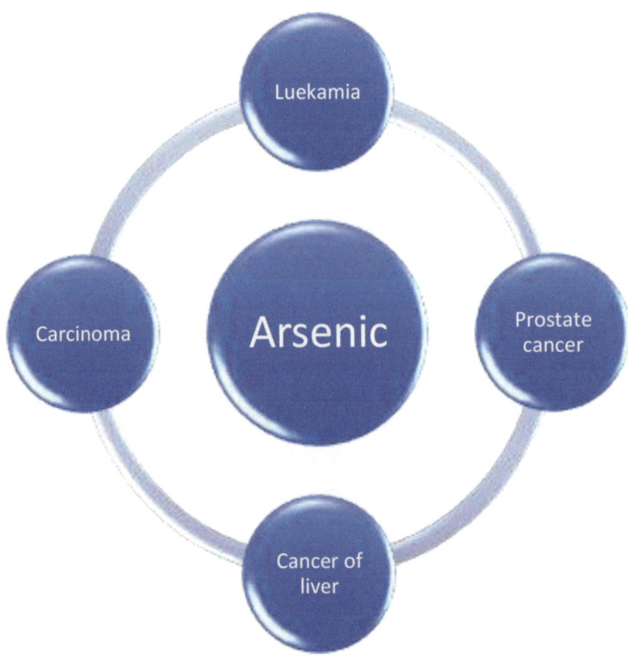

Figure 4.
Carcinogenic effect of arsenic.

the population expose to this toxic metal through atmospheric air, groundwater, and certain kind of foods. The health effects are not limited to damage to cardiovascular, endocrine, renal, and reproductive systems. In various parts of the world such as India, Pakistan, and Bangladesh it was observed that major exposure to the arsenic is through groundwater. Shahid et al. [31] reported about the sources and health effects of arsenic through the exposure of groundwater in Pakistan. This study predicts nearly 47 million people of Pakistan at risk due to arsenic contamination in groundwater. They found that over 50% of the well they studied are having higher arsenic levels than WHO recommended levels in drinking water. A recent study [32] describes the occurrence and mobilization of arsenic in the groundwater of Bangladesh. In this study, they found that intensive usage of land for agriculture and usage of agrochemicals are the major reasons for arsenic contamination in groundwater of Bangladesh. Ahmed et al. [33] reported the situation of arsenic contamination in groundwater in a village in Bangladesh. For this purpose, they discussed 20 years situation of its exposure. Based on their results they found that the cancer risk to the population who are exposing to arsenic has 40% more than the non-exposures. The Types of carcinogenic effects of arsenic toxicity was explained in **Figure 4**. Sources and health effects of all the above discussed heavy metals are summarized in **Table 2**. Arsenic has its toxic carcinogenic effect on prostate glands and cause prostate cancer, leukemia and cause lesions in hepatic regions leads to cause of cancer of the liver.

	Heavy metal	Sources	Health Effects
Essential heavy metal	Zinc (Zn)	Oil Refining Plumbing Brass manufacturing	Gastrointestinal disorders, Kidney & Liver abnormal functioning
	Copper (Cu)	Copper polishing Plating Printing	Abdominal disorders, Metabolic activity abnormalities
	Iron (Fe)	High intake of iron supplements & oral consumption	Vomiting Diarrhea Abdominal pain Dehydration & lethargy
	Cobalt (Co)	Hip alloy replacement case	Haemotological Cardiovascular Hepatic Endocrine
Non Essential heavy metal	Chromium (Cr)	Steel fabrication Electroplating Textile	Lung disorders (bronchitis,cancer), Renal and reproductive system
	Lead (Pb)	Batteries Coal combustion Paint industry	Serious effect on mental health (Alzheimer s disesase), Nervous system
	Arsenic (As)	Atmospheric deposition Mining pesticides	Highly effects dermal region (Cancer), Brain & Cardiac problems
	Mercury (Hg)	Coal combustion Fish Mining Paint industry Paper industry Volcanic eruption	Sclerosis Blindness Minamata disease Deafness Gastric problems Renal disorders
	Cadmium (Cd)	Plastic Fertilizers pesticides	Osteo related problems Prostate cancer Lung diseases Renal issues

Table 2.
Sources and health effects of heavy metals.

	US EPA			WHO			OSHA	
	Ambient Air	Drinking Water	Soil	Ambient Air	Drinking Water	Soil	Air at work place	Blood
Pb	0.15[1] μg/m^3	15[1] μg/L	400 ppm[1] (play areas); 1200 ppm non-play areas	—	15[2] μg/L	—	30[1] μg/m^3	40[1] μg/dL
Cd	6.5–130[6] ng /m^3	0.005[3] mg/L	85[4] mg/Kg	—	0.003[3] mg/L	—	5[4] μg /m^3	—
As	50[6] μg/m^3	0.01[2] mg/L	—	—	0.01[2] mg/L	—	10[5] μg /m^3	—
Hg	5[7] mg /m^3	0.002[2] mg/L	4–16[7] mg/Kg	—	0.001[2] mg/L	—	—	—

[1]*https://www.atsdr.cdc.gov/csem/csem.asp?csem=34&po=8#:~:text=EPA's%20action%20level%.20for%20lead, systems%20is%2015%20%C2%B5g%2FL.*
[2]*Ebrahimi et al. [34].*
[3]*https://www.wqa.org/portals/0/technical/technical%20fact%20sheets/2015_cadmium.pdf*
[4]*https://www.atsdr.cdc.gov/csem/csem.asp?csem=6&po=7#:~:text=OSHA%20has%20established %20workplace%20levels,people%20occupationally%20exposed%20to%20cadmium.&text = Permissible%20Exposure%20Limit%2D%20TWA%20 (PEL, %2Fm3%20(fumes).*
[5]*https://www.atsdr.cdc.gov/csem/csem.asp?csem=1&po=8#:~:text=The%20permissible%20 exposure%20limit%20for, OSHA%202,001%3B%20NIOSH%202,005%5D.*
[6]*https://ec.europa.eu/environment/archives/air/pdf/pp_as_cd_ni.pdf*
[7]*Ye et al. [35].*

Table 3.
Permissible limits of different toxic elements in environmental matrices.

The permissible limits of lead, cadmium, arsenic, and mercury in different environmental matrices suggested by various international reputed agencies such as US EPA (Environmental Protection Agency), WHO (World Health Organization), and OSHA (Occupational Safety and Health Administration) are presented in **Table 3**.

3. Conclusions

The heavy metal toxicity and their environmental impact is a global issue due to their transportation through air, soil, and water. Based on various factors such as concentration and different major sources are the possible ways of entering the heavy metals through drinking water, air and foods. In minimum traces these metals are required for cellular, metabolic and hormonal functioning in humans but if the limitation exceeds its leads to the cause of severe hazardous effects in health. The toxicity of these metals is affecting the soil vastly by killing microorganisms present in soil which are very helpful to enhance fertility and nutrition levels of soils. According to the IARC, arsenic toxic effects are the cause of cancers in prostate glands, liver, blood, and skin. Mercury is the major reason for causing carcinogenic effects on the brain, lung, skin, and colorectal parts. The adverse effects of lead are the reason for intestinal, central nervous system, and lung cancers. The toxic effects of cadmium cause gastric, breast, lung, and renal cancers in humans.

Another diagnosis was identified in china, extreme high levels of lead toxicity in children were due to the pretty dish and in women, high lead levels in the blood is due to the usage of food storage cans. Cadmium is the major cause of Alzheimer disease and due to high usage of phosphate fertilizers they are accumulating in soils and entering into food chains. WHO states that mercury is hazardous toxic metal affecting aquatic life severely and consumption such mercury affected foods by human leads to several harmful diseases such as Minamata and cause several physiological effects.

Author details

Narjala Rama Jyothi
Department of Basic Science and Humanities, Sri Padmavathi Mahila Visvavidyalayam, Tirupati, Andhra Pradesh - 517501, India

*Address all correspondence to: ramadasaradhi@gmail.com

References

[1] Jarup, L. Hazards of heavy metal contamination. British Medical Bulletin 2003, 68(1), 167-182.

[2] Duffs, J.H. "Heavy metals" - a meaningless term? Pure and Applied Chemistry 2002, 74(5), 793-807.

[3] Emslay, J. Nature's Building Blocks. Oxford university press, Oxford, 2011. ISBN: 978-0-19-960563-7.

[4] IARC. IARC monographs on the evaluation of carcinogenic risks to humans. Arsenic, metals, fibers and dusts. A review of human carcinogens. 2012, 100C.

[5] Lide, D.R. CRC Handbook of Chemistry and Physics. CRC Press, Florida. 2004. ISBN: 978-0-8493-0485-9.

[6] Jyothi, N.R., Mohamed Farook, N.A. Heavy metal toxicity in public health. 2020. IntechOpen, UK. DOI: http://dx.doi.org/10.5772/intechopen.90333.

[7] Suvarapu, L.N., Seo, Y.K., Baek, S.O. Heavy metals in the Indian atmosphere - a review. Research Journal of Chemistry and Environment. 2014, 18(8), 99-111.

[8] Suvarapu, L.N., Baek, S.O. Determination of heavy metals in ambient atmosphere : A review. Toxicology and Industrial Health 2017, 33(1), 79-96.

[9] Kim, J.J., Kim, Y.S., Kumar, V. Heavy metal toxicity : An update of chelating therapeutic strategies. Journal of Trace Elements in Medicine and Biology 2019, 54, 226-231.

[10] Giller, K., Witter, E., Mcgrath, S.P. Toxicity of heavy metals to microorganisms and microbial processes in agricultural soils: A review. Soil Biology and Biochemistry 1998, 30, 1389-1414.

[11] Yabe, J, Ishijuka, M., Umemura, T. Current levels in heavy metal pollution in Africa. Journal of Veterinary Medical Science 2010, 72, 1257-1263.

[12] Das, P., Samantaray, S., Rout, G.R. Studies on cadmium toxicity in plants : A review. Environmental Pollution 1997, 98, 29-36.

[13] Proshad, R., Kormoker, T., Mursheed, N., Islam, M.M., Bhuyan, M.I., Islam, M.S., Mithu, T.N. Heavy metal toxicity in agricultural soil due to rapid industrialization in Bangladesh: A review. International Journal of Advanced Geosciences 2018, 6(1), 83-88.

[14] Su, C., Jiang, L.Q., Zhang, W.J. A review on heavy metal contamination in the soil worldwide: Situation, impact and remediation techniques. Environmental Skeptics and Critics 2014, 3(2), 24-38.

[15] Bawa, U., Bukar, A., Abdullahi, Y. a review of lead poisoning, sources and adverse effects. ATBU journal of science, technology and Education 2015, 3(1), 71-79.

[16] Forsyth, J.E., Islam, M.S., Parvez, S.M., Raqib, R., Rahman, M.S., Muehe, E.M., Fendorf, S., Luby, S.P. Prevalence of elevated blood lead levels among pregnant women and sources of lead exposure in rural Bangladesh: A case control study. Environmental Research 2018, 166, 1-9.

[17] Ying, X.L., Gao, Z.Y., Yan, J., Zhang, M., Wang, J., Xu, J. Sources, symptoms and characteristics of childhood lead poisoning: Experience from lead specialty clinic in China. Clinical Toxicology 2018, 56(6), 397-403.

[18] Dong, C., Taylor, M.P., Gulson, B. A 25 year record of childhood blood lead exposure and its relationship to

environmental sources. Environmental Research 2020, 186, 109357.

[19] Mahuta, M.A. An overlook of sources and strategies of minimizing lead poisoning in Nigeria. Journal of Humanities and Social Sciences 2020, 20(6), 330-340.

[20] Dongre, R.S. Lead: Toxicological profile, pollution aspects and remedial solutions. 2020. IntechOpen. DOI: 10.5772/intechopen.93095.

[21] Zaltauskaite, J., Kniuipyte, I., Kugelyte, R. Lead impact on the earthworm *Eisenia fetida* and earthworm recovery after exposure. Water, Air & Soil Pollution 2020, 49, 231.

[22] Sharma, N., Singh, A., Batra, N. Impact of soil, plant-microbe interaction in metal contaminated soils. Beneficial microbes for sustainable agriculture and Environmental Management, book chapter no.9. ISBN: 13: 978-1-77188-818-9.

[23] Genchi, G., Sinicropi, M.S., Lauria, G., Carocci, A., Catalano, A. The effects of cadmium toxicity. International Journal of Environmental Research and Public Health 2020, 17(11), 3782.

[24] Bakulski, K.M., Hu, H., Park, S.K. Chapter 51: Lead, cadmium and Alzheimer's disease. The neuroscience of Dementia 2020, 2, 813-830.

[25] Suzzi, C.V., Kruse, D., Harrington, J., Levin, K., Meliker, J.R. is urinary cadmium a biomarker of long-term exposure in humans? A review. Current Environmental Health Reports 2016, 3, 450-458.

[26] Zhang, H., Reynolds, M. Cadmium exposure in living organisms: A short review. Science of the Total Environment 2019, 678, 761-767.

[27] Driscoll, C.T., Mason, R.P., Chan, H.M., Jacob, D.J., Pirrone, N. Mercury as global pollutant : Sources, pathways and effects. Environmental Science and Technology 2013, 47, 4967-4983.

[28] Yokoyama, H. Past, present and future of mercury pollution issues. In: Mercury Pollution in Minamata. Springer Briefs in Environmental Science. Springer, Singapore. 2018.

[29] Strode, S., Jaegle, L., Selin, N.E. Impact of mercury emissions from historic gold and silver mining: Global modeling. Atmospheric Environment 2009, 43(12), 2012-2017.

[30] Abdul, K.S.M., Jayasinghe, S.S., Chandana, E.P.S., Jayasumana, C., De Silva, P.M.C.S., Arsenic and health effects : A review. Environmental Toxicology and Pharmacology 2015, 40(3), 828-846.

[31] Shahid, M., Niazi, N.K., Dumat, C., Naidu, R., Khalid, S., Rahman, M.M., Bibi, I. A meta analysis of the distribution, sources and health risks of arsenic contaminated groundwater in Pakistan. Environmental Pollution 2018, 242, 307-319.

[32] Huq, M.E., Fahad, S., Shao, Z., Sarven, M.S., Khan, I.A., Alam, M., Saeed, M., Ulah, H. et. al. arsenic in groundwater environment in Bangladesh : Occurrence and mobilization. Journal of Environmental Management 2020, 262, 110318.

[33] Ahmed, S.A., Faruquee, M.H., Khan, M.H. Twenty years of arsenic contamination and Arsenicosis patients in a village of Bangladesh. American Journal of Public Health Research 2020, 8(6), 190-196.

[34] Ebrahimi, S. J. A., Eslami, A., Ebrahimjadeh, L. Evaluation of heavy metals concentration in the drinking water distribution network in Kurdistan villages in the year 2012. Research journal of pharmaceutical, biological and chemical Sciences 2015, 6(2), 55-61.

[35] Ye, B.J., Kim, B.G., Jeon, M.J., Kim, S.Y., Kim, H.C., Jang, T.W., Chae, H.J., Choi, W.J., Ha, M.N., Hong, Y.S. Evaluation of mercury exposure level, clinical diagnosis and treatment for mercury intoxication. Annals of Occupational and Environmental Medicine 2016, 28, 5.

Chapter 3

Concentrations of Heavy Metals as Proxies of Marine Pollution along Nellore Coast of South District, Andhra Pradesh

Madri Pramod Kumar, Tella Lakshmi Prasad, Kothapalli Nagalakshmi, Nadimikeri Jayaraju and Ballari Lakshmanna

Abstract

Bottom sediment samples from six stations were sampled in pre monsoon 2016, from the Govindampalli – Durgarajupatnam (GP-DP) coast. Heavy metals viz., Fe, Mn, Cr, Cu, Ni, Pb, Zn and Cd analysis was carried out by using ICP-OES, and the average concentrations are as follows Fe > Mn > Zn > Cr > Pb > Ni > Cu > Cd. Various environmental indices like Factor Analysis (FA), Geo-accumulation Index (Igeo), Enrichment Factor (EF) and Pollution Load Index (PLI) were applied to the chemical data in order to know the levels of contamination and factors contributing to the pollution. Correlation coefficient results exhibits significant positive and negative relationships among Fe, Mn, Pb, Zn, Cd. All the environmental indices suggest that heavy metals were present at higher concentrations and the impacts of anthropogenic activities are crucial that serves as source of heavy metals in the zone. Relatively, maximum number of heavy metals viz., Fe, Ni and Pb were accumulated at the brackish environment i.e., at confluence of Swarnamukhi river (GP-S Station).

Keywords: GP-DP coast, heavy metal analysis, factor analysis, geo-accumulation index, enrichment factor, pollution load index

1. Introduction

Across, the world coastal zone provides sea food to more than two billion peoples living in these areas. Apart from this, these also serve as major economic resources either directly or indirectly. From the last couple of decades, marine ecosystems have been experiencing severe and acute environmental stress, due to rapid diversifying human activities like navigation, exploration programmes for economic minerals and ores, exploration of hydrocarbons, fishing, establishment of harbors, ports and oil spills etc., In addition, to the anthropogenic activities on the terrestrial environment, natural processes like volcanism, erosion and weathering etc., also contribute to the enhancement of undesirable and unwanted chemical elements which are called pollutants in the marine environments which ultimately pose major menace to the fragile ecological system.

Among wide range of pollutants, heavy metals are the chemical elements which are generally found in low concentrations in marine and coastal environs. However, anthropogenic activities have inevitably enhanced the heavy metal concentrations which impacts not only marine ecosystem but also humans through consumption of polluted sea food [1, 2]. Heavy metals, because of their persistence and toxic nature in the natural environments, have been receiving utmost attention from the scientific community from the last couple of decades [2–14]. Several scientific studies have obtained that, once these heavy metals introduced in the marine and coastal environments by various sources, they will be redistributed both in water and sediments. Hence, it is important to determine the heavy metal concentrations in the marine ecosystem to evaluate the pollution levels in the sensitive and most fragile marine environs.

The present Investigation deals with the evaluating of heavy metal concentrations from the bottom sediments in the study area i.e., Govindampalli – Durgarajupatnam coast (GP-DP coast). Sand, silt, clay and other materials which settle down at the bottom of the water column form the sediment. Various sources like decomposition of animals, plants, erosion, weathering processes on bed rock and soils give rise to form sediment as well. It is observed from the studies of multi-elemental analysis that, apart from being habitat, sediment provides nutrients for several marine and aquatic biota. Furthermore, these sediments also play a vital role in the absorption of heavy metals in marine environments [9, 14–19]. Thus, it is imperative to determine the levels of heavy metal contaminations, their enrichment levels both in the surface and bottom sediments. Sequentially this will help to acquaint their wide spectrum of implications on the natural ecological systems.

2. Study area

The area under investigation i.e., Govindampalli to Durgarajpatnam coast (GP-DP coast) forms as a part of Nellore district (south to the Krishnapatnam port), Andhra Pradesh, South East Coast of India that lies between $14^0\ 0^1\ 10^{11}$ - $14^0\ 02^I\ 30^{\ II}$ N latitudes and $80^0\ 08^I\ 20^{II}$ - $80^0\ 19^I\ 00^{II}$ E longitudes. It falls in toposheet No.66 B3, 66 B4 & 66 C1& C5 on scale 1:50,000 of Survey of the India (SOI). The GP-DP coast is dissected by two tidal creeks and Swarnamukhi river as well (**Figure 1**) One creek exits

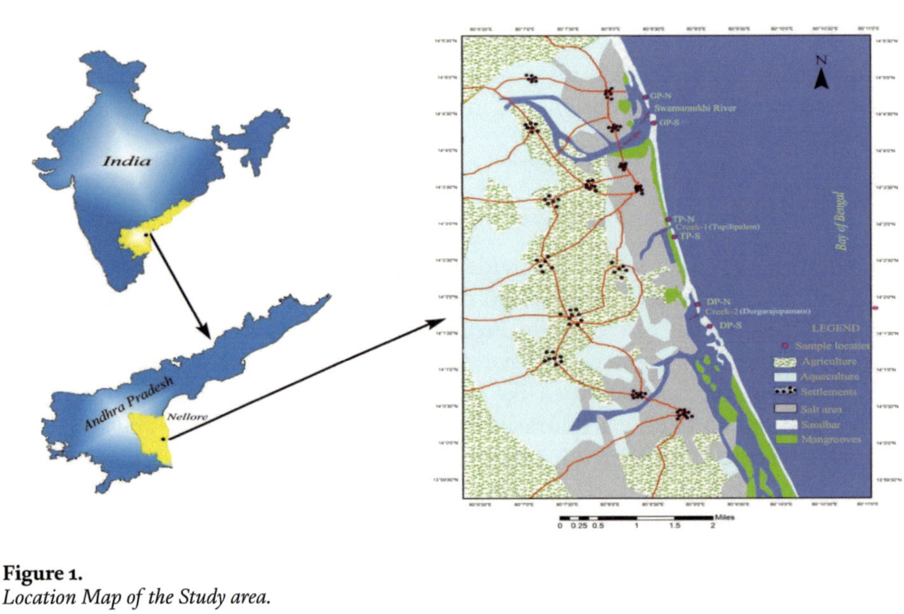

Figure 1.
Location Map of the Study area.

between Tupilipalem north (TP-N) and south (TP-S) beaches and another between Durgarajupatnam north (DP-N) and south (DP-S) beaches. The Swarnamukhi River estuary is situated in between Govindampalli north (GP-N) and south (GP-S) beaches. The total stretch of the GP-DP coast is about 7 km. The Pulicat Lake is located 120 km away towards south from the study area. Buckingham canal is present in western side of the Pulicat Lake. The Swarnamukhi River is an independent river with no tributary and serves as major source for sediment supply in this region.

2.1 Geology and geomorphology

The geology of hinter land consists of Quaternary alluvium, laterites and Triassic sandstones. The Precambrian Basement gneisses, amphibolites and

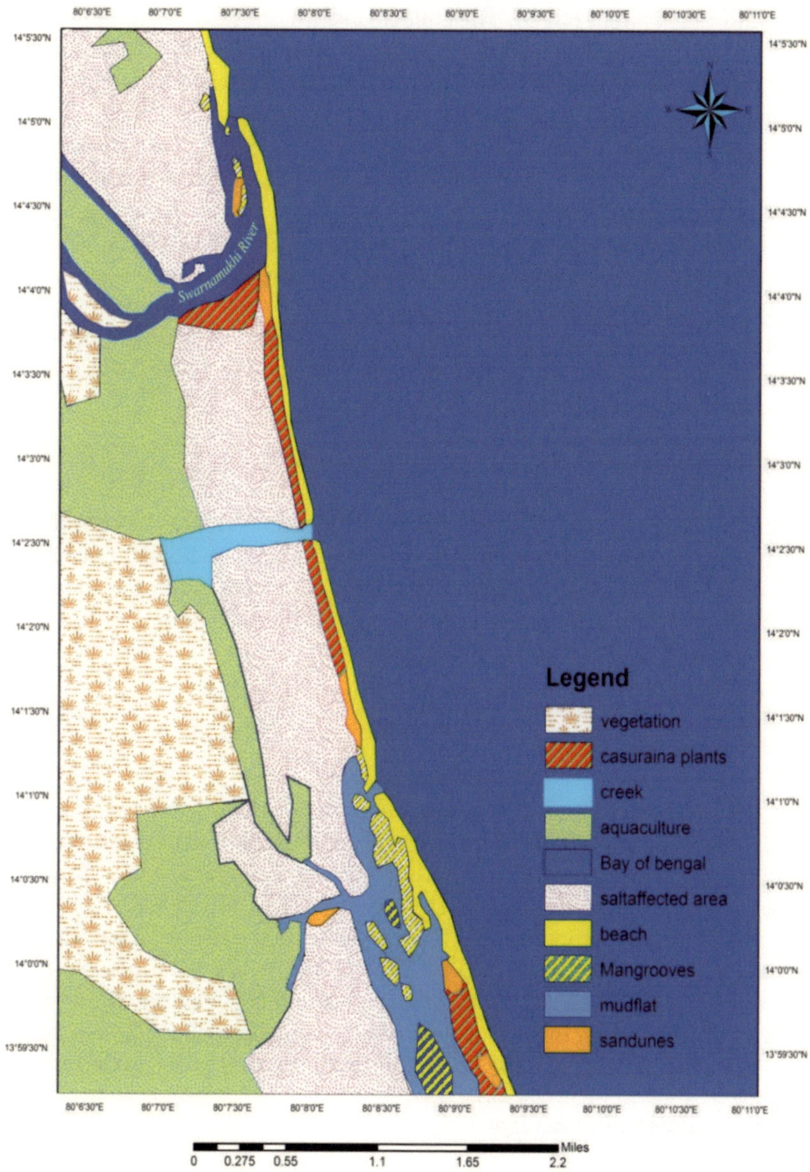

Figure 2.
Geomorphology of the study area.

migmatised pelitic schists are observed. Above these rocks recent alluvial sediments are deposited [14]. The coastal stretch, for some extent is rigorous and laterally compacted by sand. The total area is noticeable with different geomorphological features like asymmetrical sandbars/sand dunes, salt affected lands, mud flats, mangroves, aquaculture ponds etc., (**Figure 2**). The coastal fringe shows dendritic to subdendritic drainage patterns.

3. Sample collection

For evaluating the heavy metal concentrations, undisturbed bottom sediment samples were collected by using Polymerization of Vinyl Chloride (PVC) pipe from six sampling stations i.e., GP-N, GP-S, TP-N, TP-S, DP-N, and DP-S, in pre monsoon season (2016). The sampling station coordinates were recorded by using hand held Global Positioning System (GPS). The collected sediment samples were placed in polyethylene bags having zip lockers. These were pre-cleaned with double distilled water and index accordingly. Care has been taken that there was no loss or damage to polyethylene bags during transportation from field to the laboratory. Later, each sediment sample was dried at 110^0, by using pistle and mortar, the sediment samples were grounded in order to pass through 200 μ sieve. The sieved material comprises of silt and clay which may adsorb heavy metals at higher levels. About 1gm of sieved material is digested with aquaregia (HNO3: HCl) and filtered through 0.45 μ membrane and the concentrations of heavy metal [Iron (Fe), Chromium (Cr), Manganese (Mn), Nickel (Ni), Copper (Cu), Zinc (Zn), Lead (Pb) and Cadmium (Cd)] were generated by Inductively Coupled Plasma – Optical Emission Spectroscopy (ICP-OES) [20–25]. The data sets thus obtained were subjected to statistical analysis.

4. Statistical analysis (SA)

Various environmental indices like Factor Analysis (FA), Geo-accumulation Index (Igeo), Enrichment Factor (EF) and Pollution Load Index (PLI) were applied to the chemical data in XL-STAT (2013) and SPSS software's, in order to know the levels of contamination and factors contributing to the pollution. Further it is also used to unearth the levels of their (heavy metals) enrichment with respect to the natural environment.

4.1 Factor analysis (FA)

Factor analysis is very handy if the data generated, constitutes a large amount of variables. Thus, it believes redundancy among some variables, which means certain variables are interconnected with one another because of the same construct [26]. It basically provides information regarding the source of pollution and its (metals) behavior in the form of factors besides giving a glance on the controlling covariance structure among charted variables. In XL-STAT 2013 software, the Carl's Pearson coefficient matrix is transformed to diagonal matrix to attain Eigen values by using Kaiser standardization. The largest Eigen's value attributes to be Factor 1 explicates the larger variance among the datasets and the Factor 2 formulates most of the variance in general.

4.2 Geo accumulation index (Igeo)

To evaluate and to quantify the heavy metal pollution in the bottom sediments Geo accumulation Index (Igeo) was used. This was introduced by Muller [27] using the following

$$Igeo = \log_2 \left[\frac{C_n}{1.5B_n} \right] \quad (1)$$

Where, 1.5 is a constant which intends to the potential variation of the lithogenic effects if any [28] and B_n is the geochemical background value of the respective metal (n) and C_n is the (n) concentration of the element in the collected sample. Nevertheless, the background values for the Indian sediment are not available, so, the average upper continental values were used [29, 30]. To designate the sediment quality, Muller classified the Igeo values into seven categories: extremely contaminated (>5), strong to extremely contaminated (4-5), strongly contaminated (3-4), moderately to strongly contaminated (2-3), moderately contaminated (1-2), uncontaminated to moderately contaminated (0-1) and uncontaminated (<0).

4.3 Enrichment factor (EF)

In general, EF is applied to realize the degree of sediment contamination by heavy metals, other than lithogenic sources. To ascertain EF, earlier workers employed the average upper continental values of Fe and Al to normalize the determined heavy metal concentrations with respect to a background metal value [14, 31–35]. In this study, Fe is used as a conservative element to differentiate the sources of anthropogenic and natural components. Furthermore, the Fe element play a vital role in redox reaction. In the reduction phase Fe acts as sink to the heavy metals and in oxidation phase Fe has significant control on the distribution of heavy metals among the sediments [16, 36]. Owing to these facts, Fe is used as conservation element. For EF calculation the following equation was used

$$EF = M_x X Fe_b / M_b X Fe_x \quad (2)$$

Where, Fe_x is the Fe concentration in the sediments, M_b and F_b are their concentration in a suitable baseline reference material and M_x is the sediment sample concentration of heavy metals. Brich [37], classified the EF values as follow: if the EF values show >40 the sediment or soils falls in extremely high enrichment category, if EF values in range 20-40 signifies very high enrichment category, if the EF values are in range of 5-20 the sediment or soil falls in significant enrichment category, if the values are in between 2 and 5 signifies moderate enrichment and if EF values is 2 then the sediment or soil is considered as deficiency to minimal enrichment category.

4.4 Pollution load index (PLI)

The Pollution Load Index is evaluated for a zone as well as for a particular station and calculated according to Tomlinson [38]. The PLI for a particular station and for zone can be determined by the following formula

$$\text{PLI for a Station} = \sqrt[n]{CF1 \text{ X } CF2 \ldots\ldots \text{X } CF_n} \quad (3)$$

Where, CF = $C_{metal}/C_{background}$ (C_{metal} is the respective metal concentration of the sample and $C_{background}$ metal concentration of the background) and n is the number of metals and contamination factors.

$$\text{PLI for zone} = \sqrt{\text{Station1 X Station 2} \ldots\ldots \text{Station n}} \quad (4)$$

5. Results and discussion

5.1 Heavy metal concentrations

The results of the heavy metal concentrations from the bottom sediments samples of GP –DP coast were obtained by using ICP –OES (**Table 1**). The range and average concentrations (in parenthesis) in ppm were, Fe 2241.32-3715.15 (2844.79), Cr 9.14 – 11.12 (10.21), Mn 27.19-99.1 (62.02), Ni 5.2-6.67 (6.0), Cu 2.12-7.12 (4.24), Zn 12.65-19.26 (15.0), Pb 5.67-9.14 (7.0), and Cd 0.39-1.15 (0.74). The average concentration of the heavy metals arranged in decreasing as follows: Fe > Mn > zn > Cr > Pb > Ni > Cu > Cd. The concentrations of Fe, Ni, and Pb found to be higher levels at GP-S station (**Table 1**). The highest concentration of these metals is attributed mainly due to the nearness of GP-S station to the confluence of the Swarnamukhi river (the whole sediment load carried by the river drains into the Bay of Bengal). In addition, intense oblique onshore currents coupled with SW winds aid in the accumulation of heavy metals at the GP-S station [39].

The anomalous concentration of Fe metal in the whole study area is may be due to the anthropogenic activities at the Krishnapatnam port, which is present at the northern side of the investigation area. The Zn and Cd metals were found at higher levels at DP_N station. The presence of Zn at higher concentration in the sediment samples may be artributed to the basement rocks underlain by the Quaternary alluvium, amphibolites, migmatised politic schists and Triassic sandstones [14, 40, 41]. The Cd at DP-N station is contributed by the wide range of anthropogenic activities viz., ignition or burning of urban and domestic wastes at the neighborhood area of the investigation area and combustion of fossil fuels for navigation purposes [42]. Heavy metals like Cu, Cr and Mn were found in higher concentrations at stations GP – N, DP – S and TP – N respectively, contributed by irregular and untreated dissemination of industrial waste in to the drainage system.

To know the degree of linear affiliation among heavy metals, correlation coefficient was carried out (**Table 2**). Moreover these results help in understanding how closely two variables (metals) move in relation to one another. Positive correlation observed among Fe, Mn, Pb Zn and Cd, which signifies commonality in mutual dependence and identical behavior during transportation. Significant negative correlations were noticed between Fe – Cu, Cr-Cd, Cu-Pb, and Cu-Zn.

The obtained results were compared with the results of Sreenivasulu [14] (**Table 3**) who studied the geochemistry of bottom sediments in pre-monsoon seasons 2015. By comparison, it is evident in this study that the concentrations of

Sample ID	Fe	Cr	Mn	Ni	Cu	Zn	Pb	Cd
GP-N	3462	10	99	5	5	13	7	1
GP-S	3715	11	93	7	2	16	9	1
TP-N	2241	9	49	7	7	13	6	1
TP-S	2312	11	27	6	4	14	6	0
DP-N	3215	9	61	6	2	19	7	1
DP-S	2726	11	41	6	4	13	6	1
Min	2241	9	27	5	2	13	6	0
max	3715	11	99	7	7	19	9	1
avg	2845	10	62	6	4	15	7	1

Table 1.
Heavy metal concentrations (ppm) from the study area (Pre-monsoon) 2016.

Variables	Fe	Mn	Cr	Cu	Ni	Pb	Zn	Cd
Fe	1.0							
Mn	**0.9**	1.0						
Cr	0.3	0.1	1.0					
Cu	-0.7	-0.3	-0.4	1.0				
Ni	0.0	-0.1	-0.1	0.0	1.0			
Pb	**0.8**	**0.6**	0.3	-0.8	0.4	1.0		
Zn	0.4	0.1	-0.3	-0.7	0.4	**0.6**	1.0	
Cd	0.2	0.2	-0.8	0.0	0.1	-0.1	**0.5**	1.0

Bold values are indicate strong positive relationship with its corresponding elements viz., Fe-Mn, Fe-Pb, Mn-Pb, Pb-Zn and Zn - Cd.

Table 2.
Pearsons correlation coefficient matrix of the heavy metal concentration.

Sample ID	Fe ppm	Cr ppm	Mn ppm	Ni ppm	Cu ppm	Zn ppm	Pb ppm	Cd ppm
GP-N	3268	10	101	4	3	15	7	1
GP-S	3612	12	90	7	3	15	8	1
TP-N	2031	7	52	6	5	14	7	1
TP-S	2048	9	27	5	3	15	4	0
DP-N	3057	11	60	7	3	18	4	1
DP-S	2555	10	43	7	3	14	7	1

Table 3.
Heavy metal concentrations (ppm) from surface sediment samples of GP-DP coast (pre-monsoon) 2015: Sreenivasulu et al. [14].

heavy metals viz., Fe, Cr, Cu, were increased at all stations and the metal like Mn, Ni, Pb, Zn and Cd shows decreasing trend.

5.2 Hierarchical cluster analysis (HCA)

To examine the possible sources and associated heavy metals contamination, hierarchical cluster analysis was performed by using XLSTAT software. The obtained result (dendrogram) was shown (**Figure 3**). The degree of association among heavy metals represents by the distance cluster. Low distance cluster suggests a significant relationship and long distance cluster insignificant relationship among heavy metals.

Two clusters were formed i.e., TP- S to DP-S and DP –N to GP –S. Cluster 1 is formed by stations TP –S, TP – N and DP – S and cluster 2 is formed by DP-N, GP–N and GP-S. Cluster1 and 2 showed high and moderate concentration of heavy metals respectively. The high and moderate contaminations are attributed to the sediment load (which was brought down by the Suwarnamukhi River and the creeks) and proximity of these sampling stations to the inlets.

5.3 Factor analysis (FA)

Factor analysis is carried out to the heavy metal data to ascertain potential sources of contamination. A total of three factor loadings emerged with total variance of 81.72% (**Table 4**), of which factor 1 accounts 42.46%, factor 2, 25.57% and factor 3,

13.68% variance (**Figures 4–6**). Factor 1 was represented by Fe (0.926), Pb (0.91) and Mn (0.738). Therefore it may be termed as Fe-Pb-Mn assemblage. Cr (0.956) and Mn (0.663) show a significant contribution to the factor loading of 2 and 3 respectively signifies Factors controlling the sediment contamination in the coastal zone.

5.4 Geo accumulation index (Igeo)

The Geo accumulation Index is performed to know the existing environmental conditions. It also aids to determine the amount of heavy metal contamination in the natural environments. The results of the Geo accumulation Index (Igeo) which were calculated by using Eq. (1) were presented (**Table 5** and **Figure 7**).

The range of the Igeo values of the heavy metals in the study area is as follows: Fe (26-27), Mn(14-16), Cr (9),Cu (5-7), Ni (8), Pb (6-7), Zn (10) and Cd (−4 to −2). This reveals that the GP-DP coast has been extremely polluted by Fe, Mn, Cr, Cu, Ni, Pb, Zn metals (**Table 5**). Applying the Muller's classification, the results of the

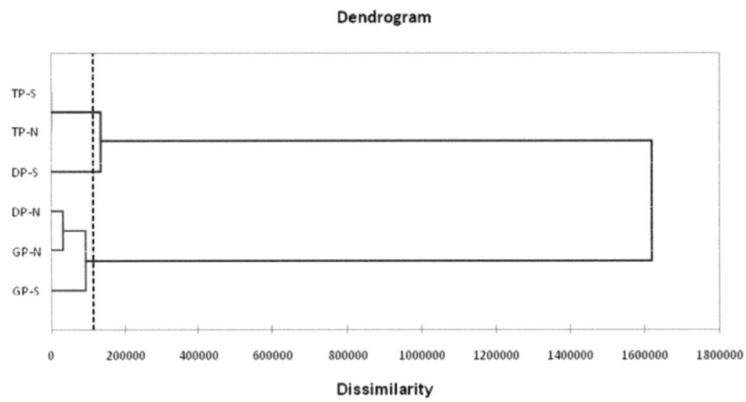

Figure 3.
Dendrogram of grouping of sediment sampling stations.

Parameters	F1	F2	F3
Fe	**0.93**	0.15	0.35
Mn	**0.74**	0.06	**0.66**
Cr	0.20	**0.96**	-0.13
Cu	-0.86	-0.11	0.38
Ni	0.18	-0.20	-0.33
Pb	**0.91**	0.11	-0.15
Zn	0.65	-0.58	-0.48
Cd	0.18	-0.88	0.30
Total	2.94	-0.48	0.60
% of variance	42.46	25.57	13.68
cumulative %	42.46	68.04	81.72

Bold values are show higher factor score loading/matrix. This indicates the higher pollution levels of respective elements.

Table 4.
Factor analysis results of heavy metals.

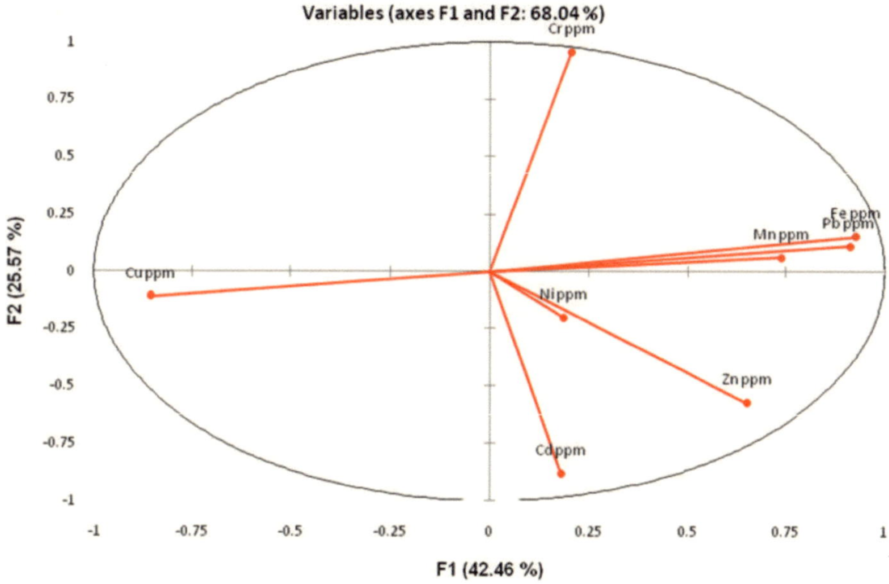

Figure 4.
Factors 1 & 2 assemblage of the heavy metals.

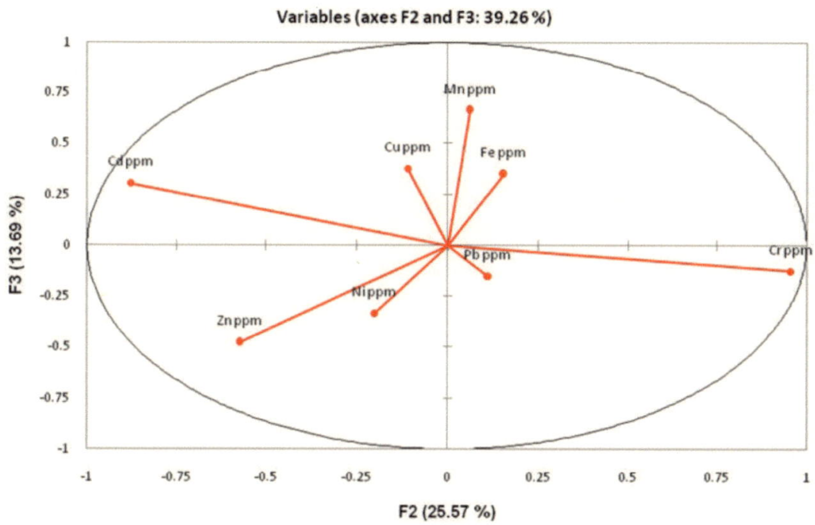

Figure 5.
Factors 2 & 3 assemblage of the heavy metals.

Igeo falls under extremely contaminated environment (>5). The GP-DP coast was unpolluted by the Cd heavy metal (Igeo results of Cd shows negative values).

5.5 Enrichment factor (EF)

In general, Enrichment Factor (EF) is deployed to understand the contribution of metals other than lithogenic origin. In the present study EF was calculated by using Eq. (2) and categorized according to Sutherland's classification [43], which states that the EF value less than or equal to 2 for metals signifies that the elements are from natural weathering processes and for the metals whose EF values are greater than 2. Those elements came from contaminated natural processes [44].

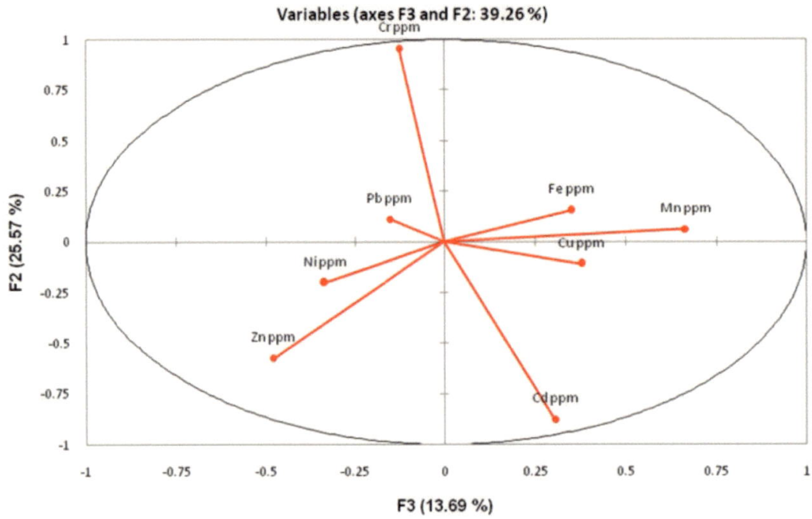

Figure 6.
Factors 3 & 2 assemblage of the heavy metals.

Sample ID	Fe	Mn	Cr	Cu	Ni	Pb	Zn	Cd
GP-N	27	16	9	6	8	6	10	-3
GP-S	27	16	9	5	8	7	10	-3
TP-N	26	15	9	7	8	6	10	-2
TP-S	26	14	9	6	8	6	10	-4
DP-N	27	15	9	5	8	7	10	-2
DP-S	26	14	9	6	8	6	10	-3

Table 5.
Geo accumulation Index for the heavy metals from sediments.

The calculated results of the EF for the sediment samples present study are given (**Table 6** and **Figure 8**).

EF values (ppm) ranges from 0.6 – 1.6 for Mn, 1.5-2.4 for Cr, 1-5.3 for Cu, 1 – 2 for Ni, 4.4 – 6.4 for Pb, 1.8 - 3 for Pb, and 23.5 – 63.2 for Cd. Average values of EF are in the order of Cd > Pb > Cu > Zn > Cr > Ni > Mn > Fe. It is evident that Fe, Mn and Ni values are <=2 which fall under deficiency to minimal enrichment category (**Table 6**). In 67% stations Cd exists falls under very high enrichment category and 33% of the stations showing extremely high enrichment. The Zn metal shows moderate enrichment in all stations except GP -N which exhibits deficiency to minimal enrichment. About 67% of the stations significant enrichment of Pb and the remaining exhibit moderate enrichment. About 50% of the stations fall under moderate enrichment and 17% of the stations come under significant enrichment category for Cu. Stations viz., TP–N, TP-S and DP–S exhibits moderate enrichment to Cr metal. The anomalous behavior of Cd metal enrichment in the sediment samples reflect anthropogenic activities like, possible burning of fossils fuels, incineration of domestic and urban wastes in the vicinity of the study area.

5.6 Pollution Load Index (PLI)

Pollution Load Index (PLI) is an empirical and quick tool proposed by Tomlinson [38] to assess pollution severity and variations along a particulate site as

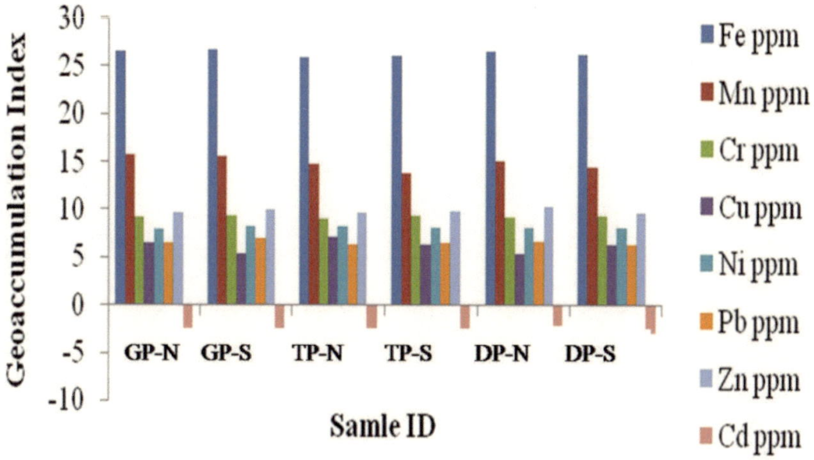

Figure 7.
Graphical representation of Igeo for heavy metals.

Sample ID	Fe	Mn	Cr	Cu	Ni	Pb	Zn	Cd
GP-N	1.0	1.6	1.5	2.2	1.0	4.4	1.8	36.0
GP-S	1.0	1.4	1.6	1.0	1.2	5.7	2.1	23.5
TP-N	1.0	1.2	2.1	5.3	2.0	6.1	2.9	63.2
TP-S	1.0	0.6	2.4	3.1	1.7	6.4	3.0	26.3
DP-N	1.0	1.0	1.5	1.1	1.3	5.3	2.9	55.7
DP-S	1.0	0.8	2.1	2.7	1.5	4.9	2.3	35.4

Table 6.
Enrichment Factor (EF) for the heavy metals.

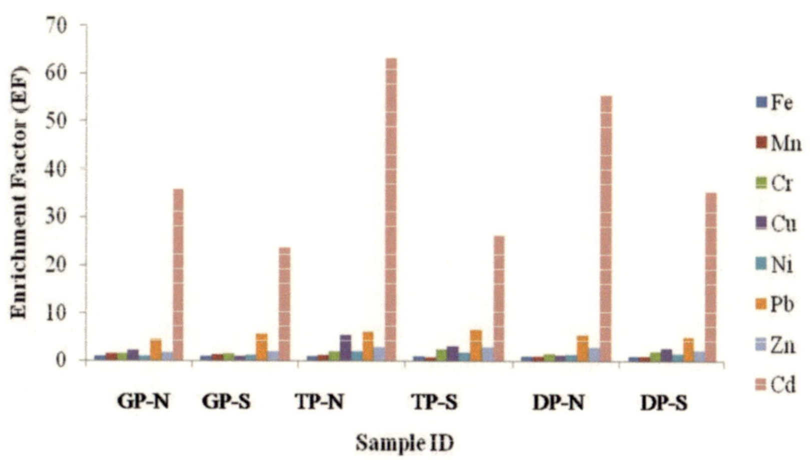

Figure 8.
Graphical representation of Enrichment Factor (EF) for heavy metals.

well as along different sites, besides providing a comparative study based on temporal basis. The Eq.((3) & (3.1)) were used to determine the PLI for a particular station and for specific zone respectively, which in turn aids to decide the sampling sites contains pollution or not. According to Tomlinson classification, if the PLI

Heavy metals	Fe	Cr	Mn	Ni	Cu	Zn	Pb	Cd
PLI	0.25	0.34	0.26	0.30	0.37	0.39	0.58	1.52

Table 7.
Pollution Load Index for zone.

Sample ID	GP-N	GP-S	TP-N	TP-S	DP-N	DP-S
PLI	0.32	0.31	0.31	0.26	0.31	0.29

Table 8.
Pollution Load Index for sampling stations.

value <1 indicates no pollution and if the PLI values show >1 meaning the zone/stations, is/are polluted. The obtained results for zone and for sampling stations were shown (**Tables 7** and **8**).

The value of the PLI of the whole study area ranges from 0.25 – 1.52 (**Table 7**). Except Cd, remaining metals i.e., Fe, Mn, Cr, Cu, Ni, Pb, and Zn were recorded below the baseline indicating no pollution when compared to the world wide sediment. The unusual concentration (1.52) of Cd metal is attributed to direct external sources like, burning of fossil fuels, industrial activities, contaminated agricultural soils, mining waste, municipal sewage effluents and the erosion of hydrothermal mineralized rocks and the black shale deposit [45]. Comparing PLI values of one sampling station with the other ranges from 0.26 to 0.32 (**Table 8**). According to Tomlinson [38], all the sampling stations were showing less than one, signifying that all the heavy metals at all stations were within threshold values. The variations in the indices are an outcome of the difference in sensitivity of these indices towards the sediment pollutants [14, 46].

6. Conclusion

The present paper deals with the assessment of pollution status by determining the heavy metals concentrations in the bottom sediment samples. By and large heavy metals are the chemical elements which occur in low concentrations in fragile coastal and marine environments. But, subsequent anthropogenic activities have inevitably enhanced concentrations in the marine ecosystem. Thus, in the present study (pre-monsoon 2016) Heavy metals (Fe, Mn, Cr, Cu, Ni, Pb, Zn and Cd) analysis was carried out by using ICP-OES, from the bottom sediments along GP-DP coast. Results show that the average concentrations are as follows Fe > Mn > Zn > Cr > Pb > Ni > Cu > Cd. Fe and Cd were recorded highest and lowest concentrations at all stations respectively in the study area.

Relatively, maximum number of heavy metals viz., Fe, Ni and Pb were accumulated at the brackish environment i.e., at confluence of Swarnamukhi river (GP-S Station) and it is also ascribed to the North-Easterly winds (since the sediment samples were collected in the southwest monsoon season). Multivariate analysis like correlation coefficient and factor analysis were carried out to understand complex dynamics of the pollutants. Correlation coefficient results exhibits significant positive and negative relationships among Fe, Mn, Pb, Zn, Cd and Fe – Cu, Cr – Cd, Cu – Pb and Cu – Zn respectively. A total of three Factors with 81.72% variance explain the controlling elements of sediment contamination. Factor 1 includes Fe- Pb – Mn assemblage, Factor 2 and 3 include Cr and Mn.

To assess the pollution in bottom sediments, Geo-accumulation Index (Igeo), Enrichment Factor -and Pollution Load Index were used. The results of Geo-accumulation Index denote that the GD- DP stretch was extremely polluted with Fe, Cr, Cu, Ni, Mn, Zn and Pb metals. Computation of Enrichment Factor and categorizing the obtained results according to Sutherland's Classification, Cd metal show very high enrichment in the 67% of the stations and extremely enriched in 33% of the stations. The Zn- Cr and Pb metals showed moderate and significant enrichment respectively. The Cu exhibits 50% and 17% of the stations falls under moderate and significant enrichment categories respectively. The variations in the indices are an outcome of the difference in sensitivity of these indices towards the sediment pollutants.

Pollution Load Index values (PLI) of Cd metal showed higher value 1.52. This may be due to the influence of direct extraneous sources like burning of fossil fuels for navigation, industrial activities, contaminated agricultural soils etc., The obtained results from the present study were compared to the results of Sreenivasulu who has studied the geochemistry of bottom sediments for the two seasons. It is evident that the concentrations of heavy metals viz., Fe, Cr, Cu were increased at all stations in the period of study and the remaining metals showed decreasing trend.

According to the current study, the heavy metals were extremely harmful to marine life. They were present at various higher concentrations and the impact of anthropogenic activities are crucial that serves as source of heavy metals in the zone. Therefore, in order to prevent severe heavy metal contamination in the investigation area, it is mandatory to enforce monitoring, mitigating and remedial strategies to reduce the loadings and cumulative concentrations of heavy metals in the sediments along fragile coastal ecosystem for sustainable development of future generations. Further care has to be taken to educate and inform relevant stakeholders to avoid converting the coastal zones into sink/dustbins in future.

Acknowledgements

The authors are thankful to the Department of Science and Technology (DST), Government of India, New Delhi, for the financial support in the form of INSPIRE Fellowship (Grant No. IF131129).

Author details

Madri Pramod Kumar[1], Tella Lakshmi Prasad[1*], Kothapalli Nagalakshmi[1], Nadimikeri Jayaraju[2] and Ballari Lakshmanna[2]

1 Department of Earth Sciences, Yogi Vemana University, Kadapa 516005, A.P., India

2 Department of Geology, Yogi Vemana University, Kadapa 516005, A.P., India

*Address all correspondence to: tlpyvugeo@gmail.com

References

[1] Glasby,G.P., Stoppers, P., Walter, P., Davis, K.R., and Renner,R.M., (1988). Heavy-metal pollutioninManukauand Waitemataharbours,NewZealand. New Zealand Journal of Marine and Fresh water Research, 22(4), 595-611. http://dx. doi.org/10.1080/00288330.1988.9516329.

[2] Ho, H.H., Swennen, R., and Damme, A.V., 2010. Distribution and contamination status of heavy metals in estuarine sediements near Cua-ong- Harbour, Ha- Long Bay, Vietnam. Geologica Belgica 13 (122), 37-47.

[3] Riba, I., Delvallas, T.A., Forja, J.M., Go'mez – Parra, A., 2002. Influence of the Aznalco'llar mining spill on the vertical distribution of heavy metals in sediments from the Guadalaquivir estuary (SW Spain). Mar. Pollut. Bull. 45, 39-47.

[4] Jarup, L., 2003. Harzards of Heavy metal contamination. British Medical Bulletin, 68, 167-182, DOI; 10.1093/bmb/ldg032.

[5] De Forest, D., Bricks, K., Adams, W., 2007. Assessing metal bioaccumulation in aquatic environments; the inverse relationship between bio accumulation factors, traffic transfer factors and exposure concentration. Aquat. Toxicol. 84, 236-246.

[6] Kruopiene, J., 2007. Distribution of heavy metals in sediments of the Nemunanas river (Luthiana). Polish Journal of Environmental studies 16 (5), 715-722.

[7] Manjare, S.A., Vhanalakar, S.A., Muley, D.V., 2010. Analysis of water quality using physic- chemical parameters of Tamdalge Tank in Kolhapur district, Maharastra. International Journal of Advanced Biotechnology and Research, 1 (2), 115-119.

[8] Fu, F., Wang, Q., 2011. Removal of heavy metal ions from waste waters; A review. J.Environ. Manage. 92, 407-418.

[9] Jayaraju, N., Sundara Raja Reddy, B.C., & Reddy, K.R., 2011, Anthropogenic impact on Andaman Coast Monitoring with Benthic Foramenifera, Andaman Sea, India; Environmental Earth Science, V.62, P. 821-829.

[10] Ravichandran, R., & Manickam, S., 2012. Heavy metal distribution in the coastal sediment of Chennai coast. The IIOAB Journal 3 (2), 12-18.

[11] Ozkan, E.Y., and Buyukisik, B., 2012. Geochemical chemical and statistical approach for assessing heavy metal accumulation in the southern black sea sediments. Ekoloji 21(83), 11-24

[12] Naser, H.A., 2013. Assessment and management of heavy metal pollution in the marine environment of the Arabian Gulf; A review. Marine Pollution Bulletin – 72, 6 -13.

[13] Baran, A., and Tarnawski, M., 2015. Assessment of heavy metal mobility and toxicity in contaminated sediments by sequential extraction and a battery of bio assess. Ecotoxicology 24 (6), 1279-1293.

[14] Sreenivasulu, G., Jayaraju, N., Sundara Raja Reddy., Reddy Balam, C., Lakshmanna, B., Rajasekhar, M., Nirmal, K., and Lakshmi Prasad, T., 2018. Assessment of heavy metal pollution from the sediment of Tupilipalem coast, south east coast of India. International Journal of Sediment Research, 33, 294-302.

[15] Ajayi, S. O., and Osibanjo. O., 1981. Pollution studies in Nigerian Rivers, water quality of some Nigerian rivers. Environmental studies (series) 2, 87-95.

[16] Muchu, A.P., Teresa, M.S.D.V., and Bordalo, A., 2003. Macrobenthic community in the douro estuary: Relations with trace metals and natural sediment characteristics. Environmental pollution, 121(2), 169-180.

[17] Akinmosin, A., Osinowo, O. and Oladunjoye, M. A. 2009. Radiogenic components of the Nigerian tarsand deposits. Earth Sci.Res. J. 13(1), 64 – 73.

[18] Olubumi, F. E., and Olorunsola., 2010. Evaluation of the status of heavy metal pollution of sediment of Agbabu bitumen deposit area, Nigeria. European Journal of scientific research 41(3), 373 – 382.

[19] Akpan, I. O., and Thompson, E.A., 2013. Assessment of heavy metal contamination of sediments along the cross river channel in cross river state, Nigeria. Journal of Environmental Science, toxicology and food technology. 2(5), 20-28.

[20] Kumar, S., 1989. Heavy metal pollution in Gomti River sediment around Lucknow, Uttar Pradesh. Current Sciecne. 58(10), 557-559.

[21] Mathew, K., and Vasudevan, N.T., 2002. Trace metals in the sediments of river Pamba. Pollution. Research, 21(3), 235-242.

[22] Singh, M., Muller, G., and Singh, I.B., 2003. Geogenic distribution and baseline concentration of heavy metals in sediments of the Ganges river, India. Jouranl. Geochemical Exploration. 80, 1-17.

[23] Vivek, K.G., Sanjay, K.G., Pandey, S.D., Gopal, K., and Virendra, M., 2005. Distribution of heavy metals in sediments and water of river Gotmi. Environmental Monitoring and assessment. 102(1-3), 419-433.

[24] Varol, M. and Sen, B., 2012. Assessment of Nutrient and Heavy Metal Contamination in Surface Water and Sediments of the Upper Tigris River, Turkey. Catena, 92, 1-10. http://dx.doi.org/10.1016/j.catena.2011.11.011

[25] Wardhani, E., Roosmini, D., and Notodarmojo, S., 2017. Status of heavy metal in sediment of Saguling Lake, West Java. IOP Conf. Earth and Environemtnal Science. 60, 1-9.

[26] Bai J, Yang Z, Cui B, Wang Q, Gao H and Ding, Q., 2009 Assessment of heavy metal contamination of roadside soils in southwest China. Stoch. Environ. Res. Risk Assess. 23, 341-347.

[27] Muller, G., 1969. Index of geoaccumulation in the sediments of the Rhine River. Geojournal, 2, 108-118.

[28] Salomons, W., & Forstner, U., 1984. Metals in the hydrocycle. New York: Springer: DOI: 10.1007/978-3-642-69325-0_3.

[29] McLennan, S.M., 2001. Relationships between the trace element composition of sedimentary rocks and upper continental crust. Geochemistry, Geophysics, Geosystems. 2(4). doi: 10. 1029/2000GC0001092000GC000109.

[30] Rajmohan, N., Prathapar, S.A., Jayaprakash, M., and Nagarajan, R., 2014. Vertical distribution of heavy metals in soil profile in a seasonally water logging agriculture fields in eastern ganges basin. Environ. Monit. Assess, 186, 511-5427.

[31] Schiff, K. C., and Weisberg, S. B. 1999. Iron as a reference element for determining trace metal enrichment in Southern California coast shelf sediments. Marine Environmental Research, 48(2), 161-176.

[32] Dragovic, S.,Mihailovic, N., and Gajic, B., 2008. Heavy metals in soils:

distribution, relationship with soil characteristics and radionuclides and multivariate assessment of contamination sources. Chemosphere, 72(3), 491-495.

[33] Seshan, B. R. R., Natesan, U., and Deepthi, K., 2010. Geochemical and statistical approach for evaluation of heavy metal pollution in core sediments in southeast coast of India. International Journal of Environmental Science and Technology, 7(2), 291-306.

[34] Nirmala, K., Ramesh, R., Ambjam, N.K., Arumugam, K., and srinivasulu, S., 2016. Geochemistry of surface sediments of a tropical brackish water lake in south asia. Environmental Earth Sciences, 75, 247. http://dx.doi.org/10.1007/sl2665-015-4964-8.

[35] Tesfamariam, Z., Younis, Y.M.H., and Elsanousi, S.S., 2016. Assessment of heavy metal status of sediment and water in Mainefhi and Toker drinking water reservoirs of Asmara city, Eritrea. American Journal of Research Communication, 4(6), 76-88.

[36] Ibrahim, A., Yaseen, B.P., and Al-Hawari, Z., 2015. Assessment of heavy metal pollution of surface sediments in wadi shuayb., Jordan. Jordan journal of civil engineering, 9(3), 303-313.

[37] Birch, G. (2003). A scheme for assessing human impacts on coastal aquatic environments using sediments. In C. D. Woodcoffe & R. A. Furness (Eds.), Coastal GIS 2003. Australia: Wollongong University Papers in Center for Maritime Policy.

[38] Tomlinson, D.L., Wilson, J.G., Harris, C.R., and Jeffrey, D.W., 1980. Problems in the assessment of heavy metal levels in estuaries and the formation of a pollution index. Helgolander Meeeresuntersuchungen. 33(1-4); 556-575.

[39] Murthy, C.S., and Veerayya, M., 1985. Long shore currents and associated sediment transport in the near shore areas of Kerala and Goa, West Coast of India, Mahasagar, Bulletin of the National Institute of Oceanography, 18(2): 163-177.

[40] Loring, D.H., and Asmund, G., 1996. Geochemical factors controlling accumulation of major and trace elements in Greenland coastal and fjord sediments.Environ. Geol., 28(1): 2-11.

[41] Ahumada, R., Gonzalez, E., Medina, V., and Rudolph, A., 2007. Analysis of mayor zinc source in marine sediments of the estuarine zone in Aysen Fjord, Chile, Invest. Mar., Valparaiso, 35 (2), 5-14.

[42] Chakraborthy, A., Bhattu, D., Gupta,T., Tripathi, S.N., and Canagaratna, M.R., 2015. Real-time measurement of ambient aerosols in polluted Indian city: sources characteristics and processing of organic aerosols during foggy and non-foggy periods. J. Geophys.Res. Atmos. 120 (17), 9006-9017.

[43] Sutherland, R.A., Tolosa, C.A., Tack, F. M. G., and Verloo M. G., 2000. Characterization of selected element concentration and enrichment ratios in background and anthropogenically impacted roadside areas. Archives of Environmental Contamination and Toxicology. 38, 428-438.

[44] Feng, H., Han, X., Zhang, W., and Yu, L., 2004. A preliminary study of heavy metal contamination in Yangtze river intertidal zone due to urbanization. Mar.Pollu. Bull. 49, 910-915.

[45] GESAMP ., 1985. UN/FAO/UNESCO/WHO/WMO/IAEA/UNEP Joint group of experts on the scientific aspects of marine pollution: cadmium,

lead, and tin in the marine environment. Reports and studies No.26.

[46] Praveena, M.S., Radojevic, M., and Abdullah, M.H., 2007. The assessment of mangrove sediment quality in Mengkabong Lagoon: An index analysis approach. International Journal of Environmental Science Education, 2(3), 60-68.

Chapter 4

Heavy Metal Contamination in a Protected Natural Area from Southeastern Mexico: Analysis of Risks to Human Health

Claudia Alejandra Aguilar, Yunuen Canedo, Carlos Montalvo, Alejandro Ruiz and Rocio Barreto

Abstract

In this chapter, a little of the history of Carmen City, Mexico is addressed; this island is immersed in a Protected Natural Area and in the "Campeche Sound" an oil extraction site. Fishing natural resources were for many years the pillar of the development of the area; the most commercially important species are still shrimp, oysters and scales. Nowadays, although the volumes of capture have decreased considerably, different species of high commercial value are still extracted. The considerable development of the oil industry has brought with its economic development and a better quality of life for its inhabitants; however, the ravages of pollution, rapid population growth, and deforestation have been the unwanted factor. This chapter addresses the effects of heavy metals on human health through a risk analysis, based on the criteria of the US Environmental Protection Agency (USEPA) that was carried out for different commercial species based on carcinogenic factors and not carcinogenic; the results show that the risk from consumption of these species is "potentially dangerous" for human health, especially in those species that, due to their eating habits (mollusks, bivalves, clams) tend to bio-accumulate heavy metals, such as cadmium, which it has been considered by the International Agency for Research on Cancer (IARC) as a risk factor; for this reason, the importance of periodically evaluating and monitoring oyster extraction banks, clams and, in general, all fishery products. Mexican legislation and various international legislations dictate the maximum permissible and tolerable levels of heavy metals in fishery products; the organisms considered in this study exceeded the permissible limits in copper and nickel, which represents a risk for human consumption.

Keywords: heavy metals, pollution, marine organisms

1. Introduction

Mexico is a mega-diverse country with 90,839,521 hectares of protected natural areas, of which Terminos Lagoon, classified as "Flora and Fauna Protection Area" has 705, 016 hectares that make it one of the largest areas in the country. Within its status as a Protected Natural Area, there are fishing activities and oil and gas extraction-conduction areas. Until a decade ago, the Campeche Sound contributed nearly 95% of the crude oil and 80% of the national natural gas; today, due to recent changes in the use of fossil energy, production has decreased, although it remains one of the most important companies in Mexico.

Campeche Sound in general, and Carmen Island in particular, have been zones of abrupt changes, beginning with the exploitation of shrimp, which in the years of 1969 to 1979, promoted the economic development of the area. Since 1976, a historical production of crude oil began for Mexico, bringing with it important changes in the population, social changes and therefore, environmental changes.

According to Cuellar et al. [1] in 1979 the company "Mexican Petroleum" (PEMEX) had a large number of facilities for the extraction and processing of crude oil and natural gas on the southwestern coasts of the Gulf of Mexico, as well as a total of 200 facilities for different purposes and 185 production platforms. These changes affected the fishing production and the lives of the inhabitants as they went from being a "fishing village" to industrial zones with an increase in the population and the services that were demanded. By 1970, there were more than 800 vessels with capacities ranging from three to fifty tons to process shrimp and more than twenty freezers and packers of the fishing product in the area, as well as four shipyards for the shrimp fleet; at present, all this activity has been in considerable decline, almost disappearing [2].

When the oil boom began, the first oil spills put fishing activity at risk and there have been very few studies in the area to determine the degree of impact of the oil industry on the deterioration of the environment; certain species such as white shrimp (*Litopenaeus setiferus*) are permanently banned to avoid completely depleting the resource; however, recent data and with the current crisis of the SARS-COV2 pandemic indicate that poaching activities have increased in the area, even with the capture of protected species; [https://www.novedadescampeche.com.mx/estado/campeche/urgente-estrategia-federal-del-control-de-la-pesca-ilegal].

The main fishery resources in this area are shrimp, oyster and scale. The oyster harvest currently has the certification of the Commission for the Protection against Sanitary Risks of Campeche (COPRISCAM, by its acronym in Spanish) in the Atasta lagoon; however, its production has been diminished due to the fishing and poaching of this resource. On the other hand, the clam was the main fishing resource in the Pom lagoon for more than four decades. Currently the catch levels show a notable decrease, which has been attributed to excessive overexploitation; some studies attribute it to pollution and deforestation in the mangrove area. According to Ramos and Villalobos [3], the mangrove ecosystems of the Terminos Lagoon Flora and Fauna Protection Area have registered in recent years, a rapid transformation towards ecosystems with low productivity and biodiversity. The causes of this rapid loss are deforestation, urbanization, industrialization, agricultural, fishing and aquaculture activities; and the alteration of the hydrological regime of the Grijalva-Usumacinta river basin.

The shrimp fishery does not show a better picture. This resource, which was exploited for many years, is now only one fifth of what was obtained in the 1980s. Historical data show that in 1972 the yield of pink shrimp (*Farfantepenaeus duorarum)* was 11,904 tons and in 2000 it was only 1,409 tons [4]. With regard to the seven-bearded shrimp (*Xiphopenaeus kroyeri)* from 1993 due to its overexploitation in the coastal marine strip and with the entry into force of Mexican standards NOM-004-PESC-1993 and NOM-002-PESC-1993 (Diario Oficial de la Federación 1994., Plan de Manejo Pesquero de camarón siete barbas *Xiphopenaeus kroyeri* en las costas de los estados de Campeche y Tabasco) its fishing has been regulated by fishing bans seasons.

Overall, the development of the oil industry, urbanization and overexploitation of marine species have had a strong environmental impact, as well as in the displacement of deep-sea fishing areas. However, very few studies have been conducted in the area that show the overall impact generated on the flora and fauna of this region. Studies have been reported on the impacts on benthic communities and their relation to the presence of hydrocarbons [5]; the studies show the presence and concentration of hydrocarbons in sediments and organisms [5–9]. There are numerous factors to be considered in the deterioration of an ecosystem, among them the great quantity of organic and inorganic substances that are generated not only by oil activity, but also by the entire related industry. In the years 2000–2001 alone, a total of 104,901 tons of sulfur oxides (SOx) and 1,747 tons of nitrogen oxides (NOx) were emitted into the atmosphere [1, 10]. There are currently no recent studies to compare these levels.

Among the inorganic contaminants that cause interest due to the adverse effects they can cause to living beings, heavy metals stand out, some of which have been cataloged as serious threats to human health because of their carcinogenic risk. Regarding the studies carried out to determine the degree of impact on the Campeche Sound, we can cite Vázquez et al. [11] who carried out oceanographic campaigns and comparative studies on the levels of Cd, Cr, Ni and V in marine sediments. In their study, they highlight that oil activity, fishing and marine traffic in the area substantially modify the levels of heavy metals; they also agree that the levels of organic matter have a direct influence on the distribution of metals in sediments; they conclude that metals can interact with organic matter in different ways forming phenomena of adsorption, ion exchange, coprecipitation and complexation.

Other studies have determined the levels of heavy metals in sediments and organisms along the Terminos, Atasta and Pom lagoons and in the Palizada, Candelaria and Chumpan Rivers. Aguilar et al. [12] attributed the levels of Cd, Cr, Cu, Hg and V detected in oysters (*Crassotrea virginica*) to anthropogenic activities; additionally, they calculated the condition index of the oyster (variable that indicates the condition of health) and attributed a decrease in it to the presence of heavy metals; likewise, the levels of Cd, Cr and Cu exceeded the permissible limits established for mollusks and fishery products in the Mexican norms NOM-031-SSA1–1993.

In another study, the concentrations of Cd, Fe, Cu, Pb and Zn were evaluated in oyster (*Crassostrea virginica*), crab (*Callinectes sapidus*) and shrimp (*Litopenaeus setiferus*). The results showed that both oyster and crab are foods that present high levels of Cd, Fe, Cu and Pb in comparison with shrimp; in this study all detected levels were within the permissible limits established by the Mexican Official Standards NOM-031-SSA1–1993 [13].

Regarding sediment studies, Montalvo et al., [14] analyzed the concentration of heavy metals in sediments of the Palizada River; the results showed a high relationship between the levels of metals found with the climatic season and the texture of the sediment. Later, Canedo et al. [15] evaluated the levels of heavy metals in sediments of the Terminos Lagoon; they concluded that the spatial distribution was influenced by river discharges and that the significant correlations found between B, Ba, Co, Mn, Ni and Zn are due to natural biogeochemical inputs; they also found heavy metal levels above background concentrations in sites near the Atasta Lagoon and considered this area vulnerable to heavy metal contamination.

1.1 Effects of heavy metals

Heavy metals exert a wide range of toxic effects in humans, aquatic and terrestrial life [12]. Different strategies have been developed to study the degree of contamination of an area, such as the use of organisms called sentinels (oysters, clams) that due to their feeding habits, their little or no mobility, their little capacity to regulate the concentrations of ions in the internal fluids and their high tolerance to the metal ions absorbed above the metabolic requirements [16], make them ideal for studies of contamination by heavy metals; likewise, studies on fish have been of considerable interest to understand the toxic effects and because they are an important source of nutrients for humans and have the potential to bioaccumulate heavy metals in their tissues [17, 18]. Food contamination can come from different sources: from contamination of the aquatic environment, during harvesting, transportation, handling or packaging.

1.1.1 Mercury (Hg)

Regarding the toxicity of heavy metals, Hg is distinguished because it does not have any biological function; its presence in the environment is due to anthropogenic causes; the natural causes of contamination by this element are not significant. It is an extremely toxic metal; organisms that have been exposed have few biological mechanisms for its elimination and it accumulates progressively through the food chain [19, 20]. The most common form of organic Hg is in the form of methyl mercury (MeHg). Usually levels above tolerance limits can alter the normal functioning of the central nervous system and affect the kidneys and the immune system [21]. Studies show that the toxicity attributed to it is associated with aging and cell death. Bryan and Langston's study [22] study on the oyster *Crassostrea virginica* showed evident embryonic abnormalities at concentrations of 5 to 10 μg/L, while the survival rates of clams, copepods, shrimp and crustaceans were affected by the increase in Hg levels.

1.1.2 Cadmium (Cd)

Cadmium is an element that has no natural source of generation so its presence in aquatic systems and organisms is entirely anthropogenic [23]. Cd does not have biochemical or nutritional functions; it is highly toxic to plants and animals. The International Agency for Research on Cancer points out the Cd and its compounds as carcinogenic. Cd intake pathways in organisms are gastrointestinal and respiratory; it has severe consequences in the blood by binding to high molecular weight proteins [24]; likewise, it has been reported that it can cause different alterations in the biology of living beings, since it accumulates mainly in the liver and can have a half-life of thirty years [25]. In phytoplankton species, growth inhibition was observed at concentrations as low as 1 μg/L [22]. Other species such as *Galaxias maculatus* exposed to acute concentrations showed deficiencies in metabolic rate and deteriorating oxygen consumption; also, stress parameters and decrease in liver catalase activity were observed [26]. In the *Henanese Sinopotamon* crab, a high deterioration of enzyme activity was found in the stomach, intestines, and hepatopancreas [27]. For *Crassotrea virginica* oyster, hepatological changes of the intestine, digestive gland and other organs were presented when exposed to Cd [28]. Due to its source of origin, the activities by which it can be generated are the industrial processes of fertilizer production, by-product of the smelting of other metals and in electronic devices [24].

1.1.3 Copper (Cu)

Cu is an essential element for the growth and metabolism of many living beings; when the levels are increased, it becomes a not very tolerable element [12]. This metal can cause harmful effects in fish, showing damage such as histopathological alteration and accumulation in different organs [29]. Other studies [22] presented experimental evidence that a considerable number of species are sensitive to concentrations of 1 to 10 μg/L of Cu, while at levels of 2 μg/L, the survival rate in young scallops was reduced; likewise, oyster and mussel embryos showed abnormalities in growth and development after exposure to 5 μg/L and the isopod crustacean *Idothea baltica* showed an increase in population mortality. Calabrese et al. [30] studied the acute toxicity of Cu in embryos of *Crassostrea virginica*; the results showed that at certain concentrations there was no development in more than 50% of the individuals under study.

1.1.4 Lead (Pb)

Pb can be in the environment in particulate form or formed into lead compounds; it can be generated as a result of human activities such as oil combustion, industrial processes and solid waste combustion; there are no natural sources of lead, its presence in the environment is anthropogenic [25]. It has been reported that in humans this metal can cause alterations of the nervous system, kidney problems and is related to the development of cancer. In exposed fish, it has been shown to decrease red and white blood cells and decrease hemoglobin levels [18]. The process of Pb accumulation in fish tissues causes oxidative stress; thus, this stress induces synaptic damage and neurotransmitter malfunction and influences immune responses [31].

1.1.5 Nickel (Ni)

Ni is a non-essential and toxic metal whose main source of exposure is food, highlighting fish and vegetables that are treated with wastewater. Its introduction to the aquatic environment is anthropogenic. The effects that it causes in different organisms were studied by Martin et al., [32] in embryos of Pacific oyster (*Crassostrea gigas*), embryos of laurel mussel (*Mytilus edulis Linnaeus*) and larvae of Dungeness crab (*Cancer magister Dana*) exposed to ten metals among them Ni; the effects caused in these species are the abnormal development in more than 50% of the studied individuals. In fish such as *Colisa fasciatius*, a freshwater teleoste, exposed to 45 ppm nickel sulfate, the adverse effects observed were leukopenia due to reduced numbers of lymphocytes and polycythemia, as well as a considerable delay in the rate of erythrocyte sedimentation of dying fish [33].

The conditions of the aquatic environment have a great influence on the transport and mobility of metals such as Ni, so Tamzin et al., [34] carried out their studies in saline waters, hoping that these conditions would decrease the impact on marine biota; however, despite the speciation of the metal in these saline environments it was determined that the physiology of the organisms is the main factor in the toxic impact, finding deterioration as inhibition of breathing and promotion of oxidative stress. In other studies, the mortality rate of African catfish, *Clarias gariepinus*, showed a linear trend with increasing concentration; the researchers concluded that the depression observed in hematocrit, hemoglobin and erythrocyte decreases in this hematological study can be used as an indicator of Ni-related stress in fish [35].

2. Methodology

2.1 Study area

Terminos Lagoon is the largest lagoon-estuarine ecosystem in Mexico by area and volume. The water body and immediately surrounding shorelands are fully incorporated into a National Flora and Fauna Reserve comprised of 705,016 ha of open water and associated wetlands and upland. Terminos Lagoon consists of about 200,108 ha of open water including associated lagoons and channels, with an average depth of 4 m, surrounded by about 259,000 ha of mangrove and cattail marsh. Of the surrounding 180,000 ha of land that is in some productive use, 90% is cattle ranching, 6% is agricultural, and 4% is urbanized, principally the City of Carmen. It is separated from the Gulf of Mexico by the Carmen Island, a 37 km long, 4 km wide barrier island with two mouths, of 3.2 and 3.8 km located to the east and west, respectively.

Terminos Lagoon was declared as a Federal Flora and Fauna Protection Zone in 1994 and is considered a "critical habitat" by the Mexican Environmental Agency [http://www.paot.org.mx/centro/ine-semarnat/anp/AN19.pdf] due to its importance as a refuge for marine species, mangrove forests, sea grass and, associated fluvial - lagoon delta system. Anthropogenic pressure mainly due to urban settlements, the disposal of wastewater in the lagoon and industrial activity based on the drilling and exploration of hydrocarbons, all these activities have been identified as a continuous threat to the quality of the ecosystems within the Terminos Lagoon.

This work summarizes the results of several investigations carried out in Terminos Lagoon Natural Protected Area where the content of heavy metals in a variety of aquatic organisms was analyzed. The sampling periods and collect sites are shown below, as well as the aquatic organisms used for the determination of heavy metals.

In 2009, during two sampling campaigns (rainy and dry seasons), the oyster (*Crassostrea virginica*) was collected at the mouth of three of the rivers that flow into the Terminos Lagoon: the Palizada River, the Chumpan River and the Candelaria River. At each site, three sampling points were established and 100 organisms were obtained from each one.

In 2013–2014 three different types of organisms were analyzed: the oyster (*Crassostrea virginica*) collected in two sites, Estero Pargo and Mouth of Atasta; shrimp (*Litopenaeus setiferus*) obtained by trawling in depths of less than 5 fathoms in the Terminos Lagoon; and the crab (*Callinectes sapidus*) collected at the mouth of the Palizada River. All the organisms were donated by the fishermen's cooperatives. 60 organisms of commercial size were obtained of each species.

In 2014, samples of three species of macrophytes (*Cyperus ligularis L.*, *Lemna minor* and *Typha domingensis*) were collected and analyzed in the "Arroyo La Caleta", which is a natural water channel parallel to the coast that crosses Carmen City, with a variable extension between both banks. The main contribution of water enters through the west mouth of the Terminos Lagoon and does not present an outlet. Other contributions of water come from land and urban drainage. The system is 7.5 km long. In the case of *T. domingensis* and *C. ligularis L.*, the complete plants were cut, stored in plastic bags, and placed in refrigeration for later analysis in the laboratory. The samples of *L. minor* were collected in plastic bags in which water from the "Arroyo La Caleta" was left and, like the previous macrophytes, they were stored in refrigeration.

In 2017, the clam *Rangia cuneata* was collected at four sampling points in the Atasta Lagoon, which is a lagoon that empties into Terminos Lagoon. In total, eight composed samples were analyzed for this study. In the same lagoon (Atasta) but in 2018, during two sampling campaigns (rainy and dry seasons), catfish (*Ariopsis felis*) was obtained by fishing with cast nets eight. 30 composed samples were analyzed.

Table 1 summarizes information about of the analyzed organisms, their sampling location and year of collection, while **Figure 1** shows the Terminos Lagoon Natural Protected Area and the sampling locations of the organisms analyzed.

Species analyzed	Sampling location	Year of collection
Oyster (*Crassostrea virginica*)	Mouth of the Palizada[1], Chumpán[2] and Candelaria[3] rivers that flow into the Terminos Lagoon.	2009
Oyster (*Crassostrea virginica)*	Pargo Estuary[4] and Atasta Mouth[5]	2013–2014
Shrimp (*Litopenaeus setiferus*)	Terminos Lagoon	2013–2014
Crab (*Callinectes sapidus*)	Mouth of the Palizada River[1]	2013–2014
Macrophytes (*Cyperus ligularis L., Lemna minor* and *Typha domingensis*)	Arroyo La Caleta[6]	2014
Clam *Rangia cuneata*	Atasta Lagoon[7]	2017
Catfish (*Ariopsis felis*)	Atasta Lagoon[7]	2018

*Superscripts indicates their location on the map (**Figure 1**).*

Table 1.
Organisms analyzed, their sampling location and year of collection.

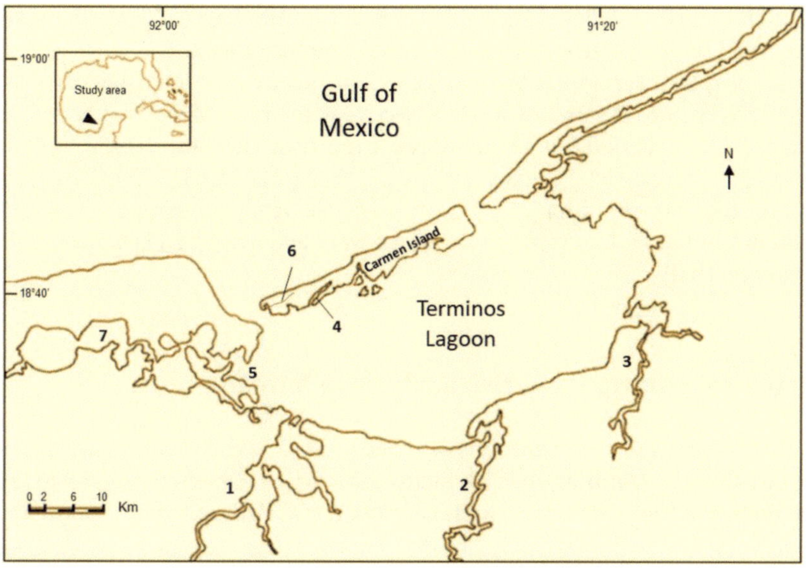

Figure 1.
Terminos Lagoon natural protected area and sampling locations of the organisms analyzed: 1) Palizada River; 2) Chumpan River; 3) Candelaria River; 4) Pargo estuary; 5) Atasta mouth; 6) arroyo La Caleta; 7) Atasta lagoon.

2.2 Sample processing and analysis

The methods used for the processing of tissues and extraction of heavy metals reported in the various studies considered for the evaluation in this work, have few variations or modifications according to "Official Mexican Standard" (NOM-117-SSA1–1994, *Test method for the determination of cadmium, arsenic, lead, copper, iron, zinc and mercury in food*) for food analysis that generally consisted of an acidic digestion of the tissues with a repetitive addition of concentrated HNO_3 and H_2O_2.

For bivalves: before extracting the tissues, they were purged during a period of 24 hours in a system with a controlled salinity of 20 psu. By so doing, the bivalves eliminated all the organic matter from their intestines that could have interfered with the results. Finally, they were shucked manually. Organisms were dried through the process of lyophilization for 24 hours and then were homogenized. Subsequently, an acid digestion was carried out, according to the official Mexican standards, as mentioned at the beginning of this section.

Bivalves samples were analyzed by Plasma Emission Spectroscopy (ICP), Perkin Elmer model 400 instrument was used, and standard solutions (J. Baker). For the evaluation of the analytical quality, the samples of oyster tissues were treated in duplicate and were analyzed in parallel with the standard certificates of "Standard reference materials oyster tissue" (SRM-1566b), with a recuperation percentage of between 84 and 94%.

For crustaceans and fish: composite samples were used for which the edible part was extracted from the organisms of each species, the tissues were homogenized with a food processor and a final sample of (20 ± 0.001 g) was taken. The digestion was carried out by adding 10 mL of HNO_3 to the tissues and placed on a heating grill at a controlled temperature, after the total destruction of organic matter, 2 mL of H_2O_2 was added to each sample in 30% solution, concentrating them up to a volume of 1 mL, finally, the concentrate was filtered through Whatman No. filter paper 32, measuring to a final volume of 20 mL for subsequent analysis. The tissue samples were analyzed in an atomic flame absorption equipment adapted with a Thermo-Scientific brand graphite furnace.

For macrophytes: samples were dried in a drying oven at a temperature of 65° C for 96 hours. The dried samples were dissected at the root, stem, and leaf. The digestion was carried out as mentioned for crustaceans and fish, following the methodology of the official Mexican standard NOM-117-SSA1–1994 (Test method for the determination of cadmium, arsenic, lead, tin, copper, iron, zinc and mercury in food). Macrophytes samples were analyzed by Plasma Emission Spectroscopy (ICP).

3. Metal concentration and risk analysis to human health

Table 2 shows the concentrations of metals found in different organisms, which are the basis for determining the risk factor analysis; as expected, organisms such as bivalve mollusks from the Candelaria, Chumpan and Palizada rivers, show the

Organism	Heavy metal levels in different organisms in μg g^{-1}					Location	Reference
	Cu	Hg	Cd	Ni	Pb		
Crassotrea virginica	56.630 (43.320–56.630)	ND	0.038 (0.020–0.038)	ND	0.137 (0.043–0.137)	Terminos Lagoon	Aguilar et al., 2014
Crassotrea virginica	60.5 (33.24–60.5)	1.1 (0.05–1.1)	2.4 (0.243–2.4)	ND	ND	Candelaria River	Aguilar et al., 2012
Crassotrea virginica	90.8 (30.50–90.8)	0.7 (0.01–0.7)	3.2 (0.23–3.2)	ND	ND	Chumpan River	Aguilar et al., 2012
Crassotrea virginica	176.5 (90.23–176.5)	0.5 (0.02–0.5)	3.0 (1.2–3.0)	ND	ND	Palizada River	Aguilar et al., 2012

Organism	Heavy metal levels in different organisms in μg g^{-1}					Location	Reference
	Cu	Hg	Cd	Ni	Pb		
Callinectes sapidus	57.800 (34.680–57.800)	ND	0.0687 0.0398–0.0687	ND	0.4253 (0.2644–0.4253	Terminos Lagoon	Aguilar Et al., 2014
Litopenaes setiferus	58.470 (41.620–58.470)	ND	0.015 (*ND-0.015)	ND	ND	Terminos Lagoon	Aguilar et al., 2014
Ariopsis felis	—	0.02934 (0.00159–0.02934)	1.2864 (0.00454–1.2864)	41.77 (0.33–41.77)	3.2097 (0.033–3.2097)	Atasta Lagoon	This Study
Rangea cuneata	308.2135 (6.1609–308.2135)	ND	0.74827 (0.2905–0.7482)	30.2055 (13.7574–30.2055)	ND	Atasta Lagoon	This Study

Table 2.
Heavy metals determined in different marine species.

National and international legislation governing heavy metal levels in fish in μg g^{-1}						
	Cu	Hg	Cd	Ni	Pb	Year
JECFA[1]	—	0.5	—	—	0.5	1989
WHO[2]	20	0.005	2	—	2	1996
USFDA[3]	—	—	—	70-80	—	1993
NOM 242[4]	—	1	0.5	—	1	2009
FAO[5]	30	0.5	0.5	—	0.5	1983
National and international legislation governing the levels of heavy metals in marine mollusks and crustaceans in μg g^{-1}						
	Cu	Hg	Cd	Ni	Pb	Year
NOM 242[6]	—	1	0.5	—	1	2009
USFDA[7]	—	—	4.0	—	1.7	1993
NAUEN[8]	32.5	—	—	—	—	1983
ISSC[9]	—	—	—	80	—	2007

[1]*JECFA, 1989 Evaluation of Certain Food Additives and Contaminants (Thirty-third Report of the Joint FAO/WHO Expert Committee on Food Additives) [meeting held in Geneva from March 21 to 30, 1988]. World Health Organization.*
[2]*WHO, 1996. Health criteria other supporting information. In: Guidelines for Drinking Water Quality p. 31–388.*
[3]*USFDA, 1993. Food and Drug Administration, Guidance for Nickel in Shellfish. DHHS/PHS/FDA/CFSAN/Office of Seafood, Washington DC, 1993.*
[4]*NOM-242-SSA-2009. Products and services. Fresh, refrigerated, frozen and processed fishery products. Health specifications and test methods.*
[5]*FAO,1983. Compilation of legal limits for hazardous substances in fish and fishery products Fish Circular 464:5–100.*
[6]*NOM-242-SSA1–2009 Products and services. Fresh, refrigerated, frozen and processed fishery products. Health specifications and test methods.*
[7]*USFDA, 1993. Guidance document for lead in shellfish. Center for Food Safety and Applied Nutrition. United States Food and Drug Administration, Washington, D. C.*
[8]*Nauen C.E, 1983. Compilation of legal limist for hazardous substances in fish and fishery products. FAO fisheries circular 764. United Nations Food and Agriculture Organization. Rome, Italy, 102 pp.*
[9]*ISSC,2007. National shellfish sanitation program. Guide for the control of molluscan shellfish. Interstate Shellfish Sanitation Conference. U.S. Food and Drug Administration, Department of Health and Human Services. Washington, D.C., 549 pp.*

Table 3.
Permissible levels of heavy metals in fishery products.

highest values in Cd, exceeding the limits established by the official Mexican Standards (NOM 242 and the USFDA standards **Table 3**). Likewise, Cu levels are high in clams *Rangea cuneata*, exceeding international specifications (**Table 3**); the Pb levels in *Ariopsis felis* exceed all the international specifications contemplated in this study. The reference values reported in national and international standards for heavy metals in marine fish and mollusks are shown in **Table 3**.

Concentrations of heavy metals in the macrophytes collected in "Arroyo La Caleta", where detected below the limit of quantification of the method (not detected ND) for B, Be, Fe, Mn and Si to 4671.29 μg g^{-1} for Fe. The average concentration of the metals analyzed presented the following order: $Be < As < V < Mn < B < Si < Fe$. Of the three species of macrophytes studied, the one that presented higher concentrations of metals was *Cyperus ligularis L.* accumulating higher percentages of Be, Fe, Mn, Si and V, followed by *Typha domingensis* which presented the highest concentration of As and average concentrations of the other metals, finally, *Lemna minor*. Regarding the structure analyzed in plants, the highest concentrations were found in the roots and to a lesser extent in the stem and leaves, which tells us about the ability to translocate (mobilize) the metals to the aerial parts as a strategy of adaptation to pollution due to heavy metals. In general, the highest concentrations found in macrophytes were related to the sites of highest anthropogenic activity [36]. It should be noted that, of all the collection sites in the Términos Lagoon Natural Protected Area analyzed in this study, the "Arroyo La Caleta" is considered one of the most contaminated because it receives wastewater from Carmen City, which is why the fish products obtained there have not been considered suitable for human consumption for several years. Due to the above, the study of the behavior of heavy metals in this location, was based on organisms with high levels of adaptation to heavy metal pollution that are not used for human consumption.

4. Estimation of the estimated daily intake, target hazard quotient, hazard index and target cancer risk

According to the USEPA [37–38] the estimate of potential risk to human health from the consumption of contaminated marine products is based on the estimated daily intake (EDI), the target hazard quotient (THQ), hazard index (HI) and target cancer risk (TR). The first parameter that was estimated is a function of the relation (EDI; μg kg^{-1} $week^{-1}$)

$$EDI = \frac{(FIR)(MC)}{BWa} \tag{1}$$

Where FIR is the food ingestion rate of fish or shellfish consumed by an adult; according to CONAPESCA, these data for Mexico, is up to 12 kg $year^{-1}$ or the equivalent of 230 g $week^{-1}$ for an adult with an average weight. For children aged 4 to 6 years, the recommended food ingestion rate is 100 grams per week (CONAPESCA: National Commission for Aquaculture and Fisheries, Fishery production statistics, www.conapesca.sagarpa.mx) or its equivalent of 5.214 kg $year^{-1}$.

According to studies by Araneda [39] the population group of adolescents is the one that shows a lower food ingestion rate of these foods; there is no data on intake in Mexico, but the recommendations indicate that an average adolescent between 14 and 17 years of age should consume between 240 and 300 grams of fish and seafood per week, the equivalent of an average of 15. 64 kg $year^{-1}$. Due to the scarcity of information in this population group, in this study it is considered that the average food ingestion rate for adolescents is 7.8 kg $year^{-1}$, which is considered a low consumption and that represents a value of 150 gr $week^{-1}$.

The parameter BWa is the reference body weight of an adult. In countries such as China, this data is 55 kg [40]; the average weight of a Mexican adult is 70 kg. The average weight for children between 4 and 6 years old is 16 kg [41] and for an average adolescent between 14 and 17 years old the weight considered is 54 kg. All the reference data are based on the characteristics and habits of the Mexican population without obesity problems. The MC parameter is the metal concentration (Cu, Pb, Ni, Hg) expressed in μg g^{-1}.

The estimated results for the EDI parameter are shown in **Table 4**. With the data from the population intake rates, we can estimate the THQ parameter which is a dimensionless amount and a relationship between the concentration of heavy metals in ingested food with other factors. According to USEPA [37–38] the THQ value should not exceed the numerical value of 1. Estimated values below 1 indicate that the contaminant levels do not cause adverse effects or potential non-carcinogenic risks in exposed persons during the estimated average life span of the Mexican population of 70 years.

The model for estimating the target hazard quotient (THQ) is determined by Eq. 2. The units were adequate for not using correction factors (**Table 5**).

$$THQ = \frac{(Efr)(EDtot)(FIR)(MC)}{(Rfd)(BWa)(Atn)} \tag{2}$$

Where Efr is the exposure frequency to the trace element, (365 days $year^{-1}$), EDtot is the exposure duration (average life span of 70 years), FIR is the food ingestion rate in grams per day for the respective food item (g day^{-1}), MC is the concentration of the trace element in the given food item (μg g^{-1}), Rfd: is the oral reference dose of the trace element (μg g^{-1} day^{-1}) (5×10^{-4} for Hg; 1×10^{-3} for Cd; 4×10^{-3} for Pb; 2×10^{-2} for Ni and 4×10^{-2} for Cu), BWa is the reference body weight (g), Atn is the averaged exposure time (Efr x EDtot).

The values obtained show a variable trend among the reference population groups, with children aged 4–6 years being those with the highest values of EDI for Cu and Ni; these values are directly related to weight.

4.1 Hazard index (HI)

The accumulated risk was evaluated by the individual sum of each of the THQ factors which represent the risk index (HI), which is shown in Eq. 3. These values, like THQ, must not exceed the numerical value of 1, otherwise it could indicate that there are considerable risk factors for the consumption of marine products reported in this study.

$$\mathrm{HI} = \sum \mathrm{THQ} \tag{3}$$

According to the estimated results (**Table 6**) no value calculated for THQ and HI exceed the parameters established to consider a risk to health by the intake of these contaminants from the consumption of fish and seafood.

The results indicate that the estimated HI values do not represent a risk for the reference population, since they do not exceed the comparison value of 1.

	Estimated daily intake EDI (μg kg^{-1} week^{-1})														
	Cu			Cd			Pb			Ni			Hg		
Organism	Adults	Children (4–6 years)	Teenagers (14–17 years)	Adults	Children (4–6 years)	Teenagers (14–17 years)	Adults	Children (4–6 years)	Teenagers (14–17 years)	Adults	Children (4–6 years)	Teenagers (14–17 years)	Adults	Children (4–6 years)	Teenagers (14–17 years)
Crassotrea virginica	186.070	353.937	156.256	0.1248	0.2375	0.10485	0.4501428	0.85625	0.37801	—	—	—	—	—	—
Crassotrea virginica	198.785	378.125	166.935	7.8857	15	6.622	—	—	—	—	—	—	3.6142	6.875	3.0351
Crassotrea virginica	298.342	567.500	250.540	10.514	20	8.8296	—	—	—	—	—	—	2.3	4.375	1.93148
Crassotrea virginica	579.928	1103.125	487.009	9.857	18.75	8.277	—	—	—	—	—	—	1.64285	3.125	1.3796
Callinectes sapidus	189.914	361.250	159.485	0.226	0.42937	0.189561	1.39741	2.6581	1.1735	—	—	—	—	—	—
Litopenaes setiferus	192.115	365.437	161.333	0.049	0.09375	0.041388	—	—	—	—	—	—	—	—	—
Ariopsis felis	143.171	313.180		4.226	1.878.04	3.54951	1.045	2.285	8.8563	137.244	261.0625	115.2542	0.096402	0.1833	0.08095
Rangea cuneata	1012.700	1926.33	850.440	2.4601	4.675	2.06392	—	—	—	99.246	188.78	83.344	—	—	—

Table 4.
Estimated daily intake (EDI), taking as a reference a population group between adults and children of different age ranges.

	Target hazard quotient (THQ)														
	Cu			Cd			Pb			Ni			Hg		
Organism	**Adults**	**Children (4–6 years)**	**Teenagers (14–17 years)**	**Adults**	**Children (4–6 years)**	**Teenagers (14–17 years)**	**Adults**	**Children (4–6 years)**	**Teenagers (14–17 years)**	**Adults**	**Children (4–6 years)**	**Teenagers (14–17 years)**	**Adults**	**Children (4–6 years)**	**Teenagers (14–17 years)**
Crassotrea virginica	0.1056	0.0884	0.0262	0.0005	0.0024	0.0007	0.0005	0.0022	0.0007	—	—	—	—	—	—
Crassotrea virginica	0.1127	0.0945	0.0280	0.0342	0.1500	0.0444	—	—	—	—	—	—	0.0291	0.1273	0.0377
Crassotrea virginica	0.1690	0.1418	0.0424	0.0457	0.2000	0.0593	—	—	—	—	—	—	0.0185	0.0820	0.0240
Crassotrea virginica	0.3287	0.2757	0.0817	0.0428	0.1875	0.0556	—	—	—	—	—	—	0.0132	0.0580	0.0172
Callinectes sapidus	0.1077	0.0903	0.0268	0.0009	0.0043	0.0013	0.0015	0.0067	0.0020	—	—	—	—	—	—
Litopenaes setiferus	0.1088	0.0913	0.0270	0.0002	0.0009	0.0003	—	—	—	—	—	—	—	—	—
Ariopsis felis	—	—	—	0.0184	0.0804	0.0239	0.0115	0.0502	0.0149	0.0299	0.1305	0.0387	0.0008	0.0034	0.0010
Rangea cuneata	0.5739	0.4815	0.1427	0.0107	0.0467	0.0139	—	—	—	0.02160	0.0944	0.0280	—	—	—

Table 5.
Estimated values of the target hazard quotient (THQ) in a reference population group.

	Hazard index values (HI)		
Organism	Adults	Children (4–6 years)	Teenagers (14–17 years)
Crassotrea virginica	0.1065	0.0930	0.0276
Crassotrea virginica	0.1467	0.2445	0.0724
Crassotrea virginica	0.2148	0.3419	0.1013
Crassotrea virginica	0.3715	0.4632	0.1372
Callinectes sapidus	0.1101	0.1013	0.0300
Litopenaes setiferus	0.1091	0.0923	0.0273
Ariopsis felis	0.0298	0.1306	0.0784
Rangea cuneata	0.5846	0.5283	0.1567

Table 6.
Estimated values of the hazard index estimated for a reference population, expressed as the sum of all the individual HI factors.

4.2 Target cancer risk

The risk from carcinogens was indicated in this study by the TR values (**Table 7**). For the estimation of these parameters, the values of Region III were taken, where the population of Mexico is included according to USEPA criteria [37–38]. The equation that represents risk for carcinogenic factors is represented by the following expression:

$$TR = \frac{(Efr)(EDI)(FIR)(MC)(Cfo)}{(BWa)(Atn)} \quad (4)$$

Where TR is the target cancer risk (dimensionless factor) and Cfo is the oral cancer slope factor; USEPA criteria [42] (μg g^{-1} bw day^{-1}). The other values are the same used for the estimation of EDI and THQ. In this study, the Cfo values used to estimate TR are 8.5×10^{-3} for Pb and 1.7 for Ni, a metal that is on the list of potent carcinogens [42]. The value of 2.59×10^{-4} was used for Cd which has been considered carcinogenic according to the International Agency for Research On Cancer (IARRC: http://www.iarc.fr/en/websites/index.php). Not all metals are considered within this classification so far; only Cd, Pb and Ni fall into this category. According to the New York State Department of Health [43] (NYSDOH) the categories of the target cancer risk (TR) are the following: TR $\leq 10^{-6}$ low risk, between 10^{-4} and 10^{-3} moderate risk, from 10^{-3} to 10^{-1} high risk and TR $\geq 10^{-1}$ very high risk that people may develop cancer at some point in their life after exposure to the metal.

The results of the target cancer risk for Cd and Pb show that the three study categories (adults, children, and adolescents) present a low risk of developing cancer from the ingestion of fish and shellfish. On the other hand, the target cancer risk calculated for Ni shows that the population group of children aged 4–6 years represents a very high risk, and a moderate risk is expected for adults and adolescents.

In certain cases, it is not advisable to limit the consumption of these marine products. It is more useful to be vigilant, as well as to limit the frequency of consumption when there is evidence of risk for the population. These actions make the difference in developed countries that have public policies and develop research

Target cancer risk (TR)									
	Cd			Pb			Ni		
Organism	**Adults**	**Children (4–6 years)**	**Teenagers (14–17 years)**	**Adults**	**Children (4–6 years)**	**Teenagers (14–17 years)**	**Adults**	**Children (4–6 years)**	**Teenagers (14–17 years)**
Crassotrea virginica	5.768E-11	2.087E-10	4.10308E-11	2.460E-08	8.920E-08	1.637E-07			
Crassotrea virginica	2.300E-07	8.325E-07	1.63668E-07	—	—	—	—	—	—
Crassotrea virginica	4.090E-07	0.004	2.90966E-07	—	—	—	—	—	—
Crassotrea virginica	3.595E-07	1.307E-06	2.55732E-07	—	—	—	—	—	—
Callinectes sapidus	1.886E-10	6.821E-10	2.37122E-07	2.371E-07	8.596E-07	1.686E-07	—	—	—
Litopenaes setiferus	8.987E-12	3.251E-11	1.34108E-10	—	—	—	—	—	—
Ariopsis felis	6.610E-08	2.391E-07	6.3933E-12	2.087E-10	5.577E-06	8.988E-05	0.458	1.658	0.325
Rangea cuneata	2.234E-08	8.086E-08	4.70213E-08	—	—	—	0.3308	1.867	0.158

Table 7.
Estimated values of the target cancer risk (TR) in a reference population.

through environmental agencies to develop models that can be applicable to different regions of the world and thus predict or estimate possible risks.

There are general recommendations in Mexico regarding the consumption of fish and seafood by children under four years of age. Certain countries such as Canada restrict consumption of species caught in rivers and lakes and recommend that consumption in the population group of children 1–4 years old be only 75 g $month^{-1}$ and in children 5–11 years old be 125 g $month^{-1}$, as well as that pregnant women should not consume more than 150 g $month^{-1}$ [44]. Some of these recommendations are in most cases based on economic interests.

Bellanger et al. [44], analyze in their studies the economic implications of exposing a population group to the toxic effects of heavy metals. In Mexico, consumption of fish and seafood is lower than that of other foods. In Mexico, fish products are governed by Mexican standards that limit the presence of heavy metals in their products; likewise, government institutions regulate the health of oyster and clam banks (due to the presence of pathogenic microorganisms), but there are no effective public policies focused on protecting the environment, stopping the deterioration of mangrove areas or monitoring and sanctioning poaching and depredation that are putting numerous species at risk.

5. Conclusions

Throughout this chapter, aspects of the region of the Campeche Sound were shown, and in particular the Terminos Lagoon, which only a decade ago produced more than 80% of the national production of crude oil, while at the same time numerous marine species of high commercial value were extracted from its waters.

Today, even though oil activity has decreased considerably, the effects of this industry, combined with population growth and the ineffectiveness of monitoring programs, still persist. The results of this study of the concentrations of heavy metals in *Crassostrea virginica*, *Rangea cuneata* and *Ariopsis felis* indicate that some of the values found are higher than those established in international and national legislation, so these fish products should not be consumed. The establishment of a monitoring program is suggested to identify the variations and conditions that favor the bioaccumulation process in exposed organisms.

Regarding the risk analysis carried out in this study, the values calculated for the target hazard quotient (THQ) and the hazard index (HI) indicate that the consumption of the studied species does not represent a risk for human health in any of the considered age groups; however, in relation to the TR that evaluates the potential risk for carcinogens, the results show worrying values, especially for the organisms that come from the Pom-Atasta lagoon system. In the two species evaluated (*Rangea cuneata* and *Ariopsis felis*), the TR values are considered "high risk" and "moderate risk", especially in the most vulnerable population group, children. For this reason, it is not recommended the consumption of these species by children under 4 years old and it is suggested to decrease their consumption in the adult and adolescent age groups. These actions are not intended to stigmatize the consumption of these products, but to have greater control and surveillance, especially in population groups of greater vulnerability.

Conflict of interest

The authors declare no conflict of interest.

Author details

Claudia Alejandra Aguilar[1*], Yunuen Canedo[1], Carlos Montalvo[1], Alejandro Ruiz[1] and Rocio Barreto[2]

1 Faculty of Chemical Sciences, Autonomous University of Carmen, Campeche, Mexico

2 Faculty of Natural Sciences, Autonomous University of Carmen, Campeche, México

*Address all correspondence to: caguilar@pampano.unacar.mx

References

[1] Cuellar J, Arreguin F, Hernández S, Lluch D. Impacto ecológico de la industria petrolera en la sonda de Campeche, México, tras tres décadas de actividad: una revisión. Interciencia; 2004. p. 310–319.

[2] Ramírez M. La pesquería de camarón en Campeche: desarrollo histórico y perspectiva. Ciencia Pesquera. 2015; 23 VOLUMEN 1 73–87. https://www.gob.mx/cms/uploads/attachment/file/123330/CP_23-1_completa..pdf

[3] Ramos J., Villalobos J, Grado de conservación del ecosistema de mangle en la laguna de Términos, Campeche. In: editors. Aspectos socioambientales de la región de la laguna de Términos, Campeche. Propuesta de políticas ambientales y acciones de restauración. Universidad Autónoma de Campeche. A 2015. p. 117–222.

[4] Wakida A, Nuñez G. Análisis de la pesquería del camarón siete barbas Xiphopenaeus kroyeri en Campeche, México. In:Memorias del III foro del camarón del golfo de México y mar caribe. 2003. p. 29–32.

[5] González Macías M. C. Análisis de la comunidad bentónica en una chapopotera del Golfo de México y sus relaciones con la presencia crónica de hidrocarburos del petróleo. [thesis]. México: UNAM.; 1997.

[6] Soto L. A, García A, Botello A. V. Study of the penaeid shrimp population in relation to petroleum hydrocarbons in Campeche Bank. In: Proc. 33rd Ann. Meet. Car. Fish. Inst. 1982. p. 81100.

[7] Botello A. V. Vigilancia de los hidrocarburos fósiles en sistemas costeros del Golfo de México y áreas adyacentes. I. Sonda de Campeche. In: *An. Inst. Cienc. Mar Limnol*. México. UNAM; 1987 *p*. 45–52.

[8] GoldB.G, Zapata-Pérez O, Noreña-Barroso E, Herrera-Rodríguez M, Ceja-Moreno V, Zavala-Coral M. Oil Pollution in the Southern Gulf of México. En Kumpf H, Steidinger K, Sherman K (Eds.) *The Gulf of Mexico Large Marine Ecosystem. Assessment, Sustainability, and Management*. Blackwell. Masachussets, EEUU. 1999. pp. 372–382.

[9] Vázquez-Botello A, Ponce-Vélez G, Díaz-González G Hidrocarburos aromático-policíclicos (PAH's) en áreas costeras del Golfo de México. *Hidrobiológica. 3*. (1994) 1–15. Available in: https://hidrobiologica.izt.uam.mx/index.php/revHidro/article/view/541.

[10] PEMEX. *Anuario Estadístico 2003.* Exploración y Producción. PEMEX. México; 200364 p.

[11] Vázquez F, Alexander H, Fraustro A. Metales pesados (Cadmio, cromo, níquel y vanadio) adsorbidos en sedimentos de la Sonda de Campeche campañaa oceanográfica SGM-9. Actas INAGEQ; 2006. 12 p.

[12] Aguilar C, Montalvo C, Rodríguez L, Cerón J, Cerón R. American oyster (Crassostrea virginica) and sediments as a coastal zone pollution monitor by heavy metals. International Journal of Environmental Science and Technology. DOI: 10.1007/s13762-012-0078-y

[13] Aguilar C, Montalvo C, Cerón J, Anguebes F. Niveles de Metales pesados en especies marinas: Ostión (Crassostrea virginica), Jaiba (Callinectes sapidus) y Camarón (Litopenaeus setiferus), de Ciudad del Carmen, Campeche, México. Revista latinoamericana de recursos naturales. 2014; 10: 9–17 p.-. Available in: http://revista.itson.edu.mx/index.php/rlrn/article/view/227

[14] Montalvo C, Aguilar C, Amador L, Cerón R, Cerón J, Anguebes F,

Córdova A. Metal Contents in Sediments (Cd, Cu, Mg, Fe, Mn) as Indicators of Pollution of Palizada River, Mexico. Environment and pollution. DOI: 10.5539/ep.v3n4p89.

[15] Canedo Y, Ruiz A, ChiA C. Diagnóstico de la contaminación por metales pesados en sedimento superficial de la laguna de Términos, Campeche: una aproximación estadística. Revista AIDIS. DOI : 10.22201/iingen.0718378xe.2014.7.2

[16] Hartwell I, Wright A, Hocutt H. Relative respiration and feeding rates ofv oyster and brackish water clam in variously contaminated waters. Mar. Poll. Bull. DOI: 10.1007/s00227-018-3351-x.

[17] Sarah R, Tabassum B, Idress N, Hashem A, Elsayed A. Bioaccumulation of heavy metals in *Channa punctatus* (Bloch) in river Ramganga (U.P.) India Saudi. J. Biol. Sci. DOI:https://doi.org/10.1016/j.sjbs.2019.02.009

[18] Dezfuly T, Aramoon A, Alishahi M, Halimi M, Rahnama R. The Effects of Lead Toxicity on the Hematological Parameters Common Carp (*Cyprinus carpio*) at Varying Salinity Levels. Iranian Journal of Toxicology. (2020)14, 1–8 p. Available in: http://ijt.arakmu.ac.ir/article-1-755-en.html

[19] Adimalla N, Wang H.Distribution, contamination, and health risk assessment of heavy metals in surface soils from northern Telangana India. Arabian J. Geosci. DOI: https://doi.org/10.1007/s12517-018-4028-y

[20] Ahmed S, Khurshid S, Qureshi F, Hussain A, Bhattacharya A. Heavy metals and geoaccumulation index development for groundwater of Mathura city, Uttar Pradesh. Desal. Water Treat. J. DOI: doi: 10.5004/dwt.2019.23322.

[21] Pandey G, Madhuri S, Shrivastav B. Contamination of mercury in fish and its toxicity to both fish and humans an overview. Int. Res. J. Pharm. (2012) 3, 44–47

[22] Bryan W, Langston J. Bioavailability, accumulation and effects of heavy metals in sediments with special reference to United Kingdom estuaries: a review. Environmental Pollution. DOI: https://doi.org/10.1016/0269-7491(92)90099-V.

[23] Corrales M. Acumulación de metales pesados en bivalvos y sus efectos tóxicos en la salud humana: Perspectivas para el estudio en Costa Rica. Ciencias naturales. Available in: http://hdl.handle.net/10669/21396

[24] Klaassen C. D, Casarett L. J, Watkins J. B, Doull J. Manual de toxicología: Casarett & Doull : la ciencia básica de los tóxicos. McGraw-Hill Interamericana; 2001. Retrieved from https://books.google.co.cr/books?id=oQJDAQAACAAJ.

[25] Scoullos J, Vonkeman H, Thornton I, Makuch, Z. Handbook for sustainable heavy metals policy and regulation. Springer science business media; 2001. DOI: https://doi.org/10.1007/978-94-010-0403-9_3.

[26] McRae K, Gaw S, Glover N. Effects of waterborne cadmium on metabolic rate, oxidative stress, and ion regulation in the freshwater fish, inanga (*Galaxias maculatus*). Aquatic Toxicol. DOI: https://doi.org/10.1016/j.aquatox.2017.10.027.

[27] Wu H, Xuan R, Li Y, Zhang X, Wang A, Wang L. Effects of cadmium exposure on digestive enzymes, antioxidant enzymes, and lipid peroxidation in the freshwater crab Sinopotamon henanense. Environ Sci Pollut Res. DOI: https://doi.org/10.1007/s11356-012-1362-6.

[28] Guzmán X, Martínez A, Rodríguez L, Gonzalez H, Vázquez A.

Tissue changes of the oyster Crassostrea virginica due to cadmium exposition and depuration. Hidrobiológica.2007; 17: 41–48. Available in: http://www.scielo.org.mx/pdf/hbio/v17s1/v17s1a5.pdf

[29] Padrilah N, Sabullah K, Shukor Y, Ahmad A. Toxicity Effects of Fish Histopathology on Copper Accumulation. Tropical Agriculture Science. (2018) 41, 519–540. Available in: http://www.pertanika.upm.edu.my/Pertanika%20PAPERS/JTAS%20Vol.%2041%20(2)%20May.%202018/02%20JTAS%20Vol%2041%20(2)%20May%202018_JTAS-0965-2016_pg519–540.pdf

[30] Calabrese A, Collire S, Nelson D.. The toxicity of heavy metals to embryos of the American osyters Crassostea virginica. Marine biology. DOI: https://doi.org/10.1007/BF00367984

[31] Wook L, Hoon C, Un-Ki H, Ju-Chan K, Yue Jai K, Kwang I, Jun-Hwan K. Toxic effects of lead exposure on bioaccumulation, oxidative stress, neurotoxicity, and immune responses in fish: A review. Environmental Toxicology and Pharmacology. DOI: 10.1016/j.etap.2019.03.010.

[32] Martin M, Kenneth O, Billig P, Neil G. Toxicities of ten metals to Crassostrea gigas and Mytilus edulis embryos and Cancer magister larvae. Marine Pollution Bulletin. DOI: 10.1016/0025-326X(81)90081-3.

[33] Shanker J, Agrawal J, Srivastra K, Chaundry H. Hematological effects of nickel toxicity in a freshwater teleost, Colisa fasciatus. Acta Pharmacologica et Toxicologica. DOI: 10.1111/j.1600-0773.1979.tb02384.x.

[34] Tamzin A, Leonard E. Mechanisms of nickel toxicity to fish and invertebrates in marine and estuarine waters. Environmental Pollution. DOI: 10.1016/j.envpol.2017.01.028.

[35] Ololade A, Oginni O. Toxic stress and hematological effects of nickel on African catfish, *Clarias gariepinus*, fingerlings. Environmental chemistry and ecotoxicology. DOI: 10.5897/JECE.9000060.

[36] GómezM. Determinación de metales pesados en macrófitas del "Arroyo la Caleta" Cd. del Carmen, Campeche [thesis]. Universidad Autónoma del Carmen. 2016.

[37] USEPA. United States Environmental Protection Agency, Regional Screening Level (RSL) Summary Table: November 2011. [Internet]. 2011. Available from: http://www.epa.gov/regshwmd/risk/human/Index.htm.

[38] USEPA. EPA Region III Risk-Based Concentration (RBC) Table 2008 Region III. 2012. Available from: 1650 Arch Street, Philadelphia, Pennsylvania 19103

[39] Araneda J, González D, Mella V, Pérez K, Quezada G, Pinheiro A. Ingesta de alimentos proteicos en adolescentes de la Ciudad de Chillán, Chile. Revista Chilena de nutrición. DOI: 10.4067/S0717-7518201900030029

[40] Yi Y, Tang C, Yi T, Zhifeng Y, Shanghong Z. Health risk assessment of heavy metals in fish and accumulation patterns in food web in the upper Yangtze River, China. Ecotoxicolol. Environ. Safety. DOI: 10.1016/j.ecoenv.2017.07.022.

[41] Soto M, Amezcua F, González R. Nonessential Metals in Striped Marlin and Indo-Pacific Sailfish in the Southeast Gulf of California, Mexico: Concentration and Assessment of Human Health Risk. Arch. Environ. Contam. Toxicol. DOI: https://doi.org/10.1007/s00244-009-9452-2.

[42] USEPA.Recommended use of BW3/4 as the default method in derivation of

the oral reference dose. EPA/100/R11/001. [Internet]. 2011. Available from: http://www.epa.gov/regshwmd/risk/human/Index.htm.

[43] NYSDOH (New York State Department of Health) Hopewell precision area contamination: appendix C-NYS DOH. Procedure for evaluating potential health risks for contaminants of concern [Internet]. 2007. Available from: http://www.health.ny.gov/environmental/investigations/hopewell/appendc.htm

[44] Bellanger M, Pichery C, Aerts D, Berlund M, Castaño A, Cejchanová M, Crettaz P, Davidson F, Esteban M, Fischer M, Gurzau A, Halzlova K, Katsonary A, Knudsen L, Kolossa M, Koppen G, Ligocka D. Economic benefits of methylmercury health exposure control in Europe: monetary value of neurotoxicity prevention. Environ Health. DOI:10.1186/1476-069X-12-3.

Chapter 5

Consumption Safety in Relation to Bioaccumulation of Heavy Metals in Periwinkles (*Tympanotonus fuscatus*) Obtained from Ogbia in the Niger Delta Region of Nigeria

Miebaka Moslen and Chioma Hope Adiela

Abstract

The study assessed human health risk and accumulation of heavy metals (Cd, Cu, Pb, Ni, Cr and Zn) in periwinkles (*Tympanotonus fuscatus*) obtained from the Niger Delta region of Nigeria. Samples were collected for six months on a monthly basis. The samples were digested according to the method described by Association of official analytical chemists and analyzed using atomic absorption spectrophotometer (AAS). Temporal variations in metal concentrations were observed with values ($mgkg^{-1}$) ranging as follows Pb (2.34–6.7), Ni (0.55–2.28), Zn (0.55–11.66), Cr (0.74–3.65), Cu (1.15–3.91) and Cd (0.22–1.06). Variation in metal concentration was significantly different ($p < 0.05$) with metals such as Pb, Ni and Cd found to be above their respective FAO/WHO permissible limits. The estimated daily intake (EDI) of all metals examined was less than their respective reference oral doses (RFD). The target hazard quotient (THQ) non-carcinogenic and the hazard index (HI) of metals were < 1 while the hazard quotient carcinogenic (HQ) ranged between 10^{-6} – 10^{-4}. The study therefore concluded gradual accumulation of metals and minimal health risk due to consumption of contaminated periwinkles in the study area.

Keywords: heavy metals, bioaccumulation, health risk, periwinkles, Niger delta

1. Introduction

Heavy metal pollution is one of the challenges of coastal waters due to human activities such as oil exploration and exploitation, construction and fabrication of marine boats, disposal of industrial and domestic wastes, sailing and illegal bunkering activities [1]. Heavy metals are generally referred to as those metals which possess a specific density of more than 5 g/cm^3 and adversely affect the environment and living organisms [2]. Nagajyoti et al. [3] and [4]; stated that metals are significant environmental pollutants and their toxicity is a problem of increasing significance for ecological, evolutionary, nutritional and environmental reasons. Heavy metal contamination in the aquatic environment is a major challenge with regards to industrialization in view of the fact that industrial and domestic wastes containing such pollutants are regularly channeled into nearby water bodies [1]. Increase in population has induced urbanization and industrialization with corresponding discharge of wastes such as heavy metals into the environment [5, 6] but Dural et al. [7] stated that heavy metals occur naturally in aquatic ecosystem but deposits of anthropogenic origin increase their levels and create environmental problems in coastal zones and rivers. Aquatic organisms like periwinkles have the ability to accumulate and bio-magnify contaminants like heavy metals, polycyclic aromatic hydrocarbons and PCB in the environment [8]. Researchers had stated that ideal bio-indicators (biomonitors) should satisfy certain conditions such as ability to accumulate pollutants without being killed by the levels encountered in the environment; sedentary in order to be representative of the study area, sufficiently long lived to allow the sampling of more than one year-class, if desired: be of reasonable size, giving adequate tissue for analysis [9, 10]. Mollusks such as *Tympanotonus fuscatus*, a deposit feeder common along the mangrove intertidal coast and estuarine swamps of the Niger Delta also satisfy most these conditions due to wide usage in biomonitoring studies [11]. Moslen [11] reported gradual bio-accumulation of heavy metals in fish (biota) with attendant minimal to moderate health risk concern but recommended regular monitoring in order to detect changes over time. *Tympanotonus fuscatus* is a mollusk (Gastropods) of high commercial and economic value in the Niger Delta region of Nigeria [11]. This study therefore, aim to evaluate consumption safety and heavy metal bioaccumulation in periwinkles (*Tympanotonus fuscatus*) obtained from Ogbia in Niger Delta, Nigeria.

2. Materials and methods

2.1 Sample collection and preparation

Periwinkle (*Tympanotonus fuscatus)* samples were obtained from landings of artisanal fishermen in Ogbia, Niger Delta, Nigeria (**Figure 1**). Samples were collected monthly for six months (February to July 2018). The samples were placed in labeled container, preserved in ice-packs and immediately taken to the laboratory for further analysis. The samples (tissue) were removed from the shell and oven

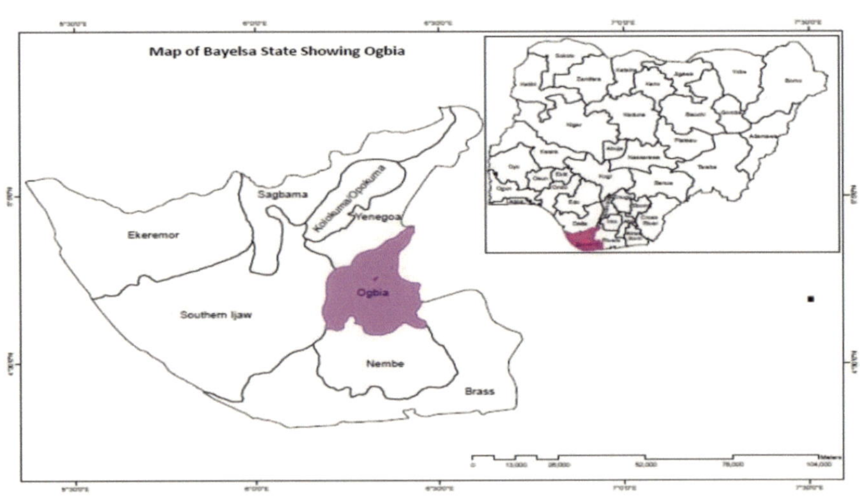

Figure 1.
Map of study area (Ogbia) with an inset map of Nigeria.

dried at 80°C for 24 hours after which, the sample was ground to powder using ceramic mortar and pestle.

2.2 Sample digestion

The homogenized sample was digested according to the method described by the Association of official Analytical Chemists [12]. Two (2 g) of the homogenized tissues was weighed into bottles and transferred into labeled boiling tube in a fume cupboard, 5 ml of 10% HCl acid was added to the sample and stirred. It was then treated with 5 ml of 10% HNO_3, and warmed on a water bath to dissolve. The digested sample was allowed to cool at room temperature and then filtered through 0.45 μm into volumetric flask. The concentrations of heavy metals in the samples were determined with an Atomic Absorption Spectrophotometer (GB Avanta PM AAS, S/N A6600 with detection limit for individual metals of study in the range of 0.001–0.02 $mgkg^{-1}$). The concentrations were blank-corrected and expressed as μgg^{-1} dry weight of sample analyzed.

2.3 Human health risk assessment

Health risk on humans was evaluated to determine possible adverse health effects due to consumption of periwinkles *(Tympanotonus fuscatus)* contaminated with heavy metals. Standard and acceptable indices were used for health risk assessment.

2.4 Exposure assessment (Estimated Daily Intake (EDI))

The Estimated Daily Intake (EDI) via consumption of heavy metal contaminated periwinkle was evaluated using Eq. 1 [13].

$$\text{Estimated Daily Intake}\left(EDI\right)=\frac{EFXEDXFIRXCfXCm}{BWXTA\left(EFXED\right)} \quad (1)$$

where EF = Exposure frequency
ED = Exposure duration
FIR = Fish ingestion rate
CF = conversion factor
Cm = heavy metal concentration in periwinkle (μg/g d-w),
BW = Adult body weight 70 kg [14, 15].
TA (EF × ED) = average exposure time [5, 6].

2.5 Assessment of non-carcinogenic health risks

The target hazard quotient (THQ) was used to evaluate the non-carcinogenic health risks associated with consumption of periwinkles contaminated with heavy metals. THQ was expressed in Eq. 2 [5, 6, 13].

$$\text{Target hazard Quotient}\left(THQ\right)=\frac{EDI}{RFD} \quad (2)$$

where EDI = Estimated Daily Intake.
RFD = Reference Oral Dose of metal.

2.6 Assessment of carcinogenic health risks

Assessment of Carcinogenic health risks with hazard quotient was done using Eq. 3.

Hazard quotient (carcinogenic) [16].

$$\text{Incremental Lifetime Cancer Risk}\left(\text{ILCR}\right) = CDI \text{ X } SF \qquad (3)$$

where CDI = Chronic daily intake (CDI) = $\frac{EDIXEFXED}{ATn}$ (4)

SF = slope factor
EDI = Estimated Daily Intake
EF = Exposure frequency
ED = Exposure duration
ATn = Average life span.

2.7 Hazard index (HI)

In view of the fact that contaminants do not act in seclusion in the environment, the HI was used to assess the total risk from various contaminant pathways. This is the sum of the target hazard quotients for all heavy metals, calculated using Eq. 5 [5, 6, 17]. HQ and HI values less than 1 were considered safe [18].

$$HI = \sum HQ = HQ_{Ni} + HQ_{Cd} + HQ_{Cr} + HQ_{Cu} + HQ_{Pb} + HQ_{Zn} \qquad (5)$$

2.8 Data analysis

Significant difference in metal concentrations between periods (months) was tested with ANOVA (General Linear Model) using the software Minitab 16. The standard deviation was calculated as the positive square root of the sample variance with number of observations as six (6) and expressed in the Eq. 6 as.

$$S_X = \sqrt{\frac{\sum_{i=1}^{n}\left(X_i - \bar{X}\right)^2}{n-1}} \qquad (6)$$

3. Result and discussion

3.1 Concentration of heavy metals in tissues of bivalves

The concentration of heavy metals in tissues of *Tympanotonus fuscatus* is presented in **Figure 2a–f** while **Table 1** compared the metal concentrations with permissible limits. Heavy metal toxicity is a major global concern due to human health risk associated with consumption of contaminated sea food, hence the need for evaluation. Mean metal concentrations (mgkg^{-1}) in the soft tissues of

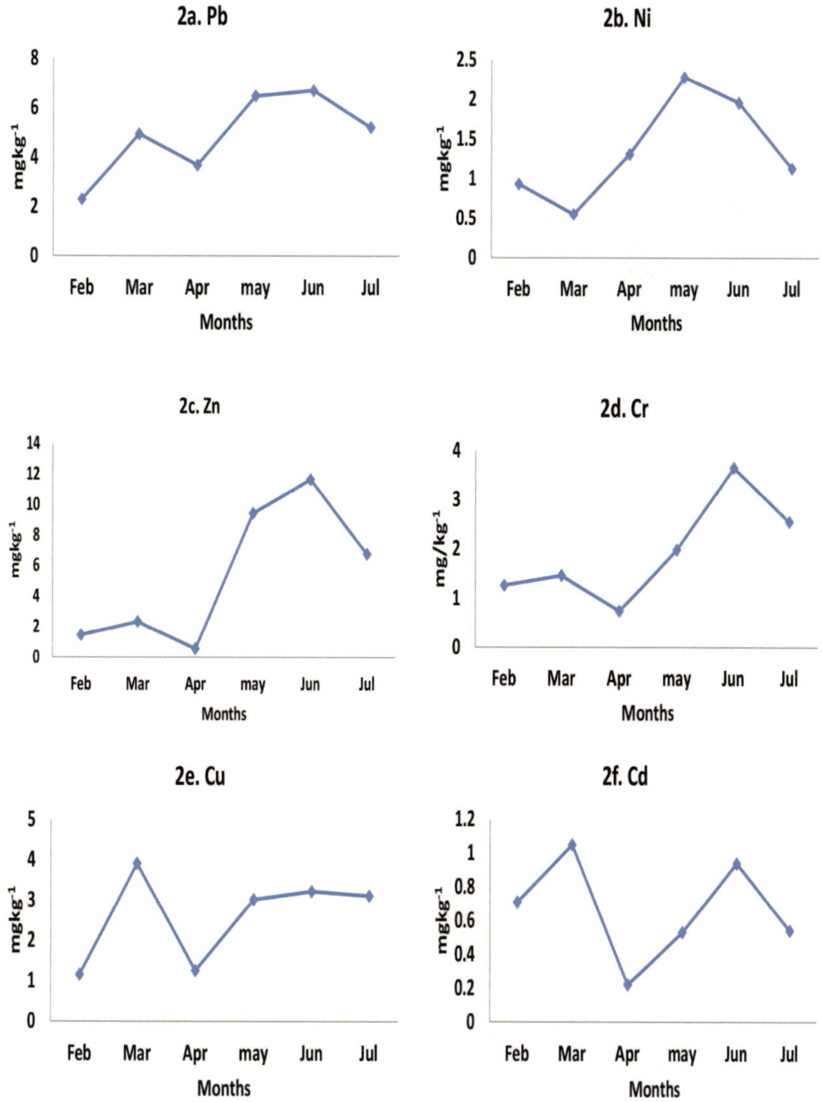

Figure 2.
(a–f) Temporal variations of heavy metal concentrations in the study area.

Heavy metals	Mean ± STDEV	Recommended limits (ppm)
Pb	4.88 ± 1.67	0.5 [19]
Ni	1.36 ± 0.64	0.2 [20]
Zn	5.36 ± 4.5	30
Cr	1.94 ± 1.04	12–13 [21]
Cu	2.61 ± 1.13	30 [22]
Cd	0.66 ± 0.3	0.1 [23]

Table 1.
Heavy metal concentrations and permissible limits.

periwinkles (*Tympanotonus fuscatus*) examined was in the order of Zn (5.36 ± 4.5) > Pb (4.88 ± 1.67) > Cu (2.61 ± 1.13) > Cr (1.94 ± 1.04) > Ni (1.36 ± 0.64) > Cd (0.66 ± 0.3). Variation in metal concentrations was significantly different ($p < 0.05$). This implies differential bioaccumulation in the concentrations of the heavy metals in gastropod tissues with respect to time, particularly those metals with elevated concentrations above regulatory limits. Consumers of periwinkle from the study area are therefore, exposed to significantly higher concentrations of heavy metals. Moslen and Miebaka [1] had reported that fish could accumulate elevated levels of heavy metals in their tissues over time without physical signs of distress but this may constitute potential health harm to consumers of such fish from presumed polluted areas. Mean values of Cr, Cu and Zn observed in this study were below their respective recommended limits of 12–13 [21, 22] , and 30 (FAO/[20]) while mean values of Ni, Cd and Pb exceeded their recommended limits of 0.2 [20], 0.1 [23] and 0.5 [19]. In other studies, Akinrotimi et al. [24] reported heavy metal concentrations ($mgkg^{-1}$) as follows Ni (1.17 ± 0.05), Cd (0.06 ± 0.01), Cr (2.44 ± 0.01), Pb (0.29 ± 0.06), Zn (5.57 ± 0.61) in bivalve Mollusks while Moslen et al. [11] reported thus Cd (0.02), Cr (1.57), Pb (0.01) and Zn (24.42) in gastropod Mollucs.

3.2 Health Risk assessment

3.2.1 Exposure assessment (Estimated Daily Intake (EDI))

Consumption of contaminated periwinkles (*Tympanotonus fuscatus*) is a key exposure route for human health risk. Estimated daily intake (EDI) was used for exposure assessment via consumption of contaminated periwinkles. EDI values ($mgkg^{-1}$) of this study ranged from 0.1×10^{-3} (Cd) – 0.84×10^{-3} (Zn) (**Figure 3**). Different researchers have reported variations in EDI values. Anaero-Nweke et al. [25] in a study of bioaccumulation of heavy metals in tissues of *Tympanotonus fuscatus* had reported EDI range of 6.57–8.81 $mgkg^{-1}$ above FAO/WHO [26] permission tolerable daily intake of 0.25 $mgkg^{-1}$ for Pb and EDI values of 0.01–0.07 $mgkg^{-1}$ within FAO/WHO [26] permissible tolerable daily intake of 0.07 $mgkg^{-1}$ for Cd. Tongo and Ezemonye [18], had reported EDI values of 0.138–0.200 $mgkg^{-1}$ in fish from the Niger Delta region while Moslen [11] also reported EDI values of fish

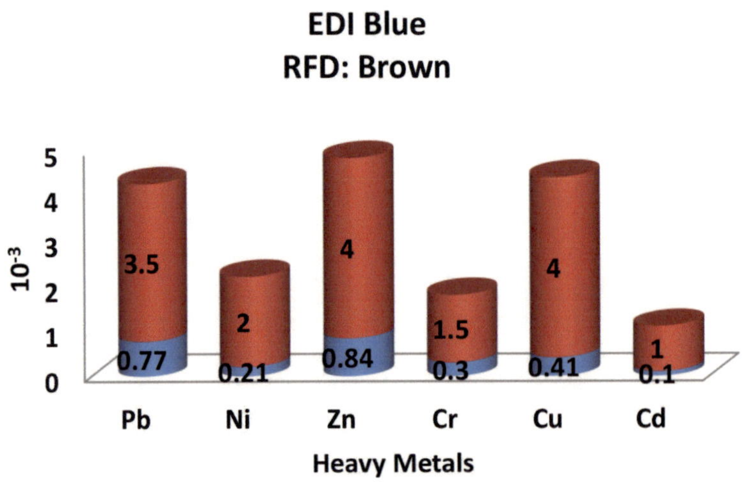

Figure 3.
Estimated daily intake of the periwinkles is presented in the study area.

tissue as follows: Cd (5.5 x 10^{-5}), Cu (6.9 x10^{-4}), Ni (3.0 x 10^{-4}), Pb (4.5 x 10^{-4}) and Ag (1.9 x 10^{-4}). The EDI values of all metals examined in the present study were less than their respective reference oral doses (RFD) implying minimal health risk due to consumption of contaminated periwinkles in the study area. However, [13] stated that RfD represents an estimation of the daily exposure of a contaminant to which the human population may be continually exposed over a lifetime without an appreciable risk of harmful effects. It is important to mention that among the different metals examined Pb, Cd, Cr and Ni are classified as chemical hazards [27, 28] meaning they could impair body functions at certain concentrations.

3.3 Target Hazard Quotient

Assessment of risk is very important because it has to do with evaluation of the amount of a substance that could lead to negative health impact for exposed persons over a particular duration. The health risk assessment of each contaminant is often based on the evaluation of the risk level and is classified as carcinogenic or non-carcinogenic health hazards [29]. The target hazard quotient was used to assess the non-carcinogenic risk. In the present study, THQ (non-carcinogenic) values ranged from 0.01 (Cu) - 0.22 (Pb) (**Figure 4**). The values of THQ (non-carcinogenic) in the present study were all <1suggeting low risk of exposure for consumers of periwinkles. Researchers have reported that THQ >1 potent risk of non-carcinogenic effect to consumers of such sea foods [5, 6, 30]. However, Khan et al. [31] stated that THQ builds some degree of alarm for carefulness but does not assess risk regarding exposure to contaminants. The THQ of the present study were also in tandem with values reported by Moslen and Miebaka [32]. Dee et al. [33] also found THQ for Cd, Cu, Mn, Pb, and Zn as 0.12, 0.06, 0.04, 0.41, and 0.03, respectively. The values of hazard quotient carcinogenic (HQ) of the present study ranged from 10^{-6} – 10^{-4} (**Figure 5**). Such values indicated low to moderate values of carcinogenic risk [34].

Metals in the environment may not act singly. The sum of the THQ (hazard index - HI) was calculated to give an indication of the total potential non-carcinogenic health impacts that could result from exposure to a mixture of heavy metals in periwinkles consumed, following EPA guidelines for health risk assessment [35, 36]. The HI value (0.56) of the present study was <1 depicting low risk of exposure for consumers of periwinkles. The HI value of the present study is less than that

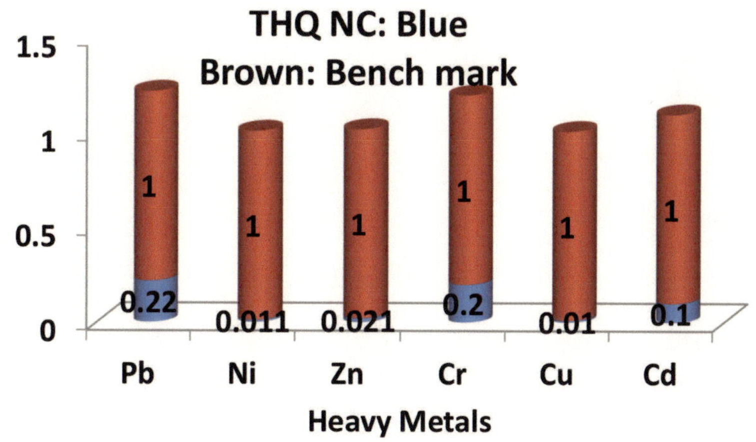

Figure 4.
The target hazard quotient non-carcinogenic of heavy metals.

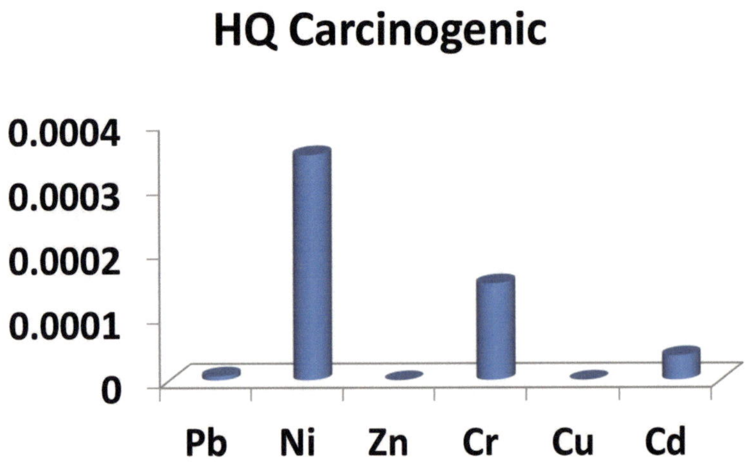

Figure 5.
The hazard quotient (carcinogenic) of heavy metals.

(1.60–5.06) reported by Denil et al. [37] for four different clams but agrees with that (0.45) in fish reported in the Niger Delta region [38]. The percentage make-up of the HI was in the order Pb > Cr > Cd> > Zn > Ni > Cu indicating the elevated input of Pb, Cr and Cd concentrations in the periwinkles examined.

4. Conclusion

The study therefore, concluded gradual bioconcentration of some heavy metals in tissues of periwinkles obtained from the study area. Both carcinogenic and non-carcinogenic indices indicated safe levels for health risk. Therefore, consumers of periwinkles in the study area have low to moderate risk of exposure for health concerns. It is import to continue monitoring to observe changes in bioaccumulation.

Author details

Miebaka Moslen* and Chioma Hope Adiela
Department of Animal and Environmental Biology, Rivers State University, Port Harcourt, Nigeria

*Address all correspondence to: moslen4c@yahoo.com

References

[1] Moslen M, Miebaka CA (2016) Temporal Variation of Heavy Metal Concentrations in Periophthalmus SP Obtained from Azuabie Creek in the Upper Bonny Estuary, Nigeria. Current Stud in Comp Edu Sci and Technol, 3(2): 136-147

[2] Järup L (2003) Hazards of heavy metal contamination. Br Med Bull. 68(1):167-182.

[3] Nagajyoti PC, Lee KD, Sreekanth TVM (2010). Heavy metals, occurrence and toxicity for plants: a review. Environ Chem Lett 8(3):199-216.

[4] Jaishankar M, Tseten T, Anbalagan N, Mathew BB, Beeregowda KN (2014) Toxicity, mechanism and health effects of some heavy metals. Interdiscip Toxicol. 7(2): 60-72.

[5] Moslen M, Miebaka CA (2017a) Heavy Metal Contamination in Fish (Callinectis amnicola) From an Estuarine Creek in the Niger Delta, Nigeria and Health Risk Evaluation. Bull Environ Contam Toxicol DOI 10.1007/s00128-017-2169-4.

[6] Moslen M, Miebaka CA (2017b). Concentration of Heavy Metals and Health Risk Assessment of Consumption of Fish (*Sarotherodon melanotheron*) from an Estuarine Creek in the Niger Delta, Nigeria. IOSR J of Environ Sci, Toxicol and Food Technol 11(3): 68-73

[7] Dural M, Goksu MZL, Ozak AA (2007) Investigation of heavy metal levels in economically important fish species captured from the Tuzla Lagoon, 102, 2007, 415-421.

[8] Davies OA, Allison ME, Uyi HS (2006) Bioaccumulation of heavy metals in water, sediment and Periwinkle (Tympanotonus fuscatus var radula) from the Elechi Creek, Niger Delta. African J of Biotechnol 5: 968-973.

[9] Langston WJ, Spence SK (1994) Metal Analysis. In: Calow P (ed) Handbook of Ecotoxicology, Blackwell Scientific, Oxford, Uk. pp: 45-78.

[10] Philips DJH, Rainbow PS (1993) Biomonitoring of trace aquatic contaminants. Environmental Management Series. Elsevier Applied Science, London. 571.

[11] Moslen M, Ekweozor IKE, Nwoka ND (2017) Assessment of Heavy Metals and Bioaccumulation in Periwinkle (Tympanotonus fuscatus var. radula (L.) obtained from the upper Reaches of the Bonny Estuary, Nigeria. J Heavy Met Tox Dis 2(2:3):1-3):5. DOI: 10.21767/2473-6457.100018.

[12] AOAC (2006) Association of Official Analytical Chemists. Official Method of Analysis (18th ed.) AOAC, 2006

[13] Akoto O, Bismark EF, Darko G, Adei E (2014) Concentrations and health risk assessments of heavy metals in fish from the Fosu Lagoon. Int J Environ Res 8(2):403-410.

[14] Moslen M, Miebaka CA, Boisa N (2019) Bioaccumulation of Polycyclic Aromatic Hydrocarbon (PAH) in a bivalve (Arca senilis- blood cockles) and health risk assessment. Toxicol Rept 6:990-997.

[15] Tongo I, Ogbeide O, Ezemonye L (2017) Human health risk assessment of polycyclic aromatic hydrocarbons (PAHs) in smoked fish species from markets in Southern Nigeria, Toxicol Rep 4:55-61.

[16] Sultana MS, Rana S, Yamazaki S, Aono T, Yoshida S, Kanan S, (Reviewing Editor) (2017) Health risk assessment

for carcinogenic and non-carcinogenic heavy metal exposures from vegetables and fruits of Bangladesh. Cogent Environ Sci 3(1) DOI: 10.1080/23311843.2017.1291107.

[17] Guerra K, Konz J, Lisi K, Neebrem C (2010) Exposure factors handbook. USEPA, (2010) Washington D.C.

[18] Tongo I, Ezemonye L (2018) Levels and Human Health Risk Assessment of Persistent organics, metalloid and Heavy metals in Fish. Nigerian J Basic Appl Sci 26(1):55-64.

[19] FAO/WHO (1989) Evaluation of Certain Food Additives and the Contaminants Mercury, Lead and Cadmium; WHO Technical Report Series (1989). No. 505.

[20] WHO (2005) Background document for development of WHO guidelines for nickel in drinking-water quality and fish. Tech. Rep. WHO (2005) /Sde/Wsh/05.08/55/.

[21] USFDA (1993) United States Food and Drug Administration. Guidance Document for Cadmium in Shellfish. US Department of Health and Human Services, Public Health Service, Office of Seafood (HFS-416), 200 C Street, SW, Washington, DC 20204, p. 44.

[22] FAO/ WHO (1983) Compilation of legal limits for hazardous substances in fish and fishery products. Fish Circ 464:5-100

[23] FAO (2003) Heavy Metals Regulations: Part 1. Legal Notice no. 66. http://www.faolex.fao. org/docs /pdf/ eri42405.pdf. (2003) Accessed 21 Apr 2012.

[24] Akinrotimi OA, Amachree D, Gentle CG (2019) Metal concentrations in some commercially important shell in Port-Harcourt, Rivers State, Nigeria. Inter J Fish Aqua Stud 7(2):134-138.

[25] Anaero-Nweke GN, Ekweozor IKE, Ugbomeh AP, Moslen M (2018) Bio-accumulation of Pb and Cd in the soft tissues of Tympanotonus fuscatus, sediment quality of the upper Bonny estuary, Niger Delta and their human health implications. Toxicol Digest 2(1):100-112

[26] FAO/WHO (2010). Summary report of the seventy third meeting of JECFA. Joint FAO/WHO expert committee on food additives, Geneva.

[27] EC (2001) Commission Regulation (EC) No. 466/2001 of 8 March 2001. Official Journal of European Communities. 1.77/1.

[28] FDA. Fish and Fisheries Products Hazards and Controls Guidance. third ed. US Food and Drug Administration: Center for Food Safety and Applied Nutrition; 2001

[29] Wongsasuluk P, Chotpantarat S, Siriwong W, Robson M (2014) Heavy metal contamination and human health risk assessment in drinking water from shallow groundwater wells in an agricultural area in Ubon Ratchathani province, Thailand. Environ Geochem Health 36:169-182.

[30] Abdou HM, Hassan MA (2014) Protective role of omega-3 polyunsaturated fatty acid against lead acetate-induced toxicity in liver and kidney of female rats. BioMed Res doi:10.1155/2014/435857.

[31] Khan S, Farooq R, Shahbaz S, Khan MA, Sadique M (2009) Health risk assessment of heavy metals for population via consumption of vegetables, World App Sci J 6:1602-1606.

[32] Moslen M, Miebaka CA. Consumption of Shrimps (Penaeus Sp) and Health Risk Evaluation Due to Heavy Metal Contamination in the Upper Reaches of the Bonny Estuary, Nigeria. Perspectives in Ecotoxicology,

ISBN: 978-93-85995-69-9, Editors: Dr. M.M. Abid Ali Khan, Murtaza Abid, Dr. M.I. Ansari and Dr. Delhi, India: Harsukh Chharang, Lenin Media Private Limited; 2018. pp. 104-113

[33] Dee KH, Abdullah F, Md Nasir SNA, Appalasamy S, Ghazi RM, Rak AE (2019) Health Risk Assessment of Heavy Metals from Smoked *Corbicula fluminea* Collected on Roadside Vendors at Kelantan, Malaysia BioMed Research International Volume Article ID 9596810, 9:1-9 pages https://doi.org/10.1155/2019/9596810

[34] NYSDOH (New York State Department of Health) (2007). Hopewell precision area contamination: appendix C-NYS DOH. Procedure for evaluating potential health risks for contaminants of concern 2007, http://www.health.ny.gov/environmental / investigations/hopewell /appendc.htm

[35] Bamuwamye M, Ogwok P, Tumuhairwe V (2015) Cancer and non-cancer risks associated with heavy metal exposures from street foods: evaluation of roasted meats in an urban setting. J Environ Pollut Hum Health 3:24-30.

[36] Huang M, Zhou S, Sun B, Zhao Q (2008) Heavy metals in wheat grain: assessment of potential health risk for inhabitants in Kunshan, China. Sci Total Environ 405:54-61.

[37] Denil DJ, Fui CF, Ransangan J (2017) Health Risk Assessment Due to Heavy Metals Exposure via Consumption of Bivalves Harvested from Marudu Bay, Malaysia. Open J Mar Sci 7:494-510.

[38] Moslen M (2017) Risk Assessment and Bioconcentration of Heavy Metals in *Mugil cephalus* (Mullet) Obtained from Azuabie Creek in Port Harcourt Nigeria. Quest J of Res Environ and Earth Sci 3 (5): 01-07

Chapter 6

Role of Heavy Metals in the Incidence of Human Cancers

Vincent Salvatore Gallicchio and Juley Harper

Abstract

There has been increased concern on many levels focused on the environmental and occupational exposure of heavy metals and their impact on disease, specifically the carcinogenic potential inducing cancer in humans. Because the impact of heavy metals on human health continues to be a major health concern, research continues to improve our understanding of the carcinogenic potential of these substances. Of particular concern have been human exposure to aluminum, arsenic, beryllium, cadmium, lead, mercury, nickel, and radium and their carcinogenic potential whether contact is via environmental or occupational exposure. This updated review focuses on the carcinogenic mechanisms heavy metals use to induce malignant transformation of cells as well as addressing the overall environmental and occupational hazards of heavy metal exposure.

Keywords: heavy metals, carcinogenesis, human exposure, toxicity, mechanisms, remediation

1. Introduction

Heavy metal exposure has long been associated with major health care concerns pertaining to human health. The metals responsible for these adverse changes in human health need to include and focus on their role in carcinogenicity. As a premise, and for obvious reasons, there has been long-standing research and clinical focus among scientists and oncologists that has produced an extensive database. Using a variety of research engines, such as the National Institute of Medicine database (PubMed) and Google Search to explore the various aspects of heavy metals and their ability to induce cancer, we have attempted to review the reported studies in this area. Importantly, the information presented in the following pages represents linking heavy metal exposure to cancer, and specific human systems most susceptible to heavy metal carcinogenesis.

1.1 Aluminum

Aluminum is unique based upon the various mechanisms of action whereby it is listed based on its carcinogenic activity. More often, human exposure to aluminum is the result of contamination of food, interestingly in the process of manufacturing vaccines for human use, and when added as a chemical salt during a variety of processes used in industry for manufactured products for commercial purposes [1, 2]. The commercial products most susceptible are those in which aluminum salts are included in the list of added ingredients such as antacid tablets and antiperspirant deodorants [1–3].

Exposure to aluminum has had a direct link to the induction of human cancer, specifically breast cancer. Experimental studies performed in mice exposed to $AlCl_3$, which interestingly is the identical form of aluminum used in the manufacture of antiperspirant deodorants for humans, demonstrated an induction of malignant transformation of epithelial cells located within mammary glands [1]. Similar results were observed following exposure to human breast tissue epithelial cells [1–3]. Aluminum has been implicated in the development of neoplasia, specifically in the development of sarcomas [4]. In the same report it was noted that one patient, following consistent chronic exposure to aluminum, developed an atypical transformation resulting in a neuroectodermal malignancy [4].

Regarding the carcinogenicity of aluminum, it was of importance to best identify the potential or possible mechanism(s) of action responsible for the induction of tumors following exposure. In research studies performed *in vitro* using human breast cells exposed to aluminum, researchers observed a reduction in the levels of the tumor suppressor gene BRCA1 mRNA [3]. This effect took place concurrent with decreased levels of other maintenance genes that regulate normal DNA levels [3]. In a complementary study, researchers exposed human breast cancer cells to aluminum and measured induced uncontrolled cell growth that was consistent [2]. Upon evaluation of these results, researchers concluded the aluminum acted as a metalloestrogen, meaning the reaction acted as an antagonist for the estrogen receptor complex on these breast cells. This kind of biochemical activity has been associated with the carcinogenesis ability following aluminum exposure [2].

When other body tissues were examined following aluminum exposure, in this specific case, the development of bladder cancer, it was revealed the bladder cancer cells had higher levels of aluminum compared to other heavy metals [5]. Although these studies were not able to directly link a cause and effect between aluminum and the induction of bladder cancer, it did provide suggestive evidence that aluminum exposure may play in the development of such cancers. This hypothesis lead to a formative therapeutic modality and that is to remove the aluminum. Use of chemical chelators has been recommended standard therapeutic procedure to be performed whenever aluminum exposure, thus poisoning, has been implicated in any physiological or cellular transformation. Physiological studies have demonstrated that when introduced into the human body, aluminum accumulates in both the soft and skeletal tissues. These are the target tissues for aluminum chelation [6]. The most common chelating agent used to detoxify aluminum exposure is desferrioxamine [6]. This chelator has proven very effective in removing the heavy metal aluminum from tissues, even though use of desferrioxamine is associated with its own level of toxicity that is associated with its clinical use in humans [6]. In order to address desferrioxamine toxicity, other chelating agents have been identified that show promise as candidates to replace desferrioxamine; however, the level of chelation associated with these agents has not yet equaled what has been demonstrated using desferrioxamine. Another option in order to reduce aluminum especially if is measured to be present in high amounts in public consumption, e.g., drinking water [7]. The method used in these conditions is reverse osmosis filtration. The procedure has been demonstrated to reduce significant aluminum levels when applied in a variety of industrial settings such as in the mining of copper and in other areas of industrial usages [7].

1.2 Arsenic

Arsenic is a heavy metal with known cytotoxicity in human tissues following exposure that can result in serious illnesses to those who are exposed. In a majority of cases, the path of exposure results from ingestion of foods and sources of drinking water contaminated with arsenic [8–11]. There are also examples of arsenic exposure that are the result of occupational exposure through environmental pollution [8–11]. Examples of occupations that provide direct risks are smelting and arsenic based pesticide industries [12]. Another well documented source of heavy metal arsenic is through contact with contaminated soil thus consumption occurs through the food chain [13].

The correlation between heavy metal arsenic exposure and human cancers is relevant because arsenic detection within tumor tissue. Specific examples of arsenic and cancer development comes from research studies demonstrating a role for arsenic in the development of bladder, lung and skin malignancies [8, 11, 12]. An additional positive correlation linking arsenic with the development of human cancers focused on the relationship between arsenic exposure and mortality rates in patients diagnosed with a variety of cancers – colon, gastric, kidney, lung and nasopharyngeal cancers [13]. Importantly, based on epidemiological data from several studies shows a clear association between the induction of both pancreatic and non-Hodgkin's lymphoma following chronic arsenic low-level exposure [14, 15].

As with all heavy metals the question is what is/are the mechanism(s) responsible for the carcinogenic activity? As it pertains to arsenic, several studies have clearly demonstrated the mechanisms responsible for arsenic induced carcinogenicity involve the formation of reactive oxygen species (ROS) that indues critical epigenetic changes leading to damaging DNA repair mechanisms [8, 9, 12]. Specifically, these important epigenetic changes induced by arsenic have included alterations in DNA methylation, histones, and miRNA, all potentially responsible for the tumorgenicity associated with arsenic exposure [9, 12]. Another postulated mechanism of action for arsenic associated carcinogenicity is arsenic's ability to induce abnormal cell growth cycles in specific cell types such as macrophages and lung epithelial [16]. This was of particular concern because in lung epithelial cells, arsenic promoted a key and significant mechanism of action inducing carcinogenesis. In this cell population arsenic was demonstrated to alter the gene expression of the tumor suppressor protein *p53,* which in turn decreased the expression of *p21,* a downstream target [17]. Thus, the result of this association between heavy metal and tumor suppression gene inactivation was increased cellular proliferation, which demonstrated the most prominent mechanism of cellular transformation. Under these conditions what develops is major oxidative stress in these cells.

In studies conducted to further understand the association between arsenic and tumor cell initiation, another important activity was attributed to arsenic. Based upon further examination, it became clear that in co-existence with changes in cell transformation, intracellular levels of glutathione, a potent ant-oxidant agent, were reduced [18]. Lowering glutathione levels thereby reduces its antioxidant activity, thus allowing altered or transformed cells to escape from being removed by suppressor T-cell lymphocytes and NK cells [18].

Another postulated mechanism of action explaining the tumorgenicity of arsenic was proposed. This alternative mechanism was identified following arsenic exposure to human bladder cells. The mechanism was attributed to the ability of chronic exposure of arsenic to inhibit proper cellular morphology attributed to altered gene expression responsible for base excision repair [19]. The key enzymatic component here is the rate limiting step catalyzed by the enzyme DNA polymerase beta, an active enzyme in the process [19]. In the presence of arsenic, the enzymatic activity was reduced in a dose-dependent manner, meaning higher concentrations of arsenic, correlating with the lack of enzymatic activity [19]. These studies demonstrated chronic exposure to arsenic influenced changes in cellular morphology and altered the gene expression for specific proteins that control cellular proliferation [20].

In order to remove arsenic from the body the use of specific chelating agents have been shown to be most effective [21]. One such example of a very effective chelating agent is 2,3-dimercaptopropanol, otherwise known as British anti-lewisite. The molecule contains 2 functional thiol groups [21]. Significant clinical data has been accumulating over the past several years demonstrating the effective chelating action of this compound. 2,3-dimercaptopropane-1-sulphonate was administered effectively with minimum side-effects to a patient diagnosed with arsenic exposure [22]. This one study provided the clear and effective use of chelators to remove excess amounts of heavy metals [22]. Based on these observations, it was proposed that incorporation of antioxidants as a component of one's dietary consumption should be recommended in order to maximize anti-cancer and reduced oxidative stress [23]. Both rice and apple juice have been found to reduce cellular stress by the presence of antioxidant compounds, in part because they contain levels of vitamin C, a potent antioxidant. Oxidative stress is a major factor leading to a number of cellular disease pathologies.

As mentioned above, safety regulators have identified apple juice and rice as two food stuffs that can often serve as source of arsenic exposure in children. The level of 5 μg/L arsenic has been set as the lowest level of toxicity exposure [24]. With these dietary links identified other alternative methods to curb the toxicity linked to food stuffs have been presented to limit arsenic uptake using genetic modifications to rice that would inhibit the absorption of arsenic [24]. Another strategy has been to use specific micro-organisms that when co-existing with arsenic in the environment reduce metal uptake [24]. Alternatively, in the cultivation of rice, use of certain watering methods in agricultural would ultimately reduce the concentration of the heavy metal when present in the environment [24, 25].

1.3 Beryllium

The heavy metal beryllium is associated with human use through its application tied to industrial processes. Thus, human consumption is linked to environmental contamination documented to most often occur from its association in power plants where it is often found in dust [26, 27]. Thus, human contact occurs via inhalation as the most common method of contact. As an environmental contaminant, it has been linked to a number of respiratory ailments including carcinogenesis of the lung [27–29]. Initially the relationship between beryllium and lung cancer was suspect, but additional studies demonstrated a clear association between exposure, especially following higher levels of beryllium exposure [28–30]. Subsequently it was shown that use of beryllium in the dental industry was another opportunity for exposure through occupational risk [29]. Thus, the intervention of personal protective equipment (PPE) had a marked ability to reduce occupational exposure related to dentistry [31]. Importantly, patients diagnosed with stage III breast cancer were found to have elevated levels of beryllium [32]. However, in this study, beryllium was not the only heavy metal to be detected thus limiting a direct cause effect situation [32]. Another cancer, osteosarcoma has also been implicated to be the result of beryllium exposure [33].

There has been a paucity of defined experimental studies conducted to determine cause and effect between beryllium and the conduction of cancer and the mechanisms involved. Much of the focus has been to address issues correlated with lung exposure. One carcinogenic mechanism studied was the link between the elevated levels of tumor necrosis factor alpha (TNF-α), which is a cytokine secreted from a specific type of T-cell ($CD4^+$) that are present in the lung [30, 34]. This factor plays an important role in the development and induction of inflammation [30, 34].

The association between TNF-α and beryllium implicates a direct link to the action of chronic inflammation exposure [30, 34].

Genetic changes are associated with beryllium exposure and have been observed to methylate the *p16* gene, which as stated previously is a known tumor suppressor gene that is activated following exposure to beryllium [30]. How to best address the removal of beryllium following exposure, in order to reduce its carcinogenetic properties, has focused on the use of chelators. Chelators are often used to remove heavy metal contamination from the body and in doing so they effectively reduce the toxic effects of exposure. Examples of effective chelators include – 4-dihydroxy-1,3-benzene disulphonic acid disodium salt [Tiron] and D-pencillamine (DPA), which demonstrated effectiveness following animal exposure [35–37]. It is of interest that a benzene derived compound would be used under any conditions because of the known carcinogenetic activity of benzene. Another chelator, meso-2,3-dimercaptosuccinic acid (DMSA) has been demonstrated to be effective when used and reported as a case-study to successfully treat a young child who was suffering from high level beryllium poisoning [38]. This experience suggested it is an effective treatment and therefore is worthy of further investigation [38]. Taken together this collective response indicates reduced exposure of beryllium will impact the overall health risks associated with its exposure [39, 40]. Addressing specific companies and other industrial sources linked to beryllium exposure should be used to support screening other methods to test employees for beryllium exposure among them. Furthermore, such companies should screen their employees using blood samples in addition to providing proper ventilation control measures in these plants and factories [40]. Along with instituting proper screening methods for employees to minimize exposure, additional strategies should be implemented, like better educating plant workers to use personal protective equipment the need arises [39, 40].

1.4 Cadmium

The heavy metal cadmium is a toxic element related to significant health consequences as an environmental contaminant. The sources of environmental exposure are generally associated with industries where it is present in their emissions. The element is used in industries such as mining, research with metallurgy, battery development, and preventing pigment precipitation when used in textiles [41]. A very serious issue regarding environmental cadmium exposure is soil contamination, as human exposure of cadmium most often is the result of ingesting contaminated food and water, inhalation and/or smoking [41, 42]. Regarding soil contamination, a specific source of cadmium contamination occurs as a result of landfills. High levels of cadmium have been found in landfills at concentrations that are much higher than recommended as tolerable in the maintenance of human health [43].Given that landfills are a major source of soil and water contamination, human exposure to cadmium more often is associated with the ingestion of contaminated foods [14, 44].

The main health issue associated with cadmium is the carcinogenicity following toxic exposure in humans, in particular, cancers of the breast, esophagus, intestines, lungs, stomach, testes [41, 45, 46], and possibly the gallbladder. The link to the gallbladder is identified in studies where gallstones have been associated as a pre-cancerous situation in many cases, when analyzed for the heavy metal contact in patients with cancer of the gallbladder [47]. When analyzed statistically significant levels of heavy metal content, cadmium and other heavy metals were found to be elevated [47]. The link between cadmium and carcinogenicity is still a significant human health concern. In other types of studies, in particular laboratory generated experiments, the results of liver cells cultured in the presence of cadmium demonstrated the oncogenic transformation of these liver cells [44]. In patients with gliomas (cancer of the brain) heavy metal analysis detected high levels of cadmium, indicating cancer of the brain may be linked to heavy metal exposure [48].

Another body organ that has also been linked to cancer following cadmium exposure is the pancreas [15, 49]. Cadmium has also been linked to the development of blood disorders, in particular, the development of chronic myeloid and lymphoblastic leukemia. When analyzed compared to controls, patients with leukemia when tested were found to have increased concentrations of cadmium in the presence of reduced levels of magnesium in both blood and serum [50]. Another significant correlation between increased levels of cadmium and carcinogenicity is the association between cadmium in urine and the development of cancer of the gastrointestinal system [51].

As was observed with other heavy metals, the overall effects correlated with the development of a variety of cancers, focused attention to determine what were the exact mechanisms involved that led to initiation of the carcinogenic processes. With respect to cadmium, the focus of the carcinogenic mechanism involved the generation of reactive oxygen species (ROS) and epigenetic changes. Both contributed to the restriction of repair mechanisms that generated altered or damaged DNA. Both also contributed to the loss of apoptosis in affected cells [41, 46, 52, 53]. Whether the exposure to cadmium is either acute or chronic, the result targets the altered signal transduction mechanisms that induce altered gene regulation, which collectively contribute to the initiation of tumor growth [44]. In this key sequence of intracellular changes that takes place following cadmium exposure, important proteins are dysregulated either via upregulation or enhanced activity or perhaps via suppression of key molecular pathways. Such an example is the inhibition of EGR-1, which is a key protein that regulates cell destructive pathways, such as transcription [44].

Adverse toxic human exposure resulting from cadmium poisoning unfortunately is not associated with any standard therapeutic measures designed to address cadmium toxicity, if presented following acute or chronic exposure [54]. With that said, research has developed compounds that upon co-administration would be effective in reducing the toxicity of cadmium exposure. Examples of compounds developed to reduce cadmium toxicity are peptide ligands that have specificity for cadmium [54]. Importantly because of their widespread availability, meaning they occur naturally are flavonoid compounds that are present widely in fruits and in fact in most plants. Collectively whether they are fruits or vegetables, they all contain flavonoids. Flavonoids are potent antioxidants, thus chemically they reduce the development of ROS and also, they can assist in cadmium chelation [55]. With that said, it is still important to more fully understand how flavonoids, specifically via their structure inhibit the development of cadmium toxicity [55].

There is experimental evidence exploring whether the use of stem cells would be effective in ameliorating the cellular damage associated with cadmium toxicity. In a study performed using rats, the testicles were exposed to cadmium causing tissue damage [56]. Following the toxic exposure, animals received bone marrow derived mesenchymal stem cells. Upon clinical treatment it was observed that within the testes the levels of proteins responsible for apoptosis reached appropriate levels to restore apoptosis, thus effecting cell regulation [56]. Within the affected tissue there was evidence that the damaged tissue had been repaired. The implications of these observations suggested that the target of recovery delivered by mesenchymal stem cells was the restoration of mitochondrial apoptosis [56].

1.5 Lead

One of the most researched heavy metals, in part because of its well-established effects on human health is lead. It has long been recognized as a significant environmental pollutant. There have been a number of pathways that either singularly or in concert attribute to impairing human health especially after chronic lead exposure [57–59].

A very common method of human lead exposure is the result of environmental contamination that involves soil and water contamination, especially sources of drinking water. Lead levels accumulate in deposits and exposure is manifested through the human food chain, thus its eventual presence in consumed food [57–59]. Another common source of lead that contributed to its exposure to humans was the presence of lead added as additive to gasoline. However, since 1995 lead has been banned as additive to gasoline for use in automobiles, yet it is still added to the fuel used for aviation purposes [59]. Another alarming link to human lead exposure was the discovery that lead was present in cigarette smoke; therefore, the lead levels in blood of smokers was reported to be high, as there is no safe concentration of lead regarding impact on human health [60]. Other occupational hazards also exist such as mining that contributes to the presence of lead exposure in those workers [57].

What have been the studies conducted to determine the overall level of toxicity of lead exposure to human health? A number of epidemiological studies have been conducted to determine the impact of lead on human health that has implicated the heavy metal as a causative factor in a number of human cancers. Whether lead exposure functions in terms of a direct cause vs. effect on inducing a specific cancer type is still under investigation [61]. In particular, interest has centered on a supportive, perhaps an additive, role in the maintenance of cancer rather than an initiating agent [61]. Lead has been detected along with other heavy metals that are also known for their impact on human health especially in children where it can impact the development of myelin, thus causing impairment in neurological development. An example was the detection of very high levels of lead in the water systems of Flint, MI and along with cadmium when analyzed in patients with gliomas (brain cancer) [61]. This observation demonstrated an increased toxic consequence to human health when such heavy metal contaminants are found together in human tissue or body fluids [61].

A study of patients with kidney cancer came to the conclusion that the cancer developed associated with high levels of lead [58]. This observation was later supported by evidence linking the development of renal cell carcinoma as associated with the presence of lead in the blood [60]. A link to the development of liver disease as the result of high lead concentrations levels along with a number of other heavy metals when tested in gallstones [47] suggested there may be a correlation between lead levels and disease of the gallbladder, perhaps inducing a pre-cancerous lesion [47]. When examined in workers exposed to high levels of lead, it was clearly demonstrated there was a significant positive correlation between the heavy metal and the presence of cancer in the lungs, along with a positive correlation linking lead exposure to the development of cancer of the brain, larynx, and bladder tissues [62]. In patients detected with pancreatic cancer, increased levels of lead in addition to several other heavy metals were measured, suggesting heavy metal exposure may contribute to the overall carcinogenicity of these heavy metals [15, 61].

The scientific literature has been devoid of studies devoted to the understanding of the mechanisms of lead induced carcinogenicity; however, several potential mechanisms have been proposed. Based on the current understanding of how lead can be carcinogenetic, one hypothesis has implicated lead as effectively disrupting internal genetic processes that result in the inability of tumor regulatory genes to function, inducing damage to DNA, and at the same time inhibiting repair of DNA damage [63]. In animal studies using mice exposed to lead, they showed that the heavy metal was capable of inducing reactive oxygen species (ROS) and by doing so the exposure effectively altered the sequence of specific gene function [63]. Another critical observation related to the ability of lead to disrupt normal cellular physiological processes were the results showing lead was effective in normal reactions controlling transcription. The reaction that mediates this transition was the substitution of lead for zinc that serves as a metal catalyst for several key enzymatic reactions that control DNA transcription [63]. Along with this observation was the important association of calcium in these enzymatic reactions based on epidemiological studies showing an increase in calcium correlated with a lower risk level for developing renal cell cancer. Consequently, as pointed out by the investigators, it clearly showed the need to have a clinical trial to determine the overall significance when these important cations and heavy metals come into contact with one another [60].

For clinical cases where heavy metals such as in lead poisoning are implicated in disease etiology and pathology, the therapeutic remedy recommended is chelation [64]. The most common chelators being used for reducing elevated lead levels are British, Anti-Lewsite, calcium disodium ethylenediaminetetraacetic acid (EDTA), D-penicillamine and Meso-2,3-dimercaptosuccinic acid. The use of any specific chelator depends on the individual clinical case [64]. Unfortunately, several of these chelating agents are associated with their own level of toxicity. Thus, to reduce the toxicity potential of these chelating agents, substitution using garlic in the clinically was found to effectively reduce blood levels of lead when lead toxicity was at moderate levels and also restricted lead associated symptoms when used clinically [64]. With the collective results, it goes worth saying the most effective treatment is to prevent lead exposure [58]. To achieve such a goal requires that all industries known to be associated with lead toxicity must address emissions of the toxic metal to the environment as well as to reduce with the goal to completely eliminate emissions such that workers are not exposed, which implies factories need to have established quality control guidelines for limiting lead exposure [64]. It stands to reason that the best and most effective way to remove lead contamination is to eliminate the sources of lead contamination [64]. In communities such as has been the case in Flint, MI that have been impacted because of lead leaching from old water pipes, the only remedy is to completely remove the old lead-based pipes for modern substitutes.

1.6 Mercury

Another heavy metal that has shown severe health consequences in humans following exposure is mercury. A minor portion of the heavy metal is found as a mineral in trace amounts with the major portion of mercury exposure the result of the environmental exposure following industrial use [65]. There are many different areas where mercury use has caused environmental problems. Common usage includes the long-term use of mercury in thermometers, dental fillings, in the manufacture of certain types of batteries, and in the burning of medical waste [65, 66]. Burning of fossil fuels has also been identified as a source of mercury pollution [65, 66]. Another contributing factor to environmental pollution and mercury is the fact that mercury often will be vaporized thus entering the atmosphere along with the other substances that when in the atmosphere, can then be incorporated into the soils and water systems [65, 67]. Regarding foods, consumption of large amounts of seafood, e.g., tuna and shellfish has been identified as another link to environmental exposure especially methyl mercury [65, 68, 69]. Collectively these sources have contributed to the environmental contamination associated with mercury.

Regarding the association between the development of cancer and mercury, there has been suggestive evidence linking mercury exposure and kidney cancer. This association is based on the physiological role of kidney in removing toxic substances when present in the body, especially within the blood [65]. Several other cancers associated with mercury are both liver and gastric cancers [70]. Also related to liver and gastric cancers, in patients with cancer of the gallbladder, mercury has been detected in gallstones at significant concentrations [47].

As has been mentioned when discussing the other heavy metals, mercury has the potential to be associated with the development of malignancies that utilize specific mechanisms that regulate the control of tumor development. The mechanisms implicated are the capacity to generate free radicals (ROS), in addition to the disruption of DNA, whether it be related to transcription events, changes in or maintenance of its molecular structure [66]. With that said there are reported other carcinogenic mechanisms that are unique to mercury. One such mechanism that addresses the carcinogenic potential of mercury is its ability to reduce levels of glutathione [71]. As mentioned earlier, glutathione is a naturally occurring antioxidant and as such it can reduce the antioxidant activity of mercury via reactive oxidant species, by inhibiting the development of oxidative stress mediated through reactive oxidant production, thus minimizing its carcinogenic potential [71]. Cells that are exposed to oxidative stress have been demonstrated to have increased rates of peroxidation of lipids, which has been proposed as another functional mechanism inducing cancer [65]. Within cells mercury has been implicated to influence the function of microtubules, which by their very nature can disrupt cellular mitosis [66].

As was stated with the other heavy metals previously mentioned, the use of chelators has been a common therapeutic approach for removing mercury from the body. For mercury two of the most effective chelating agents are dimercaptosuccinic acid (DMSA) and dimercaptopropane (DMPS) [72, 73]. With that said, there are substances that have been untested in terms of their chelating abilities for their effect against mercury. Two of these substances, desferairox and deferiprone, were tested experimentally in rats where it was observed that the combination was able to effectively chelate mercury and reduce toxic effects of mercury [74]. An experimental chelating agent that has been postulated is thiol-modified nanoporous, a silica material [75]. When tested experimentally in animals, it was observed that this substance had the potential to chelate mercury with minimal toxicity [75].

1.7 Nickel

The heavy metal nickel originally discovered as a major component constituting the earth's core has in recent years been the focal point of investigations to determine if its exposure, occupational or environmental, is involved in any carcinogenic action that compromises human health, through occupational exposure occurring primarily in the mining and refinement of nickel ore and producing metal alloys [76–78]. Nickel pollution of the environment results in its accumulation in organs and tissues within exposed organisms. As an example, nickel can enter the food chain through fish [79]. Alternatively, another route can take occur once contamination of the soil takes place [76]. On an industrial scale, nickel is often present in emissions released from oil refineries that have been identified as significant sources of environmental exposure and pollution, thus increasing the risk of exposure to those residents living close to these refineries [80].

Nickel exposure in humans has been associated with the development of a variety of cancers. Through epidemiological studies, evidence has shown there is a correlation between nickel exposure and the induction of cancer development in the lungs and in nasal and sinus tissues [13, 17, 81, 82]. In a study performed in breast cancer patients, when blood serum was analyzed for nickel it was found to be elevated significantly suggesting a potential relationship between the high nickel levels and the induction of breast cancer [83]. The correlation between nickel exposure and cancer has also been linked to the development of acute myeloid and lymphoblastic leukemia [84]. Additionally, when the urine was analyzed in patients with childhood leukemia, elevated levels of both nickel and 8-hydroxydehydrogenase implicating a causative role for nickel in inducing this childhood disease [84]. The role of nickel as a carcinogenic agent is implicated because of its ability to induce oxidative cellular damage as a primary mechanism of action [84].

Patients with pancreatic cancer, when measured for nickel levels, demonstrated elevated levels suggesting there is a positive correlation, even though other heavy metals were detected [15]. In addition, a study came to the conclusion that there may be a link between chronic nickel exposure, along with concomitant exposure of other heavy metals, to the development of T-cell lymphoma [85] and also liver cancer [13]. Collectively, the implications of these reports suggest the carcinogenic action of nickel.

Discussion of nickel and cancer addresses the need to focus on potential mechanisms of action. Several have been implicated. One mechanism involves the ability of nickel to influence noncoding RNA expression. A study demonstrated that nickel was effective in inducing materially expressed gene regulation (gene 3 MEG3) by its ability to influence the methylation of its associated promoter element [81]. This process was an effective inhibitor of PHLPPI and up-regulator of hypoxia-inducible factor-1α. Both are proteins recognized for their effective role in the processes involved in carcinogenesis [81]. As has been reported for other heavy metals, nickel as well can induce the formation of free radicals, a known carcinogenic action [86]. Exposure to nickel has been demonstrated to influence the status of the transcription and regulation status of mRNAs and also involve microRNAs [78]. Implicated in these reactions is the ability of nickel to influence immunity and the immune response, especially when it involves inflammation and the immune response, which in itself has also been implicated as having a significant role in carcinogenicity [78]. Nickel and its role in influencing the inflammatory response has been researched using animals and in combination with human cells [82]. These studies came away with the observation that there is an association between nickel exposure and cancer [78].

In addition to nickel's association with cancer, inflammation has also been investigated when tested using both animal and human cells. After dose–response studies were conducted it was determined that exposure to nickel increased the expression of certain proteins, specifically SQSTM1 and TNF. Both are known to have specific functions in the inflammation process [82]. As was observed with other heavy metals, nickel has been suggested to induce cellular following exposure epigenetic changes, an example is alteration in DNA methylation [82]. This conclusion is suggested from results that demonstrated exposure to nickel induced histone H3K4 tri-methylation [87]. The reactions associated with nickel exposure have been correlated with faulty transcriptional activation that can be a blueprint for the development of cancer [87].

Although chelation has been widely applied as a mechanism to remove heavy metal contamination, when applied to alleviate nickel contamination has produced different results. A very effector for chelating nickel, especially the cancer-linked nickel carbonyl, sodium diethyldithiocarbamate to the extent that it is the recommended remedy in the clinical setting [88]. With respect to environmental contamination, the compound ethylene diaminetetraacetic acid (EDTA) was shown to decrease the uptake of nickel when exposed to soil [89], indicating the potential for EDTA to be considered as an effective remedy for experimental exposure. The chelating compound CaNa($^{2+}$)-EDTA effectively removed nickel [90].

1.8 Radium

The heavy metal radium has had a long association with negative effects on human health. The harmful fact associated with radium is its radioactivity. Radium releases ionizing radiation through the decaying of radium into a toxic gas [91]. Radon contamination in the form of ionizing radiation can be associated through environmental and occupational exposure. Occupational exposure to radium is often associated with coal mining [92]. Coal mining exposure also implies radium contamination of any water or liquid residue used in the mining process [93]. The occupational exposure of radium can be associated with exposure through contact with building materials, soil and water systems. An Italian study demonstrated radium can accumulate when associated with confined space, such as in buildings, basements and other storage facilities [91]. Another overlooked substance that can contribute to the increased presence and concentration of radium in confined spaces is cigarette smoke [93]. This observation clearly implicates smoking and radium exposure that collectively could synergistically impact human health [93].

The development of several types of cancer have linked to radium, thus labeling it as a known carcinogen. Because the main occupational exposure of radium comes from occupations where inhalation is the primary method of exposure, the predominate form of cancer is lung cancer [91]. As a significant agent responsible for inducing cancer following radium exposure is the release of the ionizing radiation. With that said, when under controlled conditions, radium is used in the clinical treatment for human ankylosing spondylitis [94]. However, careful administration is critical because injection of radium has been associated with the development of several types of leukemia [94]. In animals, radium injections were demonstrated to induce the formation of osteosarcomas [94]. In a clinical case report, a patient being treated with radium-223 developed a cutaneous squamous cell carcinoma indicating such patients need to be followed clinically by a dermatologist [95].

2. Carcinogenicity effects on human cancer cells

2.1 Aluminum

Aluminum is known for its genotoxic profile in cosmetics, especially underarm anti-perspirant products [96]. Aluminum prevents perspiration by blocking the sweat ducts; it also absorbs through the skin. This environmental carcinogen accumulates in the human breast, transforming MCF-10A human mammary epithelial cells and inducing DNA double strand breaks (DSB). These effects have been exhibited *in vitro* with similar concentrations of aluminum to those measured in the human breast [97]. The concentrations of aluminum in the culture medium transform the MCF-10A human mammary epithelial cells, therefore enabling them to produce tumors that can metastasize [98].

To repair DSB is intrinsically mutagenic; once aluminum was removed from the culture medium, however, DSB were not reversible, therefore suggesting that mammary epithelial cells cultured in the presence of aluminum acquire mutations.

In addition, *in vitro* studies have shown that aluminum increases the migratory and invasive properties of MCF-7 or MDA-MB-231, human breast cancer cells [97].

Aluminum is a metalloestrogen, a type of inorganic xenoestrogen that is capable of binding to cellular estrogen receptors and mimicking the actions of physiological oestrogens [99]. The most commonly used aluminum-based compounds in underarm cosmetic products (UCP) are aluminum chloride and aluminum chlorohydrate. Not only do aluminum salts trigger DNA DSB, they can lead to oxidative stress, proliferation, and interference in estrogen action before and with metastasis.

A 1:1 age-matched hospital-based case–control study was performed to examine the impacts that self-reported UCP use had on breast cancer. Between a large series of breast cancer patients (aged 20–85 years) and healthy individuals, the aluminum concentrations in their breast tissue were measured and compared. The study participants were interviewed about their UCP application; their answers were categorized under "never", "1-4 times per month", "2-6 times per week", "daily" and "several times per day." A positive family history of breast cancer resulted in being the most prominent risk factor. However, self-reported use of UCP several times per day during early ages (< 30 years) showed a significant association with an increased risk of breast cancer. In addition, the aluminum in breast tissue was significantly associated with self-reported UCP use [98].

Another study showed that in an aqueous solution with a pH of 7.0, aluminum chloride and aluminum chlorohydrate yield aluminum hydroxide and are absorbed through the human skin. This suggests that with daily application of UCPs to the underarm's skin indicates a pronounced source of exposure to aluminum for the human mammary epithelium.

Aluminum has a transforming effect that is followed by the dose-dependent appearance of DNA DSB. The altered phenotype of MCF-10A cells that were cultured in the presence of aluminum chloride is not reversed by withdrawing the salt, however. These results reveal that a mutagenic effect is at least partly responsible for aluminum's transforming effect. The salt causes mutations in genes that regulate cellular proliferation, migration, metastasis and apoptosis. Mutations are also found in the genes monitoring the Max-binding protein MNT and T-lymphoma invasion and metastasis-inducing protein 2 (*Tiam2*) [1]. MNT functions as a pro-survival protein whose activity suppresses the pro-apoptotic activity of MYC, a family of proteins that contribute to oncogenesis [100]. The *Tiam2* gene serves a significant role in neuron development and human malignancies [101].

2.2 Arsenic

Arsenic is a naturally deposited metalloid that is widely distributed throughout the Earth's crust. Most arsenic-containing compounds are classified as organic and inorganic forms, with the inorganic form, specifically the trivalent arsenic (As^{3+}), being much more toxic and carcinogenic. Studies have shown that As^{3+} is an environmental etiological factor for a certain number of human cancers. There has shown to be a significant correlation between human lung cancer and environment As^{3+} exposure, either from drinking water contamination or air pollution. When As^{3+} is ingested through drinking water, it is absorbed into the bloodstream; its metabolic products, especially the methylated As^{3+}, is potentially deposited in the lung tissues due to the high partial pressure of oxygen [102].

The exact pathophysiological mechanism through which arsenic induces carcinogenesis is still to be determined; however, the increasing of oxidative stress, chromosome abnormalities (with uncontainable growth), and abnormal immune developments, are likely mechanisms. Reactive oxygen species, 8-Hydroxy-2-deoxyguanosine, is a major form of oxidative DNA damage that was acquired from the urine and skin tissue of individuals exposed by arsenic. DNA strand breaks, micronuclei in cord blood, and nitrative DNA damage were some of the early genetic effects discovered in the arsenic exposed patients. Studies have shown that arsenic also affects DNA repair machinery, which therefore causes oxidative DNA damage and mutations by the impairment of nucleotide excision repair, DNA ligase, DNA base excision repair, and DNA strand break rejoining.

Arsenic additionally affects epigenetic regulations. Chanda *et al.* claims that DNA hypermethylation of the crucial promoter region of the *p53* and *p16* genes was present in the DNA from arsenic-exposed individuals [103]. Since high exposure of arsenic is related to DNA hypermethylation of *p53* and *p16* genes, this suggests the notion that epigenetic silencing of these key tumor suppressors genes may be a notable mechanism by which arsenic induces cancer initiation [104].

Recent evidence has been reported to show that arsenic can alter miRNA expression patterns in *in vitro* and *in vivo* models of arsenic-induced carcinogenesis. Dysregulated miRNAs contribute to cancer development and progression, with the potential of acting as a novel class of oncogenes or of tumor suppressor genes. microRNAs are significant in tumorigenesis; for example, the overexpression of miR-504 negatively regulates the *p53* gene, decreasing the *p53*-mediated apoptosis, in addition to negatively regulating the cell cycle arrest in response to stress [105]. Production of reactive oxygen species (ROS) is one of the most reviewed mechanisms in arsenic carcinogenicity; as ROS reacts with DNA and induces structural DNA damage, genetic defects result, and the overexpression of antioxidant enzymes will desensitize cells to apoptosis. Arsenic can inflict oxidative stress through two different routes: direct Fenton-type reactions to produce ROS, or indirect depletion of critical antioxidants [106].

In immortalized human keratinocytes (HaCaT cells), miR-21, miR-200a, and miR-141 are overexpressed after a 4-week treatment with 500 nM sodium arsenic. For miR-21 and miR-141, these microRNAs have exhibited strong associations with the majority of human tumors. The miR-200 family has been reported to have a role in the epithelial-mesenchymal transition and cancer progression. For lung cancer development, the overexpression of miR-155 in normal cells has been a leading cause. Results indicate that urothelial human cancer is induced by miR-200 family members; the expression of miR-200a, miR-200b, and miR-200c was down-regulated in arsenic-exposed human urothelial cells (HUC1) in comparison to nonexposed HUC1 cells. The levels of these miR-200 family members in the urine of arsenic-exposed patients were also decreased [105].

2.3 Beryllium

Beginning in 1952, a collection of case reports in the Beryllium Case Registry at the Massachusetts General Hospital and cohort studies established the basis for several overlapping epidemiological reports on how beryllium induces cancer. Elevated ratios of lung cancer were shown among workers who had experienced acute berylliosis; however, the results were not similar in workers with chronic berylliosis [107]. Acute beryllium disease is mostly considered an irritative chemical phenomenon associated with high exposures; on the other hand, chronic beryllium disease is an immune-mediated granulomatous reaction to beryllium [108]. Studies showed that the increased cancer death started to occur 15 years after the onset of beryllium exposure.

Experiments were conducted by injecting zinc beryllium silicate in rabbits intravenously. Results indicated that the administration produces consistently metastasizing osteosarcomas in the long bones. Outcomes parallel to these results were obtained with the injection of beryllium oxide, beryllium phosphate, and beryllium metal into the medullary cavity of bones. This route of administration was the only route that led to the formation of osteosarcomas. Splenectomy was additionally shown to increase carcinogenicity with the IV-injected beryllium in bones; the spleen, being an important storage organ, most likely allowed the retention of a higher proportion in the reticuloendothelial system and bone.

Exposing the rats to beryllium sulfate, beryllium phosphate, beryllium fluoride, zinc beryllium manganese silicate, and beryl ore, through inhalation also produced carcinogenic properties. Throughout the duration of a 35-hour week exposure schedule, 10 micrograms of $BeSO_4$ was determined to be threshold for the induction of pulmonary adenocarcinoma in rats. The majority of malignancies were adenocarcinomas with a predominantly alveolar pattern.

In Chinese hamster V79 cells (lung fibroblasts) and in Chinese hamster ovary (CHO) cells, the induction of 8-azaguanine-resistant mutants by $BeCl_2$ and by $BeSO_4$, respectively, has demonstrated beryllium's ability to inflict gene mutations in cultured mammalian cells. $BeSO_4$ did not cause chromatid or chromosomal aberrations in Chinese hamster lung cells. In CHO cells and cultured human lymphocytes, however, $BeSO_4$ produced chromosomal breaks and sister-chromatid exchanges [107].

With a soluble beryllium compound and upon incubation of a continuous human cell line, there was shown to be a reduction of the expression of messenger RNA coding for DNA repair proteins. This observation was suggested to be a relevant mechanism for potential carcinogenicity of beryllium. To further study this claim, the DNA of rat primary hepatocytes was purposely damaged by incubation with a known DNA damaging agent, 2-acetylaminofluorene. In addition, the DNA was co-incubated with beryllium metal extracts. In the results, there was a reduction in DNA repair synthesis with the beryllium metal extract. Beryllium metal has not been confirmed to directly damage the DNA of cells; nevertheless, there is strong evidence that the metal can cause morphological cell transformation and the inhibition of DNA repair synthesis [109].

The carcinogenic properties of beryllium have been mostly demonstrated when in its metal form, some of its alloys, and a variation of its compounds. Lung cancer induced by beryllium is a main result from pulmonary instillation or inhalation with consequent direct action on the lung. The bone tumors that beryllium stimulates, a characteristic of osteogenic sarcoma, reflects the metal's bone seeking propensities [110].

2.4 Cadmium

Cadmium is a dangerous metal for humans as the human body is limited in its response to cadmium exposure; the metal is incapable of metabolic degradation to less toxic species [111]. Cadmium is a toxic heavy metal that is commonly known as a human carcinogen. Their main sources of exposure include food, cigarette smoking, and cadmium related industry. Reactive oxygen species (ROS) are measured to be the most prominent mechanism in cadmium-induced carcinogenesis. The intracellular oxidative stress that reactive oxygen species induce potentially damage macromolecules and eventually grow responsible in the formation of cancer.

There are two stages referred to when discussing cadmium-induced carcinogenesis. In the first stage, normal cells transition into transformed cells. The reactive oxygen species contribute in the malignant cell transformation of BEAS-2B (human bronchial epithelial) cells in their exposure to cadmium. For the second stage, morphologically transformed cells advance into tumorigenesis. Cadmium-transformed cells, *p62* and Nrf2, are activated and their downstream antioxidants and anti-apoptotic proteins are elevated, therefore causing a reduction in ROS, apoptosis resistance (permitting cancer cells to persist and not die), and tumorigenesis. The decrease in ROS generation in the second stage provides an optimal environment for transformed cells to survive and engage in tumorigenesis [112].

Cadmium exposure is shown to induce consistent low levels of ROS production, which causes endoplasmic reticulum stress that causes defective autophagy, which protects cadmium exposed damaged cells and encourages malignant transformation in prostate carcinogenesis. In order to maintain the quality of intracellular components, autophagy, a highly complex lysosomal-mediate degradation process, is accountable for the removal and recycling of damaged organelles. This deficient form of this activity assists in cancer cell survival as autophagy protects the cells from hypoxia and oxidative damage, in addition to promoting chemoresistance [113].

The *p62* protein performs several cellular functions for autophagy, apoptosis, ROS signaling, and cancer. The protein has been found to accumulate in autophagy-deficient cells, and the overall accumulation of *p62* due to autophagy dysfunction encourages cell survival and tumorigenesis through the activating of nuclear factor, κB. The *p62* protein is highly expressed in human lung cancer. As *p62* accumulates, it activates Nrf2 and Nrf2 target gene expression. Autophagy deficiency results in the up-regulation of *p62,* which therefore leads to the transcriptional activation of the Nrf2-dependent genes, involving antioxidant enzyme genes [114].

Similar to metal arsenic, cadmium is weakly genotoxic and mutagenic. To determine whether cadmium exposure induces properties analogous to cancer stem cells, researchers exposed immortalized human pancreatic ductal epithelial (HPDE) cells to low dose cadmium for 29 weeks. Using suspension culture spheroid formation assay, the chronic cadmium-exposed HPDE cells exhibited significantly higher levels of molecular markers for cancer stem cells, yielding 3-fold more suspension spheres than the controlled cells [115].

Cadmium does not form adducts with DNA; however, it is capable of inflicting oxidative stress that could indirectly attack DNA. This process is not instigated through participation in Fenton type chemical reactions [111]. The Fenton reaction is defined by a redox pair of ferrous ion and hydrogen peroxide (H_2O_2) that ultimately generates a reactive hydroxyl radical [116]. The potential mechanisms for cadmium-carcinogenesis include aberrant gene activation and signal transduction, suppressed apoptosis and disruption of E-cadherin-mediated-cell–cell adhesion, and altered DNA repair [111].

2.5 Lead

Lead is a metal that can be classified as an environmental pollutant and is commonly known for its usage in many industrial settings worldwide. With high lead exposure, health effects can include damage to the brain and nervous system, gastrointestinal problems, anemia, liver and kidney damage, fertility problems, and developmental delays. Inorganic lead is also suggested to be a carcinogen; epidemiological evidence for carcinogenicity in industrial workers that have been exposed to inorganic lead indicates a significant relationship with cancers of the stomach, lung, kidney, brain, and meninges.

The two primary routes through which lead enters and accumulates in the body is inhalation and oral ingestion. With this being said, even though lead has the capacity to enter the bloodstream and impact other organs of the body, the lungs and stomach are what first come into contact with lead. Due to lead's ability to pass through the blood–brain barrier, the brain and nervous system are especially vulnerable to the potential toxic effects of lead. The mechanisms that lead uses in playing a role in carcinogenesis include oxidative damage, induction of apoptosis, altered cell-signaling pathways, inhibition of DNA synthesis and repair of damage, and interaction with DNA-binding proteins [117].

In one study, results provided support for an association between occupational lead exposure and brain cancer risk. Among industrial workers who were potentially exposed to lead, the brain cancer mortality rates were greater as compared to unexposed subjects, with indications of an exposure-response trend [118]. Results, however, of many studies have showed inconsistency in determining the relationship between lead exposure and brain tumors. For results that support the association, the results suggest that lead can cross the blood–brain barrier and concentrate in the brain parenchyma due to its ability to replace calcium ions. Once the lead is absorbed, it is generally allocated to plasma, the nervous system, and soft tissues, therefore potentially developing micronucleus formation, chromosomal aberrations, and DNA damage in most mammals.

Lead's mechanism in which it causes brain cancer remains unclear; nevertheless, studies suggest the most probable mechanism is the metal's inhibiting of DNA synthesis and repair and the interacting with binding proteins that eventually hinder tumor suppressor proteins [119].

2.6 Mercury

Mercury is one of the most toxic heavy metals due to its persistence in the environment. Mercury inflicts oxidative stress and induces apoptosis. Methylmercury (MeHg) is a metalloestrogen, a small ionic metal that activates the estrogen receptor. Studies indicate that once metalloestrogens activate the estrogen receptor, there is an increase in transcription and expression of estrogen-regulated genes, therefore inducing proliferation of estrogen-dependent breast cancer [120].

The phases of cancer development are initiation, latency, promotion, and then progression. In the promotion phase, mercury has shown to cause an imbalance in the reactive oxygen species homeostasis through selectively inhibiting selenocysteine antioxidant enzymes. Mercury fulfills both the capacity to induce an inhibition of the gap junction intercellular communication and the proinflammatory cytokine release. These two mechanisms have potential to isolate cells from tissue-specific homeostasis, promoting their proliferation. In addition, they have potential to overcome the immune system defenses, checkmating the entire organism. The International Agency Research Cancer (IARC) does not classify mercury as an identified carcinogen to humans; nevertheless, if the toxic compound inhibits the gap junction intercellular communication, mercury is suggested to be a potential cancer "promoter" [121].

Animal experiments were performed to investigate the carcinogenic effects that methylmercury had on mice. They were fed with 10 mg/kg of methylmercury, and as a result, chronic kidney failure, adenoma, and carcinoma were observed. With these results, rodents that were exposed to methylmercury were reported to show a higher incidence of kidney cancer. The International Agency for Research on Cancer claims there is a satisfactory amount of evidence for methylmercury's impact in cancer on experimental animals, only classifying it as a possible carcinogenic to humans. On the other hand, the U.S. Environmental Protection Agency (EPA) judges that evidence of methylmercury's carcinogenic potential in humans was insufficient and the justification of the carcinogenicity in experimental animals was limited. Therefore, they classified methylmercury as a Group C material (possible human carcinogen) [122].

Mercury can affect multiple organ systems, especially the nervous and renal systems. One particular study wanted to determine mercury's capacity to induce centrosome amplification. Centrosomes, microtubule organizing centers of the cell, play a crucial role in cell division; they aid in the proper segregation of chromosomes into the resulting daughter cells. When metals induce cellular and genotoxic stress, however, this can interfere with the strict coordination between the centrosome and DNA cycles that ensures the cell to enter mitosis with only two chromosomes. This disrupted linkage stimulates centrosome amplification, potentially resulting in chromosome segregation and aneuploidy. For the aneuploid cells that survive, they can eventually lead to tumor formation and cancer. The study reported that methylmercury, but not inorganic mercury, prompted both a mitotic arrest and centrosome amplification in mitotic cells, therefore suggesting a possible carcinogenic mechanism [123].

2.7 Nickel

Nickel is considered a major carcinogenic heavy metal, mainly through the mechanism of DNA damage. Demonstrated by *in vitro* and *in vivo* studies, nickel destructs DNA processes through direct DNA binding and reactive oxygen species (ROS) stimulation. Nickel's carcinogenic properties also include their repressing of DNA damage repair systems through direct enzyme inhibition and downregulation of DNA repair molecule expression. Studies have shown that Ni^{2+} has potential to induce DNA damage in certain human cell systems; some include hepatocellular carcinoma (HepG2), human TK6, Chinese hamster lung fibroblast, A375, and HCT-116 cells [124, 125].

With reactive oxygen species, when they excessively attack the DNA, this results in genomic instability, a promoter of tumorigenesis. This oxidative stress or genomic instability, being a major driving force of oncogenesis, is the basic toxicological mechanism of nickel overexposure [124]. Oxidative stress is known to occur as a result of overproduction of reactive oxygen and nitrogen species through endogenous and exogenous insults. The production of these reactive oxygen species is enabled by nickel's capacity to bind with amino acids, peptides, and proteins [125].

The metal has the ability to dissolve in the human body, releasing ionic nickel, an active and occasionally genotoxic carcinogenic form of nickel. When a carcinogen is classified as 'genotoxic', this refers to chemicals that are capable of directly altering genetic material, opposed to 'non-genotoxic' carcinogens that produce cancer through indirect or secondary mechanisms. Most of the chemical carcinogens that induce direct DNA damage are therefore categorized as 'genotoxic' in their carcinogenic mechanisms. Nickel's carcinogenic potential also originates from its capacity to raise the intracellular concentration of nickel ions [126]. The nickel ions exhaust intracellular iron by hindering the membrane ion transporters, in addition to displacing iron from the active site of dioxygenase enzymes. This all leads to the inhibition of their catalytic activity [127].

DNA hypermethylation and subsequent silencing of tumor suppressor genes potentially serve as an epigenetic mechanism responsible for nickel's carcinogenicity. Promoter hypermethylation induced by nickel was observed *in vivo* as nickel sulfide was injected into *p53* heterozygous mice to induce tumor formation. Malignant fibrous histiocytomas advanced in both wild type and *p53* heterozygous mice, with all tumors exhibiting promoter hypermethylation of *p16* (a tumor suppressor gene). Additionally, Wistar rats exhibited muscle tumors that displayed DNA hypermethylation in the promotor regions of *RARβ2*, *RASSF1A* and *p16* genes, following intramuscular injection of nickel sub-sulfide [128].

2.8 Radium

Along with X-rays, radium has a carcinogenic effect of ionizing radiation in humans. The danger of ionizing radiation involves the risk of developing cutaneous squamous cell carcinoma. Additionally, studies suggest that radium treatment for the benign skin lesions may only increase the risks of sarcoma of the bone. For example, in one particular case, a patient developed a mixed tumor of carcinoma and sarcoma at the specific site where she had received radium treatment; a malignancy that developed in the same location supports the notion that the previous radium treatment caused it [129].

At elevated concentrations, naturally occurring dissolved radium can potentially be classified as carcinogenic to the human body. Following digestion, the radium can become deposited within the body where its radioactive characteristic threatens human health through cell damage, therefore increasing the overall risk of cancer [130].

Other experiments show that intra-uterine radium application or X-irradiation of the uterus can induce rat malignant uterine tumors, usually endometrial adenocarcinomas. One rat subject's uterus was exposed to direct X-irradiation and a composite endometrial tumor, also classified as an adeno-sarcoma, was produced. The tumor was not structurally similar to the mixed endometrial tumors seen in women; nevertheless, the composite structure and the potential that the tumor may also exhibit carcinomatous areas, implies that it may strongly represent the rat counterpart of the human neoplasm. Results of the experiment strengthened the suspicion that pelvic radiation can lead to an increase in long-term incidence of uterine cancer, particularly mixed tumors [131].

3. Legalization and the national and international permissible levels for these heavy metals

In recent years the legalization of producing marijuana (cannabis and cannabis-derived products) especially in certain specific states within the United States has caused a level of alarm in part because of the presence of heavy metals within these products. As the result of the expansion in the commercialization of these products, has created the challenge to now measure heavy metals in cannabis and cannabis-infused commodities. Marijuana is now legal and approved for both medical and recreational use in 33 states within the United States and the District of Columbia (Washington, DC) [132]. However, the raw materials (cannabis and hemp plants) are known to be hyperaccumulators of contaminants such as heavy metals that may be present in the medium used to cultivate the plants, whether it is the soil, the fertilizers used, and in any other growth promoting substances used to supply needed nutrients. With that said, the alarm has been sounded to critically monitor the levels of heavy metal contaminants present in any part of the growing process to ensure that the marijuana-cannabis material and its food-associated products are safe to consume [133].One of the major remaining issues, at least in the United States, is the lack of federal government oversight regarding measuring contaminants in marijuana (cannabis and cannabis-prepared food products) produced in the United States. The U.S. federal government has removed itself from this oversight and in doing so they have delegated regulatory issues to the individual states to regulate the use of marijuana cannabis and cannabis-prepared products. This adds a financial burden to states that are often financially stressed to meet these demands.

What individual states have emphasized has been to focus on the manufacturers of these marijuana (cannabis and cannabis-prepared food products) to show regulation by measuring for the following four major heavy metals: lead (Pb), arsenic (As), cadmium (Cd), and mercury (Hg). The levels of metals must be below maximum limits, based mainly on regulations set by the pharmaceutical industry in USP Chapter 232 and ICH Q3D guidelines [134, 135]. The state of California is usually the state that places severe restrictions on levels allowed. This policy is considered to be the gold standard in regulating cannabis and hemp. It determines the levels allowable in both the oral (edibles) and inhaled (vapes) cannabis products to be to safe to consume only if these four heavy metals are present at levels below those shown in **Tables 1** and **2**, based on typical consumption of 10 g/day of cannabis material [137].

For analytical measurements of heavy metal contaminants, the state of California requires that at least half a gram of sample must be used for testing purposes. The analytical testing methodology recommends that inductively coupled plasma mass spectrometry (ICP-MS) serve as the method of choice [136]. ICP-MS is a sophisticated multi-element analytical technique, capable of measuring levels to parts per trillion (ppt) using mass spectrometry to identify and measure positively charged ions. The testing methods occur in an extremely energetic argon plasma at approximately 6,000-7,000°C [138]. However, this method requires a solution technique, meaning any solid samples must be dissolved/digested before being analyzed. Most cannabis-related samples are solid materials, powders, concentrates and extracts, which invites several challenges. In addition, cannabinoid oils, which are mainly hydrophobic (not miscible with water), must also be digested prior to analysis.

As mentioned previously in the United States within the federal government exists the Environmental Protection Agency (EPA). The function of the EPA is to set federal standards for a variety of compounds and substances in terms of determining their presence in the environment, which includes the air, soil, ground water, lakes, and rivers. Over the past several years, based upon political influences, the minimum acceptable levels for a variety of substances such as heavy metals have been increased for no other reason than to reduced regulations without factoring the environmental impact. These changes come at the expense of potentially reducing the overall quality of air, soil, ground water, lakes and rivers, thus imposing potential harm to people (children and adults). As mentioned earlier in the discussion of sources of lead contamination, recent incidence of dramatically higher levels of lead present in the drinking water of those living in the Flint, Michigan area is an example of political incompetence when the politicians in the community changed the source of the community drinking water from Lake Michigan to the Detroit River, which was highly contaminated because of age of the lead pipes used to pump the water from the river. The rationale for the change was to reduce the overall costs of providing usable water for the community. Based upon the excessive negligence involved in this case, compensation costs to the citizens of the community has been in the hundreds of millions of dollars not to mention the total costs involved to completely remove and install a new water delivery system devoid of metal pipes of

Element	Maximum limit (edibles) mg/m	Maximum limit (inhaled) mg/g
Arsenic	1.5	0.2
Cadmium	0.5	0.2
Lead	0.5	0.5
Mercury	3.0	0.1

Table 1.
Heavy metal limits (cannabis & cannabis-hemp) by state of California [136].

Element	Maximum limit (soil) mg/kg	Maximum limit (plant) mg/kg
Cadmium	0.8	0.02
Chromium	100	1.3
Lead	85	2
Lead (water)	0.01/children, 0.015/adult	
Mercury	50	200
Nickel	35	10

Table 2.
Heavy metal limits according to the World Health Organization (https://www.who.int/ceh/capacity/heavy_metals.pdf?ua=1).

any kind. The long-term consequence of this unfortunate and unnecessary change on the overall sustained health of the community is yet to be determined.

In Europe the European Environment Agency (EEA), controls the level of pollutants such as heavy metals [139]. Regarding heavy metal emissions, across the 33 European countries the following is a short summary of recent achievements in the EU with respect to reducing heavy metals concentrations: (a) Since 1990 across all 33 countries, lead admissions decreased by 93%, mercury by 72%, and cadmium by 64%; (b) Reductions in levels of lead have occurred by 2004 due in part to the removal lead from gasoline; (c) Reductions in levels of mercury have occurred as the result of changes in energy use both in industries and other processes used in industry; and (d) Reductions in levels of cadmium are attributed to operational changes in industries across the board.

A remaining issue in a select set of areas of Europe is the continued presence of unacceptable levels of arsenic, cadmium, lead, mercury and nickel as the result of the presence of these substances still in the atmosphere [140]. The collective set of excessive metal excesses due cause local health issues, in part, because of the presence of localized industrial plants that release emissions of the pollutants. With that said, even though the emissions are concentrated in localized areas, this does not limit nor restrict the impact on health concerns because the pollutants are able to enter the food chain through ground soil and water contamination. Across the EU member countries there is the political will to do what is necessary and needed to sustain the momentum to continue to reduce pollutants in the environment.

4. Mitigation of the negative effects of these heavy metal materials

One could deduce that the presence of heavy metals in the environment combined with occupational exposure is a problem for human health. A pertinent question to ask then is "what can be done to reduce human heavy metal exposure?" Several remedies or actions can be considered that have been shown to be effective, they are:

a. Antioxidants - consume foods high in Vitamin C. Fruits and vegetables high in vitamin C can reduce the damage caused by heavy metal toxins by acting as an antioxidant. Vitamin C helps to convert toxins into a water-soluble form that can be easily eliminated from the body [141].

b. Porphyra. A logical approach would be to use naturally occurring organisms to monitor and remove toxic metals from aquatic systems. Such organisms could be harvested at regular intervals, dried and disposed of as contaminated solid waste or used to recover valuable heavy metals. It is the feasibility and future optimization of this approach using marine macroalgae that forms the basis of the proposed studies. Bioavailability of heavy metals is highly dependent upon several environmental factors. Biomonitors utilizing plants growing under "natural" conditions where biotic and abiotic factors are intercalated reduces the need for making assumptions regarding bioavailability of metals. Plants themselves can alter the microenvironment around them, thus altering the amount of metals that are biologically available. Bulk water analysis may not measure the conditions at the membrane level where changes occur. Benthic plants can provide valuable information regarding past environmental conditions over weeks and months. This is particularly important in plants growing within the intertidal zones where the metal content of water may fluctuate continuously [142].

c. Integrated Processes. Addressing heavy metal pollution is one of the productive areas of environmental research. Despite natural existence, various anthropomorphic sources have contributed to an unusually high concentration of heavy metals in the environment. The central problem is often these metals are characterized by their long persistence in natural environment leading to serious health consequences in humans, animals, and plants even at very low concentrations (only 1 or 2 μg in some cases). Failure of restrict regulations by government authorities is also to be blamed for heavy metal pollution. Several individual treatments, namely, physical, chemical, and biological are being implied to remove heavy metals from the environment; but they all face challenges in terms of expensiveness and *in-situ* treatment failure. Hence, integrated processes are gaining popularity as it is reported to achieve the goal effectively in various environmental matrices and will overcome a major drawback of large-scale implementation. Integrated processes are the combination of two different methods to achieve a synergistic and an effective effort to remove heavy metals. Many of the articles published so far have focused on what individual methods are most effective to remove specific heavy metals concentrating on environmental exposure. Although integrated processes are being used in mediation of heavy metal extraction, there is still the need to determine the advantages and disadvantages of each integrated process. With that said, regardless of the method more research is necessary to determine what remediation method is most effective in reducing heavy metal concentrations in the environment [143].

d. Phytoremediation. Soil heavy metal pollution has become a worldwide environmental problem that has attracted considerable public attention. This attention stems largely from the increasing concern regarding the overall security of agricultural products. In this area, heavy metals refer to several metals and metalloids that possess toxicity on biological systems. The heavy metals of most concern are arsenic, cadmium, chromium, lead, and mercury. These metals pollutants enter the soil agricultural ecosystem through natural processes derived from parent materials and anthropogenic activities. As stated previously, heavy metal pollution poses a great threat to the health and well-being of all organisms not just human beings due to the risk of increased accumulation potential that takes place through bioaccumulation via the food chain. Remediation from heavy metal exposure using chemical, physical, and biological methods has been recommended to best solve the overall problem of toxic exposure in the environment. Phytoremediation has proven to be a promising alternative to conventional approaches as it is cost effective, environmentally friendly, and esthetically pleasing. According to studies conducted, based on the natural ability of extraction, approximately 500 plants and other organisms have been identified as hyperaccumulators of one or more the heavy metals. In addition, further research integrating biotechnological approaches with comprehensive multidisciplinary research is needed to improve plant tolerance and reduce the accumulation of toxic metals in soils [144].

e. Other. As stated, heavy metals endanger overall human health. Of importance are the conditions especially when as the result of testing identifies heavy metal levels to be significantly above required standards for each. What still remains as an important factor regarding overall human health is that sustained elevated levels of heavy metals are indeed carcinogenic. The majority of studies performed designed to determine the pathway of heavy metal exposure that results in the carcinogenic effect of heavy metals in human exposure takes place via heavy metal contamination the overall food chain thereby impacting the quality agriculture, specifically the generation of agricultural products such as food and food by-products. In addition to the exposures that account for heavy metal contamination, there are additional factors that account for human exposure. This additional exposure can occur through the use of pesticides directly contaminating soil and also through waste-water run-off contamination. There are natural remediation methods that can help remediate areas of heavy metal contamination such as the presence of geological specific rock formations. With that said, it is still necessary to employ methods that address the heavy metal remediation especially when the sites of contamination are present in food products - fruits and vegetables. In addition, remediation of soil areas and water may also must be considered because these factors also contribute to heavy metal food contamination. Thus, it is imperative heavy metal remediation methods be used constructively in order to maintain overall public health [145].

5. Conclusion

The heavy metals have been shown to be responsible for a variety of human illnesses. These illnesses develop as the result of unwanted exposure whether by internal or external processes. One of the major health problems associated with heavy metal exposure is the development of a variety of cancers. The most common risk factors for developing cancer are exposure to heavy metals in the form of industrial based carcinogens, in cigarette smoke, and through foods consumed, thus via the diet. The toxicity associated with heavy metal poisoning can vary from minor conditions to major diseases, such as cancer. Both are capable of compromising overall human health. It is a fact that the major pathway responsible for human exposure, more often than not, is linked to both environmental and occupational exposure. Comparable studies have demonstrated higher levels of heavy elements, such as arsenic, aluminum, cadmium, lead, and nickel were present in cancerous tissue when compared and measured against non-heavy metal exposure in tissue from controls. Thus, limiting human exposure to heavy metals is sound public health policy; however, successful health policy must include cooperation from local, regional and national government agencies to develop, approve, implement, and then enforce those policies in order to reduce the links between heavy metal exposure and the major health concerns associated with exposure. The challenge seen for local, state, national and/or federal governments is to take these health concerns seriously and devise suitable and cost effective remedies to reduce the overall impact of heavy metals on the health consequences for its citizens.

Author details

Vincent Salvatore Gallicchio* and Juley Harper
Department of Biological Sciences, College of Science, Clemson University, Clemson, South Carolina, USA

*Address all correspondence to: vsgall@clemson.edu

References

[1] Mandriota SJ, Tenan M, Ferrari P, Sappino AP. Aluminum chloride promotes tumorigenesis and metastasis in normal mammary gland epithelial cells. International Journal of Cancer. 2016;**139**(12):2781-2781. DOI:1002/ijc.30393

[2] Darbre PD. Aluminum and the human breast. Morphologie 2016;**100**(329):65-74. DOI:10.1016/j.morpho.2016.02.001

[3] Farasani A, Darbe PD. Effects of aluminum chloride and aluminum chlorohydrate on DNA repair in MCF10A immortalized non-transformed human breast epithelial cells. Journal of Inorganic Biochemistry. 2015;**152**:186-189. DOI:10.1016/j.inorgbio.2015.08.003

[4] Roncati L, Gatti AM, Capitani F, Barbolini G, Mairoana A, Palmieri B. Heavy metal bioaccumulation in an atypical primitive neuroectodermal tumor of the abdominal wall. Ultrastructural Pathology. 2015;**39**(4):286-292. DOI:10.3109/01913123.2015.1013655

[5] Abdel-Gawad M, Elsobky E, Shalaby MM, Abd-Elhameed K, Abdel-Rahim M, Ali-El-Dein B. Quantitative evaluation of heavy metals and trace elements in the urinary bladder. Comparison between cancerous, adjacent non-cancerous and normal cadaveric tissue. Biological Trace Element Research. 2016;**174**(2):280-286. DOI:10.1007/s12011-016-0764-6

[6] Yokel RA. Aluminum chelation: Chemistry, clinical, and experimental studies and the search for alternatives to desferrioxamine. Journal of Toxicology and Environmental Health. 1994, **41**(2):131-174. DOI:10.1080/15287299409531834

[7] Ambiado K, Bustos C, Schwarz A, Bórquez R. Membrane technology applied acid mine drainage from copper mining. Water Science and Technology. 2016;**75**(3):705-715. DOI: 10.2166/wst.2016.556

[8] Martinez VD, Vucic EA, Becker-Santos DD, Gil L, Lam WL. Arsenic exposure and the induction of human cancers. Journal of Toxicology. 2011;**201**:1-13. DOI:10.1155/2011/431287

[9] Bjørklund G, Aaseth J, Chirumbolo S, Urbina MA, Uddin R. Effects of arsenic toxicology beyond epigenetic modifications. Environmental Geochemistry and Health. 2017;**39**:1-11. DOI: 10.1007/s10653-017-9967-9

[10] Chakrabotri D, Das B, Rahman MN, Nayak B, Pal A, Sengupta MK, Ahamed S, Hossain MA, Chowdhury UK, Biswas BK, Saha KC, Dutta RN. Arsenic in groundwater of the Kolkata Municipal Corporation (KMC), India: Critical review and modes of mitigation. Chemosphere. 2017;**180**:437-447. DOI: 10.1016/j.chemosphere.2017.04.051

[11] Nachman KE, Ginsberg GL, Miller MD, Murray CJ, Nigra AE, Pendergrast CB. Mitigating dietary arsenic exposure: Current status in the United States and recommendations for an improved path forward. Science of the Total Environment. 2017;**581-582**:221-236. DOI: 10.1016/j.scitotenv.2016.12.12

[12] Pershagen G. The carcinogenicity of arsenic. Environmental Health Perspectives. 1981;**40**:93-100. DOI: 10.23-7/3429223

[13] Chen K, Liao QL, Ma ZW, Jin Y, Hua M, Bi J, Huang L. Association of soil arsenic and nickel exposure with cancer mortality rates, a town-scale ecological study in Suzhou, China. Environmental Science and Pollution

Research. 2014;**22**(7):5395-5404. DOI: 10.1007/s11356-014-3790-y

[14] Satarug S, Vesey DA, Gobe GC. Kidney cadmium toxicity, diabetes and high blood pressure: The perfect storm. The Tohoku Journal of Experimental Medicine. 2017;**241**(1):65-87. DOI: 10.1620/tjem.241.65

[15] Amaral AF, Porta M, Silverman DT, Milne RL, Kogevinas N, Cantor KP, Jackson BP, Pumarega JA, López T, Carrato A, Guarner L, Real FX, Malata N. Pancreatic cancer risk and levels of trace elements. Gut. 2011;**61**(11):1583-1588. DOI: 10.1620/tjem.241.65

[16] Cui J, Xu W, Chen J, Li H, Dai L, Frank JA, Peng S, Chen G. M2 polarization of macrophages facilitates arsenic-induced cell transformation of lung epithelial cells. Oncotarget. 2017;**8**(130):21398-21409. DOI: 10.18632/oncotarget.15232

[17] Park YH, Kim D, Dai J, Zhang Z. Human bronchial epithelial BEAS-2B cells, an appropriate *in vitro* and *in vivo* model to study heavy metals induced carcinogenesis. Toxicology and Applied Pharmacology. 2015;**287**(3):240-245. DOI: 10.1016/j.taap.2015.06.008

[18] Hall MN, Niedzwiecki M, Liu X, harper KN, Alam S, Slavkovich V, Ilievski V, Levey D, Siddique S, Parvez F, Mey JL, van Geen A, Graziano J, Gamble MV. Chronic arsenic exposure and blood glutathione and glutathione disulfide concentrations in Bangladeshi adults. Environmental Health Perspectives. 2013;**121**(9):1068-1074. DOI: 10.1289/ehp.1205727

[19] Sykora P, Snow ET. Modulation of DNA polymerase beta-dependent base excision repair in cultured human cells after low dose exposure to arsenic. Toxicology and Applied Pharmacology. 2008;**228**(3):385-394. DOI:10.1016/j.taap.2007.12.019

[20] He J, Wang F, Luo F, Chen X, Jiang W, Huang Z, Lei J, Shan F, Xu X. Effects of long term low- and high dose sodium arsenite exposure in human transitional cells. American Journal of Translational Research. 2017;**9**(2):416-428

[21] Harper LK, Bayse CA. Modeling the chelation of As(III) in lewisite by dithiols using density functional theory and solvent-assisted proton exchange. Journal of Inorganic Biochemistry. 2015;**153**:60-67. DOI: 10.1016/j.jinorgbio.2015.10.004

[22] Lu PH, Tseng JC, Chen CK, Chen CH. Survival without peripheral neuropathy after massive acute arsenic poisoning: Treated by 2,3-dimercaptopropane-1-sulphonate. Journal of Clinical Pharmacy and Therapeutics. 2017;**42**(4):506-508. DOI: 10.1111/jcpt.12538

[23] Mandal P. Molecular insight of arsenic-induced carcinogenesis and its prevention. Naunyn-Schmiedeberg's Archives of Pharmacology. 2017;**390**(5):443-455. DOI: 10.1007/s00210-017-1351

[24] Stanton BA, Caldwell K, Congdon CB, Disney J, Donohue M, Ferguson E, Flemings E, Golden M, Guerinot ML, Highman J, James K, Kim C, Lantz RC, Marvinney RG, Mayer G, Miller D, Navas-Acien A, Nordstrom DK, Postema S, Rardin L, Rosen B, Sengupta A, Shaw J, Stanton E, Susca P. MDI biological laboratory arsenic summit: Approaches to limiting human exposure to arsenic. Current Environmental Health Reports. 2015;**2**(3):329-337. DOI: 10.1007/s40572-015-0057-9

[25] Spanu A, Daga L, Orlandoni AM, Sanna G. The role of irrigation techniques in arsenic bioaccumulation in rice (*Oryza sativa* L.). Environmental Science & Technology.

2012;**46**(15):8333-8340. DOI: 10.1021/es300636d

[26] Nogaj E, Kwapulinski J, Misiolek M, Golusiński W, Kowol J, Wiechula D. Beryllium concentration in pharyngeal tonsils in children. Annuals of Agricultural and Environmental Medicine. 2014;**21**(2):267-271. DOI: 10.5604/1232-1966.1108589

[27] Shay E, De Gandiaga E, Madl AK. Considerations for the development of healthbased surface dust cleanup criteria for beryllium. Critical Reviews in Toxicology. 2013;**43**(3):220-243. DOI: 10.3109/10408444.2013. 767308

[28] Boffetta P, Fryzek JP, Mandel JS. Occupational exposure to beryllium and cancer risk: A review of the epidemiologic evidence. Critical Reviews in Toxicology. 2012;**42**(2): 107-118. DOI: 10.3109/10408444. 2011.631898

[29] Hollins DM, McKinley MA, Williams C, Wiman A, Fillos D, Chapman PS, Madl AK. Beryllium and lung cancer: A weight of evidence evaluation of the toxicological and epidemiological literature. Critical Reviews in Toxicology. 2009;**39**(1):1-32. DOI: 10.1080/10408440902837967

[30] Beryllium and Beryllium Compounds. International Agency for Research on Cancer. 2012;**100C**:95-120

[31] Stark M, Lerman Y, Kapel A, Pardo A, Schwarz Y, Maier L, Fireman E. Biological exposure metrics of beryllium-exposed dental technicians. Archives of Environmental & Occupational Health. 2013;**69**(2):89-99. DOI: 10.1080/19338244.2012.744736

[32] Benderlu Cihan Y, Sönzen S, Oztürk Yildirim S. Trace elements and heavy metals in hair of stage III breast cancer patients. Biological Trace Element Research. 2011;**144**(1-3):360-379. DOI: 10.1007/s12011-011-9104-z

[33] Sandhu R, Lal H, Kundu ZS, Kharb S. Serum fluoride and sialic acid levels in osteosarcoma. Biological Trace Element Research. 2009;**144**(1-3):1-5. DOI: 1007/s12011-009-8382-1

[34] Radauceanu A, Grzebyk M, Edmé JL, Chérot-Kornobis N, Rousset D, Dziurla M, DeBroucker V, Hédelin G, Sobassek A, Hulo S. Effects of occupational exposure to poorly soluble forms of beryllium on biomarkers of pulmonary response in exhaled breath of workers in machining industries. Toxicology Letters. 2016;**263**:26-33. DOI: 10-1016/j. toxlet.2016.10.013

[35] Shukla S, Sharma P, Johri S, Mathur R. Influence of chelating agents on the toxicity and distribution of beryllium in rats. Journal of Applied Toxicology. 1998;**18**(5):331-335. DOI: 10.1002/(SICI)1099-1263(1998090) 18.5-5<331:AID-JAT5517>3.0.CO;2.0

[36] Johri S, Shukla S, Sharma P. Role of chelating agents and antioxidants in beryllium induced toxicity. Indian Journal of Experimental Biology. 2002;**40**(5):575-582

[37] Sharma P, Johri S, Shukla S. Berylllium-induced toxicity and its prevention by treatment with chelating agents. Journal of Applied Toxicology. 2000;**20**(4):313—318. DOI: 10.1002/1099-1263(200007/08)20.4: AID-JAT660>3.0CO;2-J

[38] Crinnion WJ, Tran JQ. Case report: Heavy metal burden presenting as Bartter syndrome. Alternative Medicine Review. 2010;**15**(4):303-310

[39] Mayer AS, Brazile WJ, Erb SA, Miller CM, Mroz MM, Maier LA, Van Dyke MV. Developing effective health and safety training materials for workers in beryllium-using industries. Journal of Occupational and Environmental Medicine. 2013;**55**(7):746-751. DOI: 10.1097/JOM.0b013e3182972f1b

[40] Thomas CA, Deubner DC, Stanton ML, Kriess K, Schuler CR. Long-term efficacy of a program to prevent beryllium disease. American Journal of Industrial Medicine. 2013;**56**(7):733-741. DOI: 10.1002/ajim.22175

[41] Bertin G, Averbeck D. Cadmium: Cellular effects, modifications of biomolecules, modulation of DNA repair and genotoxic consequences (a review). Biochimie. 2006;**88**(11):1549-1559. DOI: 10.1016/j.biochi.2006.10.001

[42] Chunhabundit R. Cadmium exposure and potential health risk from foods in contaminated area, Thailand. Toxicological Research. 2016;**32**(1):65-72. DOI: 10.5487/TR.2016.32.1.065

[43] Alimba CG Gandhi D, Sivanesan S, Bhanarkar MD, Naoghare PK, Bakare AA, Krishnamurthi K. Chemical characterization of stimulated landfill soil leachates from Nigeria and India and their cytotoxicity and DNA damage inductions on three human cell lines. Chemosphere. 2016;**164**:469-479. DOI: 10.1016/j.chemosphere.2016.08.093

[44] Cartularo L, Laulicht F, Sun H, Kluz T, Freedman JH, Costa M. Gene expression and pathway analysis of human hepatocellular carcinoma cells treated with cadmium. Toxicology and Applied Pharmacology. 2015;**288**(3): 399-408. DOI:10.1016/j.taap.2015.08.011

[45] Larsson SC, Orsini N, Wolk A. Urinary cadmium concentration and risk of breast cancer: A systemic review and dose-response meta-analysis. American Journal of Epidemiology. 2015(5):375-380. DOI: 10.1093/aje/kwv085

[46] Bishak YK, Payahoo L, Osaidrahimi A, Nourazarian A. Mechanisms of cadmium carcinogenicity in the gastrointestinal tract. Asian Pacific Journal of Cancer Prevention. 2015;**16**(1):9-21. DOI: 10.7314/APJCP.2015.16.1.9

[47] Mondal B, Maulik D, Mandal M, Sarkar GN, Sengupta S, Ghosh D. Analysis of carcinogenic heavy metals in gallstones and its role in gallbladder carcinogenesis. Journal of Gastrointestinal Cancer. 2016;1-8. DOI: 10.1007/s12029-016-9898-1

[48] Arslan M, Demir H, Arslan H, Gokalp AS, Demir C. Trace elements, heavy metals and other biochemical parameters in malignant glioma patients. Asian Pacific Journal of Canver Prevention. 2011;**12**(2):447-451

[49] Garcia-Esquinas E, Pollan M, Tellez-Plaza M, Francesconi KA, Goessler W, Guallar E, Umana JG, Yeh J, Best LG, Navas-Acien A. Cadmium exposure and cancer mortality in a prospective cohort: The strong heart study. Environmental Health Perspectives. 2014;**122**(4):363-370. DOI: 10.1289/ehp.1306587

[50] Khan N, Afridi HI, Kazi TG, Arian MB, Bilal M, Akhtar A, Khan M. Correlation of cadmium and magnesium in the blood and serum samples of smokers and non-smokers of chronic leukemia patients. Biological Trace Element Research. 2016;**176**(1):81-88. DOI:10.1007/sl12011-016-0816-y

[51] Ostadrahimi A, Payahoo I, Somi MH, Khajebishak Y. The association between urinary cadmium levels and dietary habits with risk of gastrointestinal cancer in Tabriz, Northwest of Iran. Biological Trace Element Research. 2016;**175**(1):72-78. DOI: 10.1007/s12011-016-0764-6

[52] Xiao CI, Liu Y, Tu W, Xia YJ, Tian KM, Zhou X. Research progress of the mechanisms underlying cadmium-induced carcinogenesis. Zhonghua Yu Fang, Yi Xue Za Zhi. 2016;**504**(4):380-384. DOI: 10.3760/cma.j.issn.0253-9624.2016.04.021

[53] Inglot P, Lewinska A, Potocki I, Oklejewicz B, Tabecka-Lonczynska A, Koziorowski M, Bugno-Poniewierska M, Bartosz G, Wnuk M. Cadmium-induced changes in genomic DNA-methylation status increase aneuploidy events in a pig Robertsonian translocation model. Mutation Research/Genetic Toxicology and Environmental Mutagenesis. 2012;**747**(2):182-189. DOI: 10.1016/j.mrgentox.2012.05.007

[54] Knight AS, Zhou EY, Francis MB. Development of peptoid-based ligands for the removal of cadmium from biological media. Chemical Science. 2015;**6**(7):4042-4048. DOI: 10.1039/C55C00676G

[55] Li X, Jiang X, Sun J, Zhu C, Li X, Tian L, Bai W. Cytoprotective effects of dietary flavonoids against cadmium-induced toxicity. Annals of the New York Academy of Sciences. 2017;**1398**:5-19. DOI: 10:111/nyas.13344

[56] Wang YJ, Yan J, Zou XI, Guo KJ, Zhao Y, Meng CY, Yin F, Guo L. Bone marrow mesenchymal stem cells repair cadmium-induced rat testis injury by inhibiting mitochondrial apoptosis. Chemico-Biological Interactions. 2017;**271**:39-47. DOI: 10.1016/j.cbi.2017/04.024

[57] Orisakwe OE, Dagur EA, Mbagwu HO, Udowelle NA. Lead levels in vegetables from artisanal mining sites of Dilimi River, Bukuru and Barkin Ladi North Central Nigeria: Cancer and non-cancer risk assessment. Asian Pacific Journal of Cancer Prevention. 2017;**18**(3):621-627. DOI: 10.22034/APJCP. 2017.18.3.621

[58] Rashidi M, Alavipanah S. Relation between kidney cancer and soil leads in Isfahan Province, Iran between 2007 and 2009. Journal of Cancer Research and Therapeutics. 2016;**12**(2):716-720. DOI: 10.4103/0973-1482.154936

[59] McCumber A, Strevett KA. A geospatial analysis of soil lead concentrations around regional Oklahoma Airports. Chemosphere. 2017;**167**:62070. DOI: 10.1016/j.chemosphere. 2016.09-127

[60] Southard EB, Roff A, Fortugno T, Richie JP, Kaag M, Chinchilli VM, Virtamo J, Albanes D, Weinstein S, Wilson RT. Lead, calcium uptake, and related genetic variants in association with renal cell carcinoma risk in a cohort of male Finnish smokers. Cancer Epidemiology Biomarkers & Prevention. 2011;**21**(1):191-201. DOI: 10.1158/1055-9965.EPI-11-0670

[61] Silbergeld EK. Facilitative mechanisms of lead as a carcinogen. Mutation Research/Fundamental and Molecular Mechanisms of Mutagenesis. 2003;**553**(1-2);121-133. DOI: 10.1016/6j.mrfmmm.2003.07.010

[62] Steenland K, Barry V, Anttila A, Sallmén M, Mcelvenny D, Todd AC, Straif K. A cohort mortality study of lead-exposed workers in the USA, Finland and the UK. Occupational and Environmental Medicine. 2017;May 2017. DOI: 10.1136/oemed-2017-104311

[63] Silbergeld EK, Waalkes M, Rice JM. Lead as a carcinogen: Experimental evidence and mechanisms of action: American Journal of Industrial Medicine. 2000;**38**(3):316-323. DOI: 10.1002/1097-0274(200009)38:3<316:AID-AJIM11>3.0CO;2-P

[64] Kianoush S, Sadegehi M, Balali-Mood M. Recent advances in the clinical management of lead poisoning. Acta Medica Iranica. 2015;**53**(6):327-336

[65] Rice KM, Walker EM, Wu M, Gillette C, Blough ER. Environmental mercury and its toxic effects: Journal of Preventive Medicine & Public Health. 2014;**47**(2):74-83. DOI: 10.3961/ipmph.2014.47.2.74

[66] Crespo-Lòpez ME, Macêdo GL, Pereira SI, Arrifano GP, Picanço-Diniz DL, Nascimento JL, Herculano AM.

Mercury and human genotoxicity: Critical considerations and possible mechanisms. Pharmacological Research. 2009;**60**(4):212-220. DOI: 10.1016/j.phrs.2009.02.011

[67] Branco V, Caito S, Farcina M, Rocha J, Aschner M, Carvalho C. Biomarkers of mercury toxicity: Past, present, and future trends. Journal of Toxicology and Environmental Health, part B. 2017;**20**(3):119-154. DOI: 10.1080/10937404.2017.1289834

[68] Junqué E, Gari M, Arce A, Torrent M, Sunyer J, Grimalt JO. Integrated assessment of infant exposure to persistent organic pollutants and mercury via dietary intake in a central western Mediterranean site (Menorca Island). Environmental Research. 2017;**156**:714-724. DOI: 10.1016/j.envres.2017.04.030

[69] Alcala-Orozco M, Morillo-Garcia Y, Caballero-Gallardo K, Olivero-Verberl J. Mercury in canned tuna marketed in Cartagena, Columbia and estimation of human exposure. Food Additives & Contaminants: Part B. 2017;**10**:1-7. DOI: 10.1080/19393210.2017.1323803

[70] Yuan W, Yang N, Li X. Advances in understanding how heavy metal pollution triggers gastric cancer. BioMed Research International. 2016;**2016**:1-10. DOI: 10.1155/2016/7825432

[71] Valko M, Morris H, Cronin MT. Metals, toxicity and oxidative stress. Current Medicinal Chemistry. 2005;**12**(10):1161-1208. DOI: 10.2174/0929867053764635

[72] Kosnett MJ. The role of chelation in the treatment of arsenic and mercury poisoning. Journal of Medical Toxicology. 2013;**9**(4):347-354. DOI: 10.1007/s13181-013-0344-5

[73] Sears ME. Chelation: Harnessing and enhancing heavy metal detoxification – A review. The Scientific Word Journal. 2013;**2013**:1-13. DOI: 10.1155/2013/219840

[74] Iranmanesh M, Fatemi SJ, Golbafan MR, Balooch FD. Treatment of mercury vapor toxicity by combining deferasirox and deferiprone in rats. BioMetals. 2013;**26**(5):783-788. DOI: 10.1007/s10534-013-9656-9

[75] Sangvanich T, Morry J, Fox C, Ngamcherdtrakul W, Goodyear S, Castro D, Fryxell GE, Addleman RS, Summers AO, Yantasee W. Novel oral detoxification of mercury, cadmium, and lead with thiol-modified nanoporous silica. ACS Applied Materials & Interfaces. 2014;**6**(8):5483-5493. DOI: 10.1021/am5007707

[76] Ngole-Jeme VM, Fantke P. Ecological and human health risks associated with abandoned gold mine tailings contaminated soil. Plos One. 2017;**12**(2): DOI: 10.1371/journal.pone.0172517

[77] Pavela M, Uitti J, Pukkala E. Cancer incidence among copper smelting and nickel refining workers in Finland. American Journal of Industrial Medicine. 2016:**60**(1);87-95. DOI: 10.1002/ajim.22662

[78] Gölz L, Buerfent BC, Hofmann A, Rühl H, Fricker N, Stamminger W, Oldenburg J, Deschner J, Hoerauf A, Nöthen NM, Schumacher J, Hübner MP, Jäger A. Genome-wide transcriptome induced by nickel in human monocytes. Acta Biomaterialia. 2016;**43**:369-382. DOI: 10.1016/j.actbio.2016.07.047

[79] Plavan G, Jitar O, Teodosiu C, Nicoara M, Micu D, Strungaru SA. Toxic metals in tissues of fishes from the Black Sea and associated human health risk exposure. Environmental Science and Pollution Research. 2017;**24**(8):7776-7787. DOI: 10.1007/s11356-017-8442-6

[80] Harai R, Harari F, Forastiere F. Environmental nickel exposure from oil refinery emissions: A case study in Ecudor. Annali dell'Istituto Superiore di Sanità. 2016;**52**(4):495-499. DOI: 10.4415/ANN_16_04_06

[81] Zhou C, Huang C, Wanf J, Huang H, Xie Q, Zhu J, Li Y, Zhang D, Zhu Q. Lnc RNA MEG# downregulation mediated by DNMT3b contributes to nickel malignant transformation of human bronchial epithelial cells via modulating PHLPP1 transcription and HIF-1α translation. Oncogene. 2017;**36**:3878-3889. DOI: 10.1038/onc.2017.14

[82] Huang H, Zhu J, Li Y, Zhang L, Gu J, Xie Q, Jin H, Che X, Huang C, Chen LC, Lyu J, Gao J, Huang C. Up regulation of SQSTM1/p62 contributes to nickel-induced malignant transformation of human bronchial epithelial cells. Autophagy. 2016;**12**(10):1687-1703

[83] Yu M, Zhang J. Serum and hair nickel levels and breast cancer: Systematic review and meta-analysis. Biological Trace Element Research. 2017:**175**(2):1-7. DOI: 10.1007/s12011-017-0949-7

[84] Yang Y, Jin X, Yan C, Tian Y, Tian J, Sen X. Urinary level of nickel and acute leukemia in Chinese children. Toxicology and Industrial Health. 2008;**24**(9):603-610. DOI: 10.1177/0748233708100091

[85] Tilakaratne D, Sidhu S. Heavy metal (monoclonal) bands: A link between cutaneous T-cell lymphoma and contact allergy to potassium dichromate, nickel and cobalt? Australasian Journal of Dermatology. 2014;**56**(1):59-63. DOI: 10.1111/ajd.12182

[86] Zambelli B, Uversky VN, Ciurli S. Nickel impact on human health: An intrinsic disorder perspective. Biochimica Et Biophysica Acta (BBA) – Proteins and Proteomics. 2016;**1864**(12):1714-1731. DOI: j.bbapap.2016.09.008

[87] Ma L, Bai Y, Pu H, Gou F, Dai M, Wang H, He J, Zheng T, Cheng N. Histone methylation in nickel-smelting industrial workers. Plos One. 2015;**19**(10). DOI: 10.1371/journal.pone.0140339

[88] Sunderman FW. Chelation therapy in nickel poisoning. Annals of Clinical and Laboratory Science. 1981;**11**(1):1-8

[89] Atma W, Larouci M, Meddah B, Benabdeli K, Sonnet P. Evaluation of the phytoremediation potential of *Arundo donax* L. for nickel-contaminated soil. International Journal of Phytoremediation. 2017;**19**(4):377-386. DOI: 10.1080/15226514.2016.12252591

[90] Gopal R, Narmada S, Vijayakumar R, Jaleel CA. Chelating efficacy of CaNa2 EDTA on nickel-induced toxicity in *Cirrhinus mrigala* (Ham.) through its effects on glutathione peroxidase, reduced glutathione and lipid peroxidation. Comptes Rendus Biologies. 2009;**332**(8):685-696. DOI: 10.1016/j.crvi.2009.03.004

[91] Di Loreto G, Sacco A, Felicioli G. Radon in workplaces, a review. Giornale Italiano Di Medicina Del Lavoro Ed Ergonomia. 2010;**32**(4):251-254

[92] Chalupnik S, Wysocka M, Janson E, Chmielewska I, Wiesner M. Long term changes in the concentration of radium in discharge waters of coal mines and Upper Silesian rivers. Journal of Environmental Radioactivity. 2017;**171**:117-123. DOI: 10.1016/j.jenrad.2017.01.007

[93] Abdel Ghany HA. Enhancement of radon exposure in smoking areas. Environmental Geochemistry and Health. 2007;**29**(3):249-255. DOI: 10.1007/s10653-007-9082-4

[94] Wick RR, Nekolla EA, Gaubitz M, Schulte TL. Increased risk of myeloid leukemia in patients with ankylosing spondylitis following treatment with radium-224. Rheumatology. 2008;**47**(6):855-859. DOI: 10.1093/rheumatology/ken060

[95] Benejegerdes KE, Brown SC, Housewright CD. Focal cutaneous squamous cell carcinoma following rdium-223 extravasation. Proceedings (Baylor University Medical Center). 2017;**30**(1):78-19

[96] Darbre PD. Underarm antiperspirants/deodorants and breast cancer. *Breast Cancer Res*. 2009;11 Suppl 3(Suppl 3):S5. doi: 10.1186/bcr2424 [doi].

[97] Mandriota SJ. A case-control study adds a new piece to the aluminium/breast cancer puzzle. *EBioMedicine*. 2017;22:22-23. doi: S2352-3964(17)30260-8 [pii].

[98] Linhart C, Talasz H, Morandi EM, et al. Use of underarm cosmetic products in relation to risk of breast cancer: A case-control study. *EBioMedicine*. 2017;21:79-85. doi: S2352-3964(17)30233-5 [pii].

[99] Darbre PD. Metalloestrogens: An emerging class of inorganic xenoestrogens with potential to add to the oestrogenic burden of the human breast. *J Appl Toxicol*. 2006;26(3):191-197. doi: 10.1002/jat.1135 [doi].

[100] Link JM, Hurlin PJ. The activities of MYC, MNT and the MAX-interactome in lymphocyte proliferation and oncogenesis. *Biochim Biophys Acta*. 2015;1849(5):554-562. doi: S1874-9399(14)00081-9 [pii].

[101] Chen JS, Su IJ, Leu YW, Young KC, Sun HS. Expression of T-cell lymphoma invasion and metastasis 2 (TIAM2) promotes proliferation and invasion of liver cancer. *Int J Cancer*. 2012;130(6):1302-1313. doi: 10.1002/ijc.26117 [doi].

[102] Li L, Bi Z, Wadgaonkar P, *et al.* Metabolic and epigenetic reprogramming in the arsenic-induced cancer stem cells. *Semin Cancer Biol*. 2019;57:10-18. doi: S1044-579X(18)30106-8 [pii].

[103] Huang HW, Lee CH, Yu HS. Arsenic-induced carcinogenesis and immune dysregulation. *Int J Environ Res Public Health*. 2019;16(15):2746. doi: 10.3390/ijerph16152746. doi: 10.3390/ijerph16152746 [doi].

[104] Lu G, Xu H, Chang D, *et al.* Arsenic exposure is associated with DNA hypermethylation of the tumor suppressor gene p16. *J Occup Med Toxicol*. 2014;9(1):42-5. eCollection 2014. doi: 10.1186/s12995-014-0042-5 [doi].

[105] Cardoso APF, Al-Eryani L, States JC. Arsenic-induced carcinogenesis: The impact of miRNA dysregulation. *Toxicol Sci*. 2018;165(2):284-290. doi: 10.1093/toxsci/kfy128 [doi].

[106] Chen QY, DesMarais T, Costa M. Metals and mechanisms of carcinogenesis. *Annu Rev Pharmacol Toxicol*. 2019;59:537-554. doi: 10.1146/annurev-pharmtox-010818-021031 [doi].

[107] Magos L. Epidemiological and experimental aspects of metal carcinogenesis: Physicochemical properties, kinetics, and the active species. *Environ Health Perspect*. 1991;95:157-189. doi: 10.1289/ehp.9195157 [doi].

[108] Cummings KJ, Stefaniak AB, Virji MA, Kreiss K. A reconsideration of acute beryllium disease. *Environ Health Perspect*. 2009;117(8):1250-1256. doi: 10.1289/ehp.0800455 [doi].

[109] Strupp C. Beryllium metal II. a review of the available toxicity data. *Ann Occup Hyg*. 2011;55(1):43-56. doi: 10.1093/annhyg/meq073 [doi].

[110] Kuschner M. The carcinogenicity of beryllium. *Environ Health Perspect*. 1981;40:101-105. doi: 10.1289/ehp.8140101 [doi].

[111] Arita A, Costa M. Epigenetics in metal carcinogenesis: Nickel, arsenic, chromium and cadmium. *Metallomics*. 2009;1(3):222-228. doi: 10.1039/b903049b [doi].

[112] Wang Y, Mandal AK, Son YO, *et al.* Roles of ROS, Nrf2, and autophagy in cadmium-carcinogenesis and its prevention by sulforaphane. *Toxicol Appl Pharmacol*. 2018;353:23-30. doi: S0041-008X(18)30259-X [pii].

[113] Kolluru V, Tyagi A, Chandrasekaran B, Ankem M, Damodaran C. Induction of endoplasmic reticulum stress might be responsible for defective autophagy in cadmium-induced prostate carcinogenesis. *Toxicol Appl Pharmacol*. 2019;373:62-68. doi: S0041-008X(19)30138-3 [pii].

[114] Son YO, Pratheeshkumar P, Roy RV, *et al.* Nrf2/p62 signaling in apoptosis resistance and its role in cadmium-induced carcinogenesis. *J Biol Chem*. 2014;289(41):28660-28675. doi: 10.1074/jbc.M114.595496 [doi].

[115] Wang Z, Yang C. Metal carcinogen exposure induces cancer stem cell-like property through epigenetic reprograming: A novel mechanism of metal carcinogenesis. *Semin Cancer Biol*. 2019;57:95-104. doi: S1044-579X(18)30153-6 [pii].

[116] Choi J, McGill M, Raia NR, Hasturk O, Kaplan DL. Silk hydrogels crosslinked by the fenton reaction. *Adv Healthc Mater*. 2019;8(17):e1900644. doi: 10.1002/adhm.201900644 [doi].

[117] Liao LM, Friesen MC, Xiang YB, *et al.* Occupational lead exposure and associations with selected cancers: The shanghai men's and women's health study cohorts. *Environ Health Perspect*. 2016;124(1):97-103. doi: 10.1289/ehp.1408171 [doi].

[118] van Wijngaarden E, Dosemeci M. Brain cancer mortality and potential occupational exposure to lead: Findings from the national longitudinal mortality study, 1979-1989. *Int J Cancer*. 2006;119(5):1136-1144. doi: 10.1002/ijc.21947 [doi].

[119] Ahn J, Park MY, Kang MY, Shin IS, An S, Kim HR. Occupational lead exposure and brain tumors: Systematic review and meta-analysis. *Int J Environ Res Public Health*. 2020;17(11):3975. doi: 10.3390/ijerph17113975. doi: 10.3390/ijerph17113975 [doi].

[120] Gaudet HM, Christensen E, Conn B, Morrow S, Cressey L, Benoit J. Methylmercury promotes breast cancer cell proliferation. *Toxicol Rep*. 2018;5:579-584. doi: 10.1016/j.toxrep.2018.05.002 [doi].

[121] Zefferino R, Piccoli C, Ricciardi N, Scrima R, Capitanio N. Possible mechanisms of mercury toxicity and cancer promotion: Involvement of gap junction intercellular communications and inflammatory cytokines. *Oxid Med Cell Longev*. 2017;2017:7028583. doi: 10.1155/2017/7028583 [doi].

[122] Hong YS, Kim YM, Lee KE. Methylmercury exposure and health effects. *J Prev Med Public Health*. 2012;45(6):353-363. doi: 10.3961/jpmph.2012.45.6.353 [doi].

[123] Holmes AL, Wise JP. Mechanisms of metal-induced centrosome amplification. *Biochem Soc Trans*. 2010;38(6):1687-1690. doi: 10.1042/BST0381687 [doi].

[124] Guo H, Liu H, Wu H, *et al.* Nickel carcinogenesis mechanism: DNA

damage. *Int J Mol Sci*. 2019;20(19):4690. doi: 10.3390/ijms20194690. doi: 10.3390/ijms20194690 [doi].

[125] Cameron KS, Buchner V, Tchounwou PB. Exploring the molecular mechanisms of nickel-induced genotoxicity and carcinogenicity: A literature review. *Rev Environ Health*. 2011;26(2):81-92. doi: 10.1515/reveh.2011.012 [doi].

[126] Stannard L, Doak SH, Doherty A, Jenkins GJ. Is nickel chloride really a non-genotoxic carcinogen? *Basic Clin Pharmacol Toxicol*. 2017;121 Suppl 3:10-15. doi: 10.1111/bcpt.12689 [doi].

[127] Scanlon SE, Scanlon CD, Hegan DC, Sulkowski PL, Glazer PM. Nickel induces transcriptional down-regulation of DNA repair pathways in tumorigenic and non-tumorigenic lung cells. *Carcinogenesis*. 2017;38(6):627-637. doi: 10.1093/carcin/bgx038 [doi].

[128] Sun H, Shamy M, Costa M. Nickel and epigenetic gene silencing. *Genes (Basel)*. 2013;4(4):583-595. doi: 10.3390/genes4040583 [doi].

[129] Chakari W, Pilt AP, Lock-Andersen J. Sarcomatoid tumor following radium treatment. *Case Rep Dermatol*. 2018;10(1):13-16. doi: 10.1159/000486476 [doi].

[130] Schrag JM. Naturally occurring radium (ra) in home drinking-water wells in the sandhills region of South Carolina, USA: Can high concentrations be predicted? *Geohealth*. 2017;1(4):138-150. doi: 10.1002/2017GH000069 [doi].

[131] Bird CC, Willis RA. The possible carcinogenic effects of radiations on the uterus. *Br J Cancer*. 1970;24(4):759-768. doi: 10.1038/bjc.1970.91 [doi].

[132] Marijuana Policy by State: https://www.mpp.org/states/

[133] Measuring Heavy Metal Contaminants in Cannabis and Hemp: A Practical Guide; R. J. Thomas, CRC Press, Boca Raton, FL, US, ISBN: 9780367417376, September, 2020.

[134] United States Pharmacopeia General Chapter 232 Elemental Impurities – Limits: First Supplement to USP 40–NF 35, 2017, https://www.usp.org/chemical-medicines/elemental-impurities-updates

[135] ICH Q3D Step 5 Guidelines, ICH Website: http://www.ich.org/products/guidelines/quality/article/quality-guidelines.html (Q3D)

[136] Lackerman B, Peters J, and Thomas R. Heavy Metals and Cannabis. *New Food-Food Integrity 2020,* April 23, 2000. *https://www.newfoodmagazine.com/article/109084/heavy-metals-and-cannabis/*

[137] California Bureau of Cannabis Control, Medical Cannabis Regulations, https://bcc.ca.gov/law_regs/cannabis_order_of_adoption.pdf

[138] Practical Guide to ICP-MS: A Tutorial for Beginners, CRC Press, Boca Raton, FL, US, ISBN: 978-1-4665-5543-3, 2014

[139] https://www.eea.europa.eu/data-and-maps/indicators/eea32-heavy-metal-hm-emissions-1/assessment-10#fieldsetlegend-e3c9ab4866564bbfa78e8a428782d6cd-0

[140] https://www.eea.europa.eu/publications/air-quality-in-europe-2018

[141] Cooper AM, Felix D, Alcantara, F, Zaslavsky I, Work A, Watson PL, Pezzoli K, Yu Q, Zhu D, Scavo AJ, Zarabi Y, and Schroeder JI. Monitoring and mitigation of toxic heavy metals and arsenic accumulation in food crops: A case study of an urban community garden. *Plant Direct.* 2020;4:1-12. doi: 10.1002/pld3.198

[142] https://seagrant.unh.edu/project/research/monitoring-and-mitigation-heavy-metals-porphyra

[143] Selvi A, Rajasekar A, Theerthagiri J, Ananthaselvam A, Kuppusamy S, Kuppusamy S, Madhavan J, and Pattanathu, KSMR. Integrated Remediation Processes Toward Heavy Metal Removal/Recovery from Various Environments-A Review. Front. Environ. Sci., 22 May 2019 | https://doi.org/10.3389/fenvs.2019.00066

[144] Li C, Zhou K, Qin W, Tian C, Yan X, and Han W. A Review on Heavy Metals Contamination in Soil: Effects, Sources, and Remediation Techniques. *Soil and Sediment Contamination: An International Journal,* 2019; 28: 4, 380-394, DOI: 10.1080/15320383.2019.1592108

[145] Akün, ME. Heavy Metal Contamination and Remediation of Water and Soil with Case Studies from Cyprus. In: Heavy Metal Toxicity in Public Health; 2020, DOI: 10.5772/intechopen.90060

Chapter 7

Occurrence and Impact of Heavy Metals on Some Water, Land, Flora and Fauna Resources across Southwestern Nigeria

Olufemi Akinnifesi, Femi Adesina, Germaine Ogunwole and Sylvanus Abiya

Abstract

Rapid urbanization and industrialization in communities of Nigeria contribute significantly to environmental pollution. Amongst the diversity of these environmental contaminants are heavy metals, a rarely biodegradable and toxic class of metals. Heavy metals are known to be harmful to plants, aquatic species, and subsequently endanger human health through bioaccumulation or biomagnification. Even at low concentrations, heavy metals may affect key soil microbial processes; inhibit plant metabolism and growth. Toxic metals in groundwater affect water quality and potability, and their presence in aquatic systems also facilitate the production of reactive oxygen species that can damage physiological processes in fishes and other aquatic organisms. This chapter highlights the occurrence and impact of heavy metals in different environmental matrices and organisms sampled across some Southwestern states in Nigeria. Various studies including those of the authors found varying levels of heavy metals, especially in concentrations that can imperil ecosystem functions. While results of studies included in this chapter may suggest heavy metal introduction through anthropogenic-urbanization means, the lack of proper implementation of environmental monitoring laws in Nigeria also clearly exist. As such, the mitigation of heavy metals amongst other pollutants demands better home-grown decentralized technologies.

Keywords: heavy metals, groundwater, lead, freshwater, bioaccumulation, mercury, food chain

1. Introduction

Nigeria is a developing and populous country in the Africa continent, with an estimated population of 200 million people making it the most populous country in the continent. It lies between latitudes 4° and 14 °N, and longitudes 2° and 15 °E. There are about 36 states and a Federal Capital Territory aggregated into six geopolitical zones. Southwest of Nigeria, which is one of these zones, comprise of six states (Ekiti, Lagos, Ogun, Ondo, Osun, Oyo). **Figure 1**, in which two (Lagos and Oyo) are among the five most populated states in Nigeria. Southwestern Nigeria has two distinct weather seasons; the dry season-from November to March; and the

Figure 1.
Map of Nigeria showing the Southwestern region and its states.

raining season-which ranges from April to October [1]. The region has an annual temperature range of 21 to 29°C, with relatively high humidity. The annual rainfall varies from 1150 mm in the northern part and 2000 mm in the southern areas [2].

Southwestern Nigeria holds major economic hub of the country, where most of the country's factories are located. The region is also endowed with rich mineral and agricultural resources that often encourage mining and farming activities. These increasing industrialization, migration, urban development, agricultural practices and human activities continue to expose southwestern Nigeria to environmental pollution. Solid waste, landfills, and inefficient urban runoff and sewage management systems are not uncommon in the region and this can further create hazards associated with heavy metal pollution in the region [3]. More so, the major cities of the region is often associated with very high vehicular activities. Automobiles are known to introduce a number of heavy metals from exhaust fumes, and in lubricants of faulty engines, which are consequently deposited on roadsides [4, 5]. or incorporated into dust particles in size range as small as 10^{-9} and 10^{-6} m [6].

Heavy metals are metallic, naturally occurring compounds with a very high density greater than 5 g/cm3 [7]. Heavy metal classification also includes some basic metals, metalloids, transition metals, lanthanides and actinides and groups III to V metals of the periodic table e.g. As, Pb, Hg, Cd, Cr, Co, Ni, Cu, Zn, Se, Al, Cs, Mn, Mo, Sr., U, Be and Bi [8]. Given that heavy metals are not usually biodegradable, they are often transported and accumulated into plants and animal tissues from soil and water matrices, which predispose long term deleterious effects on the receiving species. The inherent capacity of species to react differently to heavy metals does exist; hence, some species are more tolerant and resistant to heavy metals while others are not [9]. More so, the recalcitrant nature of heavy metals makes them persist in the environment for years and thus may pose threats to man and other organisms. With these characteristics of heavy metals and increasing pollution trends in our environment, it suffices to suggest that heavy metal is a serious threat [10].

The toxic effects of heavy metals on living organisms, especially when consumed above the recommended limits set by various regulatory bodies, are far reaching [11, 12]. Elements like arsenic, mercury and lead might be toxic even at low levels of exposure. Once consumed by humans, heavy metals continue to concentrate in vital organs like the brain, liver, bones, and kidneys, for years causing damaging health consequences [13]. Many serious health conditions may also arise from the direct intake of heavy metal contaminated food. Heavy metal intakes have often been linked with health problems such as immunological compromise, intrauterine growth defect, impaired psycho-social behavior, and nutritional defect and many more [14].

Heavy metals contamination raises serious concern, however, governments of developed countries, through various regulatory agencies continues to introduce possible methods for heavy metal pollution prevention and remediation [15, 16]. Unlike these developed economies, Nigeria still faces waste management crises. Large introduction of heavy metals into the environment is yet inevitable, especially due to the dearth of infrastructures and weak environmental laws to cater for the 25 million tonnes of municipal waste generated annually [17].

2. Heavy metals accumulation in soil

The soil is a critically essential component for all living organisms. Especially for plants, soil serves as nutrient media for the growth of plants and other microflora. However, these life-supporting functions, which include carbon cycling, essential nutrients provision, water filtering and storage, atmospheric regulation [7], predispose soil matrixes to heavy metal occurrence. Heavy metals can occur naturally in soil environment via the pedogenetic weathering of parent materials viz. weathering of minerals, erosion and volcanic activity. More significantly, the anthropogenic introduction of heavy metals in soil arise from activities such as mining, smelting, electroplating, effluent discharge, as well through bio-solids (livestock manures, composts), waste incineration, vehicle exhausts [18–20]. Contamination of soils by heavy metals through anthropogenic activities from industrial areas has been reported in Southwest of Nigeria (**Table 1**) [22, 23, 26, 27, 29, 31].

Owning to the soil adsorption processes which binds inorganic and organic pollutant, soil has long been recognized as a sink for many pollutants [32]. A number of studies have reported that topsoil and roadside soil in close proximity to urban areas with heavy traffic are suceptible to heavy metal contamination from atmospheric deposition and adsorption [33, 34]. In heavily commuted cities such as in Southwest Nigeria, metals deposit such as Cd, Cu, Pb and Zn are good indicators of contaminations from automobiles as they appear in gasoline, oil lubricants, car component and industrial incinerator emissions [32]. Numerous studies conducted in recent years found high levels of heavy metals as street dust contaminants near traffic routes in Southwest Nigeria [24, 35, 36]. The re-suspension of metal-laden road dust may also be identified as the main source of road pollution, especially along roads with intense traffic and high vehicular emission. These have made ecological risk assessment of heavy metals in polluted soils to gain more attention in recent years [37–39].

Mining is another means of heavy metal pollution in the soil. There is prepoderence of artisanal gold and gemstone mining in Southwestern Nigeria. Most of these mines are found in deep reserved locations not readily accessible by regulatory agencies, thus, make illegal mining practices to continue unchecked [22]. With lack of rules guiding artisanal mining, this have significantly contributed to soil-heavy metal pollution in the region. Very recently, the increased government interest in the mining and extractive industries may also heighten the challenges associated with heavy metal pollution, especially if environmental protection is not given top priority. Some proponents have stated that ecological risk assessment can reveal the possibility of soil being polluted, and even determine the ecological functions that may likely be altered by concerned heavy metals [40]. However, in contrast, [41]

			Heavy metals (mg/kg)									
Authors	Sites	State	Pb	Zn	Ni	Cd	CCu	Fe	As	Hg	Cr	Mn
[21]	T Trailer T Terminal	Ogun	0.20–9.60	1.60–64.55	0.65–1.65	0.20–1.40	0.76–3.13	0.65 5.46	12.0–65.0*	0.80 –43.25*	0.20–0.25	
[22]	Mining Site	Ekiti	0.1	4.95	0.17	1.53	2.71	54.7	0.01			
[23]	Mining Site A	Osun	0.216	0.7	0.08	0.13	0.629		0.09			
	Mining Site B	Osun	0.278	0.628	0.081	0.09	0.629		0.051			
[24]	Dumpsite	Ondo	18.6	148.7	17.7	1.2	17.8				45.5	
[25]	Mining site	Ogun	0.296	0.116	0.027	0.595			0.535	0.291	0.225	
[26]	Roadside	Lagos	5.57–69.20	25.87–198.32	0.94–42.73			403–1528.30			1.58–347*	3.72 -953.52
[27]	Dumpsite	Ondo	0.1	2.92			0.76	4.59				
	Induced soil	Ondo	0.20–0.70	2.04–7.09			0.07–0.39	8–12.80				
[28]	Dumpsite	Oyo	4273.8*			258.38*		7910				
[29]	Roadside	Osun	72.7									
[30]	Industrial areas	Oyo	3.5–363.9*		0.7 - 19.1	5.6–22.9*	0.0–163.5*	0.3 2.9			77.6–84.9	50.5–1685.0
WHO/FAO			100	300	50	3	100	50,000	20	4	100	2000

*Above maximum allowable limits.

Table 1.
Heavy Metals occurrence in soil reported from southwest Nigeria.

has stated that more studies and practices have shown that such results from risk assessment have little capability to reveal the real degree of the potential toxic effects of metals without primary empirical research.

The result of the baseline heavy metal content in mined soil (in Ijero Ekiti, Ekiti State, Nigeria) by [22] showed mined soil have significant amount of cadmium (1.530 ppm), chromium (0.810 ppm), iron (54.700 ppm), arsenic (0.010 ppm), zinc (4.950 ppm), lead (0.100 ppm) and nickel (0.170 ppm), except copper (2.710 ppm) which is not significant ($p > 0.05$) to control soil. Similarly, [23] studied mined soil of different depth (0–15 cm and 15–30 cm) at Ijana in Osun state, Nigeria and high levels of Zn, Cd, Pb and Cu were detected in the soil from mine site. Heavy metals in mined soil were consistently higher than in the control soil. Which also aligns with earlier studies by [42–44]. Following the submission of [23] a continuously exploited mining site would lead to increased soil degradation, higher soil acidity levels and eventual increase in heavy metals solubility in the soil of the mine site.

Also, in a study conducted by [45] who examined an illegal dumpsite in Lagos State, Nigeria, a high level of Zn, 25.48 mg/L was reported, followed by Fe, 15.63 mg/L. Cu, Mn, Pb and Cd found in soil samples were 2.11, 1.04, 0.22 and 0.09 mg/L respectively. Except Ni (0.01 mg/L) found at the lowest concentration and mercury (Hg) which was below detectable levels, all metals detected were above the permissible limits of the Nigerian Standard for Drinking Water Quality (NSDWQ), World Health Organization (WHO), and United State Environmental Protection Agency (USEPA).

Similarly, in Ibadan (Oyo State), Pb and Cd concentrations detected in top-soil of battery waste dumpsite exceeded the environmental quality standards set by National Environmental Standards and Regulations Enforcement Agency (NESREA) for soils in Nigeria [28]. Street dumpsite soil samples in Osogbo, Osun state was also examined by [29] and the highest Lead concentration detected was 72.7 mg/kg. This value for lead is comparable to that recorded in other states such as Ondo, Oyo and Lagos States. Heavy metal contents in soils varied significantly from location to location in Lagos State. Mn content in soils ranged from 3.72–953.52 mg/kg, Ni ranged from 0.94–42.73 mg/kg, Pb ranged from 5.57–69.20 mg/kg, Cr ranged from 1.58–347 mg/kg, Zn ranged from 25.87–198.32 mg/kg and Fe ranged from 403 to 1528.30 mg/kg [26].

In 2017, [27] examined soils of a cocoa industry dumpsite in Akure, Ondo State and heavy metal contents detected were Fe (4.59 mg/kg), Pb (0.10 mg/kg), Zn (2.92 mg/kg) and Cu (0.76 mg/kg). Induced soil with different concentration of spent engine oil, also contained Fe (8.00–12.80 mg/kg), Pb (0.20–0.70 mg/kg), Zinc (2.04–7.09 mg/kg) and Cu (0.07–0.39). Likewise, [24] reported the contamination indices and the potential ecological risks of toxic metals contents of soil from municipal waste in Ondo. Though the contamination index recorded was 2.8, depicting a low degree of contamination, the total ecological risk index was 285.5, which portray the site as high risk. Also, the heavy metals contamination of soils at Ogere trailer terminal along the busy Ibadan-Lagos express road in Ogun State revealed significant heavy metal presence [21].

3. Heavy metal accumulation in plants

Owing to the fact that decayed and composted wastes enhance soil fertility, it is a common practice in Nigeria that dumpsites and previously cleared landfills in urban regions are converted to farming sites [55]. Most of these wastes sites and associated soils often contain heavy metal remains in various forms and at different contamination levels [46], which can be transferred into plants parts and their produce. Heavy metals are some of the known form of contamination in food supply, especially vegetables [47]. Their uptake by plants from the soil depends on different factors, including agrochemical applications, solubility of heavy metals, soil pH, soil type, plant species and soil-to-plant translocation factors (TFs) of the metals [48, 49]. Regardless of which factor is more dominant, crops planted directly on municipal waste sites can absorb heavy metals through their roots as soluble ions present in the soil, or absorption by leaves. These absorbed metals can get transferred and bioaccumulated in the storage and vegetative plant parts. Although micronutrients such as Fe, Mn, Cu, Zn, Mo are essential for normal metabolic processes in plants, excessive quantities can be quite harmful to plants [50].

It has also been established that high heavy metal pollution causes adverse effects on the physiological processes of agricultural plants. These often caused reductions in plant growth and dry matter accumulation. Furthermore, excessive heavy metal accumulation in agricultural soils frequently leads to high heavy metal uptake by crops, and thus affects food quality and safety [51]. The direct intake of heavy metal contaminated foods by humans make heavy metals persist in the food chain [52]. Heavy metal presence in the human systems can deplete some essential nutrients in the body, tinker immunological defenses, impair psychosocial faculties, trigger malnutrition, and most often incites the prevalence of gastrointestinal disorder and cancer.

The uptake, metabolism, and effects of most heavy metals on different plant species in South-west Nigeria have been documented in literature (**Table 2**). Adesina and Akinnifesi [53] revealed that heavy metals from crude oil contamination were absorbed by tomato plant roots and translocated to other parts of the plants, thereby storing them in the tissues. The experiment also showed that fruits were also reduced in sizes compare to tomato fruit on uncontaminated soil. *Chromolena odorata* plants growing around gold mining site in Ijana, Osun State were found to contain heavy metals although within the permissible limits [23]. In a screenhouse experiment, maize seeds were planted on a mined soil from Ijero-Ekiti, Ekiti State, plant growth parameters were retarded as the soil, roots and shoots from the mine soil have higher heavy metals content than in the control soil [22].

Accumulation of heavy metals in plants was also proven by [28] where high levels of Pb, Cd and Fe in maize roots, stems, leaves, and grains were recorded. This was attributed to contaminated battery waste soil on which the maize plants were grown in Ibadan, Oyo State. Also, 30 edible plants belonging to species of *Celosia argentea*, *Corchorus olitorius* (Jute), *Colocasia esculenta* (Cocoyam), *Musa sp.* (Banana) and *Saccharum officinarum* (Sugar cane) growing on mining soil at Ewekoro, Ogun State were examined by [54]. The mean concentrations of heavy metals in plants studied were Cu (26.32 mg/kg), Pb (15.46 mg/kg), Zn (213.94 mg/kg), Cr (30.62 mg/kg), Co (0.45 mg/kg) and Ni (3.77 mg/kg). These levels detected in all the plants were all above international limits and Zn > Cr > Cu > Pb > Ni > Co. In 2018, [56] used moss plant (*Dicranum scoparium*) as bioindicator to evaluate heavy metals from different locations in Yaba College of Technology Campus, Lagos State. The presence of Zn (8.63 mg/l), Pb (1.12 mg/l), Cu (0.51 mg/l), Ni (0.43 mg/l) and Cd (0.026 mg/l) reflect 80.46%, 10.49%, 4.78%, 4.03% and 0.24% respectively. The highest amount of Zinc was attributed to emission of zinc from paint chips from the walls of buildings, use of cosmetics including perfumes, pigments, corrosion of galvanized safety fence and wearing of iron.

Many Bioaccumulation factor (hereafter BAF) and translocation factor (herafter TF) experiments have helped evaluate phytoremediation potentials of some plant Eqs. (1) and (2) [60]. While BAF is a calculated value that indicates the ability of

				Heavy metals (mg/kg)					
Authors	Plants	Sites	State	Lead	Zn	Ni	Cd	Cu	Cr
[53]	Tomato	Crude oil	Ondo	0.01–0.06	—	0.01–0.46	0.01–0.06	0.02–0.28	0.01–0.22
[23]	*Chromolena odorata*	Mining	Osun	0.13–0.20	1.02–1.21	0.08–0.09	0.01–0.02	0.59–0.60	0.12
[22]	Maize	Mining	Ekiti	Nd	0.02–0.64	0.04–0.08	0.02–0.04	0.01–0.08	Nd
[54]	30 edible plants	Mining	Ogun	0.66–109*	14.9 – 252*	0.66–29.10	—	2.59–305.6*	1.20 – 343*
[55]	*Amaranthus cruentis*	Landfill	Oyo	110.5–678.5*	—	—	2.01 - 3.20*	—	—
[56]	Moss plant (mg/L)	within college	Lagos	0.57–1.61	3.62–10.18	0.08–1.52	0.0001–0.06	0.0001–2.38	—
[57]	Basil	Industrial Dumpsite	Lagos	1.85–2.54	—	—	0.99–1.41	—	—
[58]	5 plants	Farm	Ekiti	—	247.6–784.2*	—	147.8 - 366.5*	—	—
[59]	Fruit and vegetables	Market	Lagos	0.07–1.93	0.03–0.13	0.05–0.29	0.004–0.09	0.01–0.07	—
WHO/FAO				*0.3*	*100*	*67*	*0.1*	*73*	*1.3*

**Above maximum allowable limits, nd – not detected.*

Table 2.
Heavy Metals occurrence in some plants reported in southwest Nigeria.

plants to remove metal compounds from the soil/substrate, the translocation factor indicates the ability of the absorbed compound to be transferred from plant roots to other plant organs [61].

$$\mathbf{BAF} = \frac{\mathbf{Metalconcentrationinstemorleaves}}{\mathbf{Metalconcentrationinsoil}} \tag{1}$$

$$\mathbf{TF} = \frac{\textbf{Metal concentration in stem or leaves}}{\textbf{Metal concentration in root}} \; or \; \frac{\textbf{BAF in stem or leaves}}{\textbf{BAF in root}} \tag{2}$$

Plants that have BAF and TF >1 can be used as bioaccumulators [62].

Also, plants can be deployed as phytostabilizers if they have BAF >1 and TF <1; and as phytoextractors if they have BAF <1 and TF >1 [63]. The knowledge of these values also help as guides when planting crops in areas of possible heavy metal contaminations.

4. Heavy metal accumulation in surface and groundwater

One of the targets of Millennium Development Goals (MGDs) is to globally achieve clean water before 2030. Most livelihood activities depend on the availability of water for drinking, irrigation, and industrial processes or recreational purposes. The availability of quality water depends on the physical, chemical, and biological properties surrounding the water [9]. In Southwest Nigeria, fresh surface water is usually available throughout the year, but in abundant quantities during the rainy season, which almost last for 7 months. While most rural communities are served by surface water, urban communities in Southwestern Nigeria depend largely on groundwater abstraction to meet their water need. Achieving quality water in this region has continued to be major concern, giving the spate of pollution to both surface and groundwater in the region. High industrial use, agricultural activities, poor sewage management, urban runoff and overall poor environmental monitoring has increased the pollution menace on water resources in the Southwest.

The introduction of foreign contaminants into water bodies may either pollute the water or increase nutrients for aquatic biota [64]. Water is polluted or contaminated if its physicochemical properties are altered, and these alterations affect the aquatic lives adversely [65]. Heavy metal can be introduced into water matrices through anthropogenic activities like mining, crude oil spills, and untreated waste effluents containing metals. Surface water can be polluted by contaminants washed into it either from far polluted sites as non-point source pollution or through direct introduction into water bodies, as point source pollution [66]. Akinyemi et al. [9] sampled river water beside a cassava processing factory at Ibadan, Oyo State for heavy metals (**Table 3**) and reported the mean concentration of Cu at (0.035 mg/l), Cr (0.026 mg/l), Ni (0.036 mg/l), Cd (0.026 mg/l), Zn (0.075 mg/l) and Pb (0.032 mg/l). Though the heavy metals reported were within WHO permissible limits, the discharged from the cassava factory near the river was reported to contribute to the heavy metal contamination of the river.

Furthermore, the cross interaction of water at the surface water and groundwater interface may also contribute heavy metals to water especially in regions with high water table [68] investigated some shallow groundwater wells that were in close contact with polluted streams in.

			Heavy Metals (mg/l)						
Authors	**Sites**	**State**	**Lead**	**Cd**	**Cr**	**Zn**	**Ni**	**Cu**	**Fe**
[9]	River	Ogun	0.002–0.135*	0.009–0.085*	0.013 – 0.068*	0.026–0.214	0.018 – 0.076*	0.001–0.164	—
[45]	Leachate	Lagos	0.22*	0.09*		25.48*	0.01	2.11*	15.63*
	Groundwater		0.004– 0.009	0.002–0.009*	—	0.15 – 4.10	0.002–0.007	0.01–0.023	0.02–0.06
[28]	Stream	Oyo	0.024–0.275*	0.002 – 0.025*	—	—	—	—	0.262 – 2.038
[67] [68]	Rivers Stream Groundwater	Osun Ondo	– 0.518* 0.000–0.079*	– 0.252* 0.000–0.820*	0–2100* 0.372* 0.000–0.079*	10-650* – –	– – –	0–1350* – –	– –
[69]	Lake	Oyo	0.015*	nd	nd		nd		0.0066
[70]	Lagoon	Lagos	—	—	—	0.4–0.54	0.1–0.60*	—	7.3 – 6.65*
[10]	River	Ekiti	0.16*	0.13*		4.65		0.84	5.87*
[71]	Water treatment facility	Oyo	0.07–0.36*	0.08 –0.10*	0.04–0.22*	—	—	—	—
	Tap water		0.12–0.65*	0.08 –0.11*	0.02–0.29	—	—	—	—
	Dam		0.03–0.15*	0.06 –0.08*	0.13–0.37*	—	—	—	—
[72]	Reservoir	Osun	3.05*	2.9*	3.88*	8.99*	7.26*	3.19*	176.22*
[73]	Coastal water	Ondo	0.00–0.01	0.00–0.01*	—	0.00 - 0.01	—	nd	0.06–0.43
WHO/FAO			*0.01*	*0.003*	*0.05*	*5*	*0.02*	*2*	*3*

**Above maximum allowable limits, nd – not detected.*

Table 3.
Heavy Metals occurrence in Surface and Groundwater from Southwest Nigeria.

Ondo City, Ondo State. Highest level of aluminum, cadmium, lead, and chromium concentrations in well water at 1.632 mg/l, 0.820 mg/l, 0.079 mg/l and 0.079 mg/l recorded in the rainy season exceeded WHO recommendations.

Landfill leachates also contribute toxic metals to surface water and groundwater through percolation of the subsoil. Ogunwole et al. [45] investigated the impact of leachates on groundwater resources at varying horizontal distances from some dumpsites in Lagos State. Their findings showed high concentrations of Zn (25.48 mg/l), Fe (15.63 mg/l). Cu, Mn, Pb, Cd, Ni had a concentrations of 2.11, 1.04, 0.22, 0.09 mg/l and 0.01 respectively. For the groundwater, heavy metals level ranges as Cd (0.004–0.009 mg/L) and Zn (4.10 mg/L) detected in some of the wells exceeded the permissible level to be found in drinking water. Prohibitory levels of Cd found in the groundwater closest to the dumpsite were attributed to the low pH around the dumpsite area. Mobility of metals in soil column is often determined by the pH of the subsurface area. This mostly affects the sorption/desorption of metals, precipitation/dissolution, complex formation and oxidation/reduction reactions of metals [74]. Ten rivers were examined in Osun State by [67], while most of the parameters measured fell within the threshold values for drinking water, heavy metals like iron (60–960 μg L^{-1}), chromium (0–2100 μg L^{-1}), aluminium (0–800 μg L-1) and calcium (6400–232,000 μg L-1) exceeded at some locations. For public health concerns, individuals drinking from heavy metal contaminated water are highly liable to cancerous conditions. Hence, there is need to regulate heavy metal pollutant that is discharged into these water bodies.

5. Heavy metals toxicity in freshwater fishes

Aquatic environment is one of the major receiving ends for pollutants, particularly heavy metals which are routed back into the food webs through bio-accumulation in aquatic macroinvertebrates, invertebrate fishes and finally biomagnified in human [10].

Heavy metals in aquatic ecosystems are usually found in water, sediments and associated biota [75]. They generally exist in low concentrations in water and attain considerable levels in sediment and biota. The adhesive property of sediments makes it possible to sink pollutants like pesticides and heavy metals; and also important for the remobilization of contaminants in aquatic systems [76]. The biota that inhabits contaminated sites is generally exposed to toxic concentrations of these metals because many of them process food from sediment. Most filter feeding fishes and macro-invertebrates take up heavy metals and bioaccumulate them. This can potentially threaten the health of these aquatic species and others across the food chain, especially birds, fish and humans [77].

The process by which metal accumulate in the tissues of living organisms via the food chain is called bioaccumulation, and are measured as either Bioaccumulation factor (BAF) and Bioconcentration factor (BCF). If BAF is high, it indicates that the bioaccumulation of heavy metals occurred in organisms. Bioaccumulation is often related to the fact that metals of different kind tend to accumulate differently in the tissue of aquatic species, considering the feeding, swimming, and metabolic activity of individual species. Previous documented studies reveal accumulated concentrations of heavy metals in various parts of the organism. Fish gill was reported to store the highest amount of Pb (15.5 mg/kg) and Mn (604.32 mg/kg), while liver tissue had the highest Cd (7.88 mg/kg) and Zn (11.80 mg/kg) level in fish species from Ogun River, Nigeria [78].

Olusola and Festus [79] during their study of coastal waters of Ondo state, noted that distribution of heavy metals in different fish parts were organ specific,

				Heavy metals (mg/kg)						
Authors	Species	Site	State	Lead	Cd	Zn	Fe	Ni	Cu	Cr
[78]	Fish	River	Ogun	8.69–15.5*	1.53 - 7.88*	6.78–11.80	—	15.9–24.80*	—	—
[69]	Fish	Lake	Oyo	0.053	0.064	—	7.43*	0.016	—	0.004
	Prawn			0.104	0.045	—	1.61*	0.019	—	0.001
[10]	*Clarias gariepinus*	*River*	Ekiti	0.09	0.04	0.95	1.09*	—	0.66	—
[79]	Fish	Coastal water	Ondo	1.14	3.18*	0.28 - 4.14	—	0.01	5.72*	3.41*
[82]	Catfish	Lake	Oyo	1.3–12.1*	0.3 - 4.2*	3.5–71.2	82.1 - 2303.8*	0.7–39.3*	1.3 - 40.1*	—
[83]	Fish	Lagoon	Lagos	3.19–5.88*		8.73 - 15.78	7.27–8.69*	—	2.04 - 3.31*	—
[73]	Shrimps	Costal water	Ondo	0.01–0.027	0.01	5.35–7.51	4.21–5.90*	—	1.86 - 3.06*	—
[70]	Tilipia	Lagoons	Lagos	—	—	0.16–1.30	0.10–8.02*	0.00 - 4.00*	—	—
	Catfish			—	—	0.25–1.95	0.21–8.40*	0.00 - 2.80*	—	—
[80]	Frozen fish	Storage facilities	Oyo	2.11–13.18*	0.59 - 2.18*	—	—	—	2.89 - 9.74*	—
WHO/FAO				2	0.5	30	0.5	0.02	3	0.05

**Above maximum allowable limits.*

Table 4.
Heavy Metals occurrence in aquatic animals from Southwest Nigeria.

as higher concentrations of metals were recorded in the gills and eyes compared to other organs. In contrast, [80] recorded highest concentration of Pd, Cd, Cu and Hg in liver than other parts of frozen fishes sold in storage facilities in Ibadan, Oyo State. Some of the concentrations recorded in these parts were considered to be above maximum acceptable limit by [81]. Severe exposure of aquatic animals to heavy metals through different sources could result in observable structural and/ or functional changes. This could further cause mortality, and also loss of biodiversity. Leachates collected from dumpsite were investigated for toxicity on fishes (*C. gariepinus and O. niloticus*) and it was found out that the treatment (leachates) induces biochemical alterations in the organism, as well as mortality [45].

Several researchers have reported accumulation of heavy metals in aquatic animals in Southwest Nigeria (**Table 4**). Two lagoons (Epe and Badagry) in Lagos were investigated and found to contain Zn, Ni and Fe in both water and sediment [70]. The fishes dwelling inside these studied lagoons were further examined and similar metals were detected in various parts of the fishes (tilapia and catfish). While concentrations detected in the study were considered safe for consumption, the researchers still emphasized the need for continuous monitoring to prevent adverse human health effects [78, 84, 85], especially if heavy metals detected in fishes are in elevated concentrations.

6. Conclusion

Many of the heavy metals reported in this chapter exceed the maximum limit by WHO and FAO. Despite the massive and continued documented evidences on the occurrence and adverse effects of heavy metals in the South-west region and the larger Nigeria, little is been done in terms of control measures. Many domestic and industrial wastes in the South-western region of Nigeria are still discharged into water bodies untreated. This contributes a large intermix of heavy metals in many of the natural waters in the urban areas and endangers the quality and health of aquatic environment. Deploying advanced technologies for wastewater and sewage treatment have greatly reduced the menace of toxic metal pollution in developed countries. Developing a similar but strategic approach in Nigeria should lessen the impact of heavy metal pollution in the Southwestern Nigeria. More so, with efficient water and sanitary infrastructure, larger populace will have access to clean water supply and hygiene practices that will limit their use of contaminated water.

Most inhabitants in South-west Nigeria, especially in highly commuted cities, are still constantly predisposed to metal contamination through industrial and automobile emissions. Poor combustion engines add large quantities of lead and cadmium into the air and surfaces as particulate matters. Lead in human causes cardiovascular diseases, predispose renal, immunological and reproductive dysfunctions. Also, Cadmium is carcinogenic, neurotoxic and causes skeletal disorders, and disrupts a mineral balance in the body.

As a booming revenue generator, the hydrocarbon industries in the Southwest Nigeria also contribute immensely to heavy metal pollution in the region. Large deposits of crude oil spills and other hydrocarbon exploring technologies add metal contaminants into land, surface and groundwater. Likewise, the unguarded activities of artisanal miners of gold, copper, limestone etc. are another major heavy metal contributor in this region. These crude mining methods are implicated to not only unearth natural heavy metals, but introduces others forms of heavy metals through explosives and catalyst.

With research showing that heavy metals do accumulate in agricultural produce, edible fruits and seafood, it is evident that most local populace are constantly, but insensibly taking in heavy metals through many food sources. More worrisome is that the concept of quality control of most food products in Nigeria is still very low. Hence, most people could be exposed to the long term health effect of heavy metals. The exposure to cadmium, lead, nickel and mercury compounds from consumption of contaminated plant, water and seafood is dangerous, especially during prenatal development and infancy, as it causes irreversible changes in the central nervous system.

From the reports in this chapter, and many other uncovered studies of heavy metals in south west region of Nigeria, there is enough empirical based need for urgent attention to control waste and activities that contribute heavy metals to the environment.

Author details

Olufemi Akinnifesi*, Femi Adesina, Germaine Ogunwole and Sylvanus Abiya
Department of Biology, Federal University of Technology Akure, Ondo State, Nigeria

*Address all correspondence to: olufemiakinnifesi@yahoo.com

References

[1] Akinloye L, Akinpade J, Makinde A: Analysis of climate of Southwestern Nigeria for building design. International Journal of Constructive Research in Civil Engineering, 2017; 3: 1-7.

[2] Ogunrayi OA, Akinseye FM, Goldberg V, Bernhofer C: Descriptive analysis of rainfall and temperature trends over Akure, Nigeria. Journal of Geography and Regional Planning, 2016; 9(11): 195-202.

[3] Orisakwe OE, Blum, JL, Sujak S, Zelikoff JT: Metal pollution in Nigeria: A biomonitoring update. Journal of Health and Pollution, 2014; 4 (6):40-52.

[4] Ogunfowokan AO, Asubiojo OI, Adeniyi AA, Oluyemi EA:"Trace lead, zinc and copper levels in Barbula lambarenensis as a monitor of local atmospheric pollution in Ile Ife, Nigeria," Journal of Applied Sciences, 2004; 4(3): 380-383.

[5] Okunola OJ, Uzairu A, Uba S, Ezeanyanaso CS, Alhassan Y: Distribution pattern of metals in atmospheric settling dust along roads in Kano Metropolis, Nigeria. Journal of Applied Chemistry, 2015; 739325, 12.

[6] Annergarn HJ, Moja SJ, Malahela J, Kgashane P, De-Lange B "Vanderbijlpark-Golden highway project: dust fall-out monitoring report,"AER, 2002; 22.178S GHP.

[7] Singh R, Ahirwar NK, Tiwari J, Pathak J: Review on sources and effect of Heavy Metal in soil: its bioremediation. International Journal of Research in Applied Natural and Social Sciences, 2018; 1-22.

[8] Cairney T. Contaminated Land. Blackie; 1993: London; p. 4

[9] Akinyemi LP, Odunaike RK, DE, Alausa SK: Physico-chemical parameters and Heavy Metal concentrations in Eleyele River in Oyo State, South-West of Nigeria. International Journal of Environmental Science and Toxicology Research, 2014; 2(1): 1-5.

[10] Edward JB, Idowu EO, Oso JA, Ibidapo OR: Determination of Heavy Metal concentration in fish samples, sediment and water from Odo-Ayo River in Ado-Ekiti, Ekiti-State, Nigeria, International Journal of Environmental Monitoring and Analysis, 2013: 1(1):27-33. doi: 10.11648/j.ijema.20130101.14

[11] WHO 2009: Global Framework for Action on Sanitation and Water Supply. Working Document.

[12] USEPA 2000: National Primary Drinking Water Regulations. United States Environmental Protection Agency.

[13] Norris RD, Hinchee RE, Brown R, McCarty PL, Semprini L, Wilson JT, Kampbell DH, Reinhard M, Bouwer EJ, Borden PC, Vogel TM, JThomas M, Ward CH (1993). Handbook of Bioremediation, Lewi; 1993: Boca Raton, FL.

[14] Nweke OC, Sanders WH: Modern environmental health hazards: a public health issue of increasing significance in Africa. Environ Health Perspectives, 2009; 117(6): 863-70.

[15] Ahmad K, Ejaz A, Azam M, Khan ZI, Ashraf M, Al-Qurainy F, Fardous A, Gondal S, Bayat AR, Valeem EE: Lead, cadmium and chromium contents of canola irrigated with sewage water. Pakistan Journal of Botany, 2011; 43(2): 1403-1410.

[16] Etim EU, Onianwa PC: Heavy Metal pollution of topsoil in the vicinity of an industrial estate co-located with a housing estate in Southwestern Nigeria. Journal of Environmental Protection, 2013; 4:91-98.

[17] Beatrice A, Jussi K: Municipal solid waste management problems in Nigeria: Evolving knowledge management solution", International Journal of Water Resources and Environmental Engineering, 2013; 7(6).

[18] Bilos C, Colombo JC, Skorupka CN, Rodriguez PM: "Sources, distribution and variability of airborne trace metals in La Plata City area, Argentina". Environmental Pollution, 2001; 111: 149-58.

[19] Modaihsh A, Al-Swailem M, Mahjoub M: Heavy metal contents of commercial inorganic fertilizer used in the Kingdom of Saudi Arabia. Agric. Mar. Sci. 2004, 9, 21-25.

[20] Sabiha-Javied, Mehmood T, Chaudhry MM, Tufai M, Irfan N: Heavy metal pollution from phosphate rock used for the production of fertilizer in Pakistan. Journal of Microchemistry, 2009. ; 91:94-99.

[21] Fashae OA, Olusola AO, Orekan P. (2019) Assessing Heavy Metal Distribution and Contamination of Soil in Ogere Trailer Terminal, Ogun State (Southwestern Nigeria). In: Boughdiri M, Bádenas B, Selden P, Jaillard E, Bengtson P, Granier B, Editors. Paleobiodiversity and Tectono-Sedimentary Records in the Mediterranean Tethys and Related Eastern Areas. CAJG 2018. Advances in Science, Technology & Innovation (IEREK Interdisciplinary Series for Sustainable Development). Springer, 2018: Cham; doi.org/10.1007/978-3-030-01452-0_73

[22] Abiya SE, Ogunwole GA, Odiyi BO, Olanrewaju AB: Growth behaviour and metal accumulation of two varieties of maize *Zea mays* L. sown on a soil obtained from a mining site. Brazilian Journal of Biological Sciences, 2020; 7(15): 29-38.

[23] Abiya SE, Odiyi BO, Ologundudu FA, Akinnifesi OJ, Akadiri S: Assessment of Heavy Metal pollution in a gold mining site in Southwestern Nigeria. Journal of Genetics and Cell Biology, 2018; 1(2): 30-35.

[24] Ogundele LT, Oladejo OF, Akinola AC. Concentrations, source identification and human health risk of heavy metals in the road dust collected from busy junctions in Osogbo Southwest, Nigeria. International Journal of Environmental Quality, 2020; 38, 24-36.

[25] Oyeyemi, K. D., Aizebeokhia, A. P., Anake, W. U., Sanuade, O. A., Akinsiku, A. A., Olofinnade, O. M., Jonathan H. O. and Ayara, W. A. (2019). Geospatial distribution of heavy metal contamination in Ewekoro Limestone, SW Nigeria. Journal of Physics: Conf. Series1299 (2019) 012080IOP. Doi:10.1088/1742-6596/1299/1/012080

[26] Olukanni DO, Adeoye DO: Heavy Metal concentrations in road side soils from selected locations in the Lagos Metropolis, Nigeria. International Journal of Engineering and Technology, 2012; 2(10):1743-1752.

[27] Adesina FP, Ashamo MO: Assessment of Growth of *D. bulbifera* (L.) on contaminated soil (spent engine oil) and industrial effluent. International Research Journal of Environmental Sciences, 2019; 8(2): 1-6.

[28] Afolayan AO: Accumulation of heavy metals from battery waste in topsoil, surface water, and garden grown maize at Omilende Area, Olodo, Nigeria. Global Challenges, 2018; 2 (1700090): 1-12.

[29] Olayiwola OA, Agbolade JO, Oladapo AS: Daily exposure of roadside topsoil to lead at Olaiya Roundabout, Osogbo, Nigeria. International Journal of Engineering and Science Invent., 2013; 2(5):55-60.

[30] Kolawole, T. O., Olatunji, A. S., Jimoh, M. T. and Fajemila, O. T. Heavy

Metal contamination and ecological risk assessment in soils and sediments of an industrial area in Southwestern Nigeria. Journal of Health & Pollution, 8(19): 1-16.

[31] Adegoke JA, Owoyokun TO, Amore IO: Open Land Dumping: An analysis of heavy metals concentration of an old lead-battery dumpsite. Pacific Journal of Science and Technology, 2009; 10(2): 592-595.

[32] Popoola OE, Bamgbose O, Okonkwo OJ, Arowolo TA, Odukoya O, Popoola AO: "Heavy metals content in playground topsoil of some public primary schools in metropolitan Lagos, Nigeria". Research Journal of Environmental and Earth Sciences, 2012; 4(4): 434-439.

[33] Arowolo TA, Bamgbose O, Odukoya OO: "The chemical forms of lead in roadside dust of metropolitan Lagos, Nigeria". Global Journal of Pure and Applied Science, 2000; 6(3): 483-487.

[34] Xiangdong L, Chi-Sun P, Pui Sum L: "Heavy Metal contamination of urban soils and street dusts in Hong Kong". Applied Geochemistry, 2001; 16: 1361-1368.

[35] Azeez JO, Mesele SA, Sarumi BO, Ogundele JA, Uponi AO, Hassan AO: Soil metal pollution as a function of traffic density and distance from road in emerging cities: a case study of Abeokuta, Southwestern Nigeria. Archives of Agronomy and Soil Science, 2014: 60 (2): 275-295. doi.org/10.1080/03650340.2013.792406.

[36] Famuyiwa AO, Lanre-Iyanda YA, Osifeso O: Impact of land use on concentrations of potentially toxic elements in urban soils of Lagos, Nigeria. Journal of Health and Pollution, 2018; 8(19): 180904.

[37] Pekey H, Karakas D, Ayberk S, Tolun L, Bakoglu M: Ecological risk assessment using trace elements from surface sediments of Izmit Bay, Turkey. Marine Pollution Bulletin, 2004; 48: 946.

[38] Qu J, Yuan V, Wang X, Shao P, Cong Q: Distribution of heavy metals, chemical fractions and ecological risks around a molybdenum mine in Liaoning Province, China. Vitamins & Trace Elements, 2004; 1(3):.doi.org:10.4172/2167-0390.1000104.

[39] Cao HC, Luan ZQ, Wang JD, Zhang XI: Potential ecological risk of Cd, Pb and As in agricultural black soil in Jilin Province, China. Stochastic Environmental Research and Assessment, 2009; 23(1): 57.

[40] Fairbrother A, Wenstel R, Sappington K: Framework for metals risk assessment. Ecotoxicology and Environmental Safety, 2007; 68: 145.

[41] Wu Y, Xu Y, Zhang J, Hu S: Evaluation of ecological risk and primary empirical research on heavy metals in polluted soil over Xiaoqinling gold mining region, Shaanxi, China. Trans, Nonferrous Met. Soc. 2010; 20: 688.

[42] Eludoyin AO, Ojo AT, Awotoye OO: Effects of artisanal gold mining activities on soil properties in a part of Southwestern Nigeria. **Cogent Environmental Science**, 2017; 3: 1-11.

[43] Oladipo OG, Olayinka A, Awotoye OO: Ecological impact of mining on soils of Southwestern Nigeria. **Environmental and Experimental Biology**, 2014; 12: 179-186.

[44] Olatunji OS, Osibanjo O: Determination of selected heavy metals in inland fresh water of lower River Niger drainage in North Central Nigeria. African Journal of Environmental Science and Technology, 2012; 6(10): 403-408.

[45] Ogunwole GA, Abiya SE, Otitoloju AA: Illegal dumpsites and it impact on

groundwater and fish resources (*Clarias gariepinus and Oreochromis niloticus*). International Journal of Environmental Science and Toxicology Research, 2019; 7(1): 1-13.

[46] Opaluwa OD, Aremu MO, Ogbo LO, Abiola KA, Odiba IE, Abubakar MM, Nweze NO: Heavy metal concentrations in soils, plant leaves and crops grown around dump sites in Lafia Metropolis, Nassarawa State, Nigeria. Advances in Applied Science Research, 2012, 3 (2):780-784.

[47] Chauhan G, Chauhan UK: Human health risk assessmentof heavy metals via dietary intake of vegetables grown inwastewater irrigated area of Rewa, India. International Journal of Scientific Research and Publications, 2014; 4(9): 1-9.

[48] Lubben S, Sauerbeck D: The uptake and distribution of heavy metals by spring wheat, Water, Air, & Soil Pollution, 1991; 57(1): 239-247.

[49] Rattan RK, Datta SP, Chhonkar PK, Suribabu K, Singh AK: Long term impact of irrigation with sewage effluents on heavy metal content in soil, crops and groundwater- a case studyAgriculture Ecosystem and Environment, 2005, 109, 310-322. doi: 10.106/j.agee.2005.02.025

[50] Zwolak A, Sarzyńska M, Szpyrka, Stawarczyk K: Sources of soil pollution by Heavy Metals and their accumulation in vegetables: a review. Water, Air, Soil Pollution, 2019; 230: 164. https://doi.org/10.1007/s11270-019-4221-y.

[51] Muchuweti MJ, Birkett W, Chinyanga E, Zvauya R, Scrimshaw M.D, Lister JN: Heavy Metal content of vegetable irrigated with mixture of wate water and sewage sludge in Zimbabwe: Implications for Human health. Agriculture Ecosystem and Environment, 2006; 112(1): 41-48

[52] Khan S, Cao Q, Zheng YM, Huang, YZ, Zhu, YG: Health risk of Heavy Metals in contaminated soils and food crops irrigated with wastewater in Beijing, China. Environmental. Pollution, 2007; 152(3): 686-692. Doi: 10. 1016/j.envpol.2007.06.056.

[53] Adesina FP, Akinnifesi OJ: Occurrence and uptake of heavy metals in soil and plant (tomato) associated with crude oil: Agriculture and Forestry, 2020

[54] Laniyan TA, Adewumi AJ: Evaluation of contamination and ecological risk of heavy metals associated with cement production in Ewekoro, Southwest Nigeria. Journal of Health & Pollution, 2020; 10 (25): 1-13.

[55] Ogunyemi S, Awodoyin RO, Opadeji T: Heavy metal contamination of *Amaranthus cruentis* L grown on domestic refuse landfill soil in Ibadan Nigeria. Journal of Agricultural Science, 2003; 15(2): 87-94.

[56] Ojiodu CC, Olumayede EG, Okuo JM: The level of heavy metals in the atmosphere of a macroenvironment, Lagos State, Southwestern- Nigeria using moss plant (*Dicranium scorparium*) as bioindicator. Science World Journal, 2018; 13(4): 69-74.

[57] Adeyolanu OD, Kadiri, OJ, Are KS, Oluwatosin, GA, Moral, MT: Lead and cadmium contents in a medicinal plant/ spice grown in an urban city of Nigeria. Cogent Food & Agriculture, 2016; 2: 1136016.

[58] **Adey**eye EI: Trace metals in soils and plants from Fadama farms in Ekiti State, Nigeria. Bulletin of the Chemical Society of Ethiopia, 2005; 19(1): 23-34.

[59] **Ogunkunle ATJ,** Bello OS, Ojofeitimi OS: Determination of heavy metal contamination of street-vended fruits and vegetables in Lagos state, Nigeria, International Food Research Journal, 2014; 21(5).

[60] Yoon J, Cao X, Zhou Q, Ma, LQ: Accumulation of Pb, Cu, and Zn in native plants growing on a contaminated Florida site. Science of Total Environment, 2006; 368:456-464.

[61] Mellem JJ, Baijnath H, Odhav B: Bioaccumulation of Cr, Hg, As, Pb, Cu and Ni with the ability for hyperaccumulation by *Amaranthus dubius*. African Journal of Agricultural Research, 2012; 7(4):591-596. doi:10.5897/AJAR11.1486.

[62] Usman ARA, Alkredaa RS, Al-Wabel, MI: Heavy metal contamination in sediments and mangroves from the coast of Red Sea: *Avicennia sp. marina* as potential metal bioaccumulator. Ecotoxicol Environ Saf. 2013; 97: 263-270. Doi:10.1016/j.ecoenv.2013.08.009.

[63] Sopyan, RS, Sumarni NK: Fitoakumulasi merkuri oleh akar tanaman bayam duri (*Amarantus spinosus* Linn.) pada tanah tercemar. Online Journal of Natural Science, 2014; 3(1): 31-39, http://dx.doi.org/10.2012/.

[64] Boukori Y, Bawa IM, Djaneye BG: Characterization of some Togo surface waters. Bulletin of the Chemical Society of Ethiopia, 1999; 13(1): 11-21.

[65] Eletta OAA: Determination of some trace metal levels in Asa River using AAS and XRF techniques. International Journal of Physical Sciences, 2007; 2 (3): 056-060.

[66] Oyediran IA, Aladejana JA: In: Proceeding of the environmental management conference, Federal University of Agriculture, Abeokuta, Nigeria, 2011.

[67] Titilawo Y, Adeniji A, Adeniyi, M, Okoh, A: Determination of levels of some metal contaminants in the freshwater environments of Osun State, Southwest Nigeria: A risk assessment approach to predict health threat. Chemosphere, 2018; 211: 834-843.

[68] Akinnifesi OJ, Akinneye JO: Comparative bacteriological and heavy metal toxicity evaluation of groundwater wells proximal to pollution receiving streams in Ondo City, Nigeria. The Open Microbiology Journal, 2019; 13: 315-323. doi: 10.2174/1874285801913020315,

[69] Jenyo-Oni A, Oladele AH: Heavy Metals assessment in water, sediments and selected aquatic organisms in lake **Asejire**, Nigeria. European Scientific Journal, 2016; 12:24.

[70] Olowu RA, Ayejuyo OO, Adewuyi GO, Adejoro IA, Denloye AAB, Babatunde AO, Ogundajo AL: Determination of heavy metals in fish tissues, water and sediment from Epe and Badagry Lagoons, Lagos, Nigeria. E-Journal of Chemistry, 2010; 7: 215-21.

[71] Adewuyi GO, Etchie AT, Etchie TO: Health risk assessment of exposure to metals in a Nigerian water supply. Human and Ecological Risk Assessment: An International Journal, 2014; 20(1): 29-44.

[72] Aduwo AI, Adeniyi IF: The heavy metals/trace elements contents of sediments from Owalla Reservoir, Osun State, Southwest Nigeria. Advances in Oceanography and Limnology, 2018; 9(2): 68-78.

[73] Olawusi-Peters OO, Akinola JO, Jelili AO: Assessment of Heavy Metal pollution in water, shrimps and sediments of some selected water bodies in Ondo State. Journal of Researches in Agricultural Sciences, 2017; 5(2): 55-66.

[74] McLean JE, Bledsoe BE: Ground water issue: Behavior of metals in soil, U.S. Environmental Protection Agency, 1992. No. EPA/540/S 92/018)

[75] Camusso M, Vigano L, Baitstrini R: Bioaccumulation of trace metals in Rainbow Trout. Ecotoxicological and Environmental. Safety, 2005; 31: 133-141.

[76] Rashed MN: Monitoring of environmental heavy metals in fish from Nasser Lake. Environmental International, 2001; 27: 27-33.

[77] Wright P, Mason CF: Spatial and seasonal variation in heavy metals in the sediments and biota of two adjacent estuaries, the Orwell and the Stuor in Eastern England. Science of Total Environment, 1999; 226: 139-156.

[78] Ayanda IO, Ekhator UI, Bello OA: Determination of selected heavy metal and analysis of proximate composition in some fish species from Ogun River, Southwestern Nigeria. Heliyon, 2019; 5(10): 02512. doi.org/10.1016/j.heliyon.2019.e02512

[79] Olusola JO, Festus AA: Levels of Heavy Metal in Some Selected Fish Species Inhabiting Ondo State Coastal Waters, Nigeria. Journal of Environmental Analytical Toxicology, 2015; 5: 303. doi:10.4172/2161-0525.1000303.

[80] Kareem OK, Orisasona O, Olanrewaju AN: Determination of Heavy Metal levels in some commonly consumed frozen fish in Ibadan, Southwest, Nigeria. Research Journal of Environmental Toxicology, 2016; 10 (1): 82-87.

[81] FAO/WHO.: Joint FAO/WHO food standards programme codex committee on contaminants in foods. 5th Session, The Hague Netherlands, 21-25 March, 2011, ftp://ftp.fao.org/codex/meetings/CCCF/cccf5/cf05_INF.pdf

[82] Utete B, Fregene BT: Heavy metal concentrations in surface water, sediments and fish and their potential toxicity in highly degraded multipurpose peri-urban Eleyele Lake, Ibadan City, Oyo State, South-western Nigeria. Cogent Environmental Science, 2020; 6: 1785137.

[83] Taiwo IO, Olopade OA, Bamidele NA: Heavy metal concentration in eight fish species from Epe Lagoon (Nigeria). Transylv. Rev. Syst. Ecol. Res, 2019; 21(1): 69-82. doi: 10.2478/trser-2019-0006.

[84] Al-Busaidi M, Yesudhason P, Al-Mughairi S, Al-Rahbi WAK, Al-Harthy KS, Al-Mazrooei NA, Al-Habsi SH (2011). Toxic metals in commercial marine fish in Oman with reference to national and international standard. Chemosphere, 2011; 85:67-73.

[85] Rahman MS, Molla AH, Saha N, Rahman A: Study on heavy metals levels and its risk assessment in some edible fishes from Bangshi River, Savar, Dhaka, Bangladesh. Food Chemistry, 2012; 134: 1847-1854.

Chapter 8

Environmental Impacts of Heavy Metals and Their Bioremediation

Ayşe Handan Dökmeci

Abstract

Fast consumption, increasing energy needs, unplanned urbanization, and unconscious discharge of industrial wastes cause pollution of air, soil, food and water resources. Among these pollutants, heavy metals and metalloids are not biodegradable and accumulate in compartments such as water, soil and plants, threatening human and environmental health. Monitoring studies show that heavy metals such as arsenic, lead, mercury, cadmium, nickel, zinc, copper, chromium and trace elements are in first place according to their availability in the environment. Preventive and remedial measures should be taken to reduce the effects of heavy metals. Legal regulations, monitoring studies, the use of soluble and non-toxic compounds in environmental compartments (air, water, soil and plants) in industrial processes, heavy metal-free pesticides, appropriate wastewater treatment plants and use of renewable energy sources instead of fossil fuels are among the priority measures to reduce concentrations of heavy metals in the environment. As a bioremediation approach, removing toxic wastes from the environment by using bioaccumulatory organisms such as plants or mussels maintains its importance among studies aimed at recovery. Studies have shown that integrated methods - especially the combination of suitable plants and microorganisms - are very effective in mitigating the effect of heavy metals in the environment.

Keywords: heavy metal, environmental matrix, mitigation, human health, ecosystem health, bioremediation

1. Introduction

While the 21st century offers many technological possibilities to the service of humanity, humans try to dominate nature by using science and technological facilities and change the existing balance. The deteriorating natural balance and environmental problems that occur as a result are the most important hazards that threaten health in our age. The effects of climate change increase in parallel with increasing environmental problems. With industrialization, mining, pesticide use and increasing energy needs, heavy metals come to the forefront primarily in relation to occupational diseases, and they are also among the pollutants that disrupt the ecological balance as a result of water, soil and air pollution. Under normal conditions, the proportion of heavy metals in nature is low. When the concentration ratio in the natural environment increases, heavy metals such as arsenic, mercury, cadmium, chromium and lead have toxic effects, especially on organisms, and inhibit enzymes. They create environmentally toxic effects with their non-degradable and accumulatory properties in environmental compartments [1–3]. While some metals are important as trace elements for living organisms, they may accumulate in living organisms above a certain concentration and cause toxic effects or they can convert to other compounds in the environment; during this transformation, toxic and water-soluble compounds of a metal may sometimes occur. For this reason, heavy metals and trace elements are classified as vital and non-vital according to their degree of participation in biological processes. Those defined as vital must be present at a certain concentration in the structure of the organism. Because these metals participate in biological reactions, they must be consumed regularly in foods. For example, copper is an indispensable element in red blood cells and in many oxidation and reduction processes in animals and humans. On the other hand, non-vital heavy metals can cause health problems by affecting the physiological structure even at very small concentrations. The best example of this group is mercury, which binds to sulfurous enzymes. Whether a heavy metal is necessary for living organisms depends on the organism being considered. For example, while nickel is toxic to plants, it must be present as a trace element in animals [4–6].

2. Heavy metals and their impacts

There are more than 32 metals in the periodic table. Heavy metals are metals with high density and toxic effects even at very low concentrations, with a density of 5 g/cm^3 or higher [7]. These metals enter water, nutrients, soil and air ecosystems from the earth through natural processes and through anthropogenic activities [8]. These are considered major pollutants because of their toxic effects on living organisms and their prevalence in the world. Al, Fe, Cr, Sb, As, Be, Cd, Cu, Pb, Hg, Ni, Se, Ag and Zn are classified as primary contaminant metals by the United States Environmental Protection Agency (US EPA) [9]. In addition, the Agency for Toxic Substances and Disease Registry (ATSDR) has included heavy metals on the priority chemicals list due to their potential threat to human health and their toxicity [10].

2.1 Arsenic and arsenic compounds

Inorganic arsenic (As) is a gray-colored element with a molecular weight of 74.92 g/mol found naturally in the earth's crust. However, inorganic arsenic often coexists with other elements such as oxygen, chlorine, and sulfur. Inorganic arsenic compounds include arsenic pentoxide, arsenic trioxide, and arsenic acid. The combination of carbon and hydrogen with arsenic creates organic arsenic. Organic arsenic compounds include arsanilic acid, arsenobetaine and dimethylarsinic acid. Pure arsenic is rarely found in nature and is not very toxic, but it can be oxidized easily in humid air and become toxic by conversion to arsenic anhydride. Arsenic, which has four different oxidation steps, is mostly in the form of +3 and + 5 valences in nature [1, 11, 12].

There are many uses of arsenic compounds in industry. Arsenic acid (AsO4H3), arsenic trichloride (AsCl3), arsenic triiodide (AsI3), arsenic anhydride (As2O5), copper arsenide ([Cu3 (AsO3) 2) and very toxic compounds such as calcium arsenide [Ca3 (AsO4)5], lead arsenide [Pb3AsO4) and sodium arsenate (Na2HAsO4), which are highly water soluble, are used for the production of paint (green, yellow), glass, ceramics, and semiconductors. Arsenic compounds can also be used as insecticides. Sodium arsenate (Na2HAsO4, 7H2O) is used to kill some parasites in viticulture, fruit cultivation and potato growing. Arsenic compounds have also been used as chemical weapons. Lewisite containing arsenic is a volatile liquid (chlorovinyldichloroarsine) and Adamsite (diphenylaminechlorarsine) is a solid substance used in the form of aerosol. These substances have vesicant, irritant and sternutatory properties. Arsenious anhydride or arsenic trioxide (As_2O_3) is one of the most toxic compounds among arsenic compounds. It is a white powder, dissolves quickly in alkalis, acids and hot water, and contains arsenous acid (H_3AsO_3). Since it is volatile, it easily enters through the lungs, but there is a risk of loss during toxicological searches. As_2O_3 was used as a rat poison since medieval times. Today, arsenic compounds are used in limited amounts (10%) as rodenticide. Arsine (AsH_3), or arsenous hydrogen, is a colorless gas with an unpleasant garlic odor. It is released as a result of contact between inorganic arsenic compounds with hydrogen or the hydrolysis of metallic (zinc, aluminum, etc.) arsenides with water or an acid, especially during washing of some metal parts with acids (H_2SO_4). Ethyl arsine is formed by the interaction of various fungi with arsenic compounds. The toxicity of arsenic compounds generally depends on their solubility in water: Pure metal arsenic and inorganic compounds (sulfides) are insoluble in water and theoretically lack toxic effects, though in a humid environment, they can be oxidized to toxic compounds (As_2O_3). Water soluble inorganic compounds (arsenous anhydride, arsenide, alkaline arsenates) are very toxic; and arsenic hydrogen or arsine (AsH_3) has a different toxicity than other forms of inorganic arsenic as a hemolytic poison. There is a reaction with thiol groups. Mold and dampness increase the toxicity of arsenic compounds. Organic arsenic compounds have lower toxicity, they are eliminated faster from organisms [1, 13–15].

Arsenic compounds block the thiol (-SH) groups of many enzymes that play a role in cell respiration and different metabolisms in organisms. By inhibiting the oxidative decarboxylation of pyruvic acid, they disrupt the metabolism of carbohydrates, lipids and amino acids and form a true "biochemical lesion". Since the skin and keratinized tissues are rich in sulfur, the accumulation of arsenic compounds here is directly related to their attraction to thiol groups [1]. The International Agency for Research on Cancer (IRAC) has classified inorganic arsenic as a Group 1 carcinogenic to humans [16]. The Environmental Protection Agency (EPA), on the other hand, classified it as a Group A human carcinogen. The EPA calculated the cancer slope factor for inorganic arsenic oral exposure as 1.5 mg/kg/day. No cancer inhalation studies are available in humans or animals for arsine. The EPA has not classified arsine for carcinogenicity. The EPA set limits on the amount of arsenic that can be released into the environment from industrial sources. For this, the amount of arsenic permitted in drinking water was determined as 0.01 ppm [17].

2.2 Lead and lead compounds

Lead and its derivatives are industrial toxic substances that have been widely used by humans for centuries and therefore play an important role in environmental pollution. Lead is a metal found in nature. It is a bluish gray metal that is rarely found in the earth's crust. It occurs mainly as lead ore, and the most important ore is lead sulfide (galena). It is found in a variety of compounds such as lead, lead acetate, lead chloride, lead chromate, lead nitrate, and lead oxide. While pure lead is insoluble in water, the solubility of lead compounds in water varies. The chemical symbol of lead is Pb and its atomic weight is 207.2 g/mol. Its vapor pressure is 1.77 mm Hg at 1000°C [18]. In nature, pure metal or its oxides, inorganic salts such as carbonate, sulfate, nitrate and acetate, and organic compounds such as lead tetraethyl exist. The primary use of lead is in the manufacture of batteries. In addition, it is widely used in in the production of paint, alloys and metal products (lead-antimony and lead-tin) in industry, electronic conductors, rubber, sheet lead, waterproof plates (roofing) or sound and radiation insulation, linoleum, solder, pipes, ammunition and toys. Its use for ceramic glazes, paint and pipe soldering has decreased significantly. In addition, it is used in the composition of some ointments (lead acetate), diluted lead acetate (as abortive) is used infrequently in treatment under the name white water and as an insecticide (lead arsenate) in agriculture. Tetraethyl lead was used as a gasoline additive to increase the octane rating, before the EPA banned its use in gasoline in 1996 in the USA. Leaded gasoline is still used for propeller planes and some race cars [1, 19]. Metallic lead is a bluish, gray-silver soft metal. It melts at 327°C, is quite dense and can easily stretch (can take the form of a wire). It is not affected by dilute hydrochloric acid and sulfuric acid. It forms concentrated sulfuric acid and hot lead sulfate. Lead melts even in cold temperatures, forming lead nitrate and nitrous vapors. Organic acids (acetic acid, citric acid, etc.) affect lead slowly and cause the formation of their respective salts. It emits steam at temperatures above 500°C and harmful metallic dust. Lead is slightly soluble in water and may be the source of toxicity in some drinking water. Lead oxides usually form from water-insoluble compounds. PbO "oxide" (yellow masikot) forms yellow-orange lead oxide (lilac) when heated and is used for glass, enamel, accumulators, etc. in industry. PbO2 "bioxide" is the red oxide of lead and is an energetic oxidizer. Pb3O4 "minium" (cinnabar, plombi orthoplumbus) is a red-colored compound added to antirust paints and used in the varnish, enamel, crystal and battery industries. It is formed by heating lead oxide in air flow at 500°C. Lead sulfide is the substance that forms the mineral galena in its natural state. It is insoluble in water. It is found in the combination of yellow-colored varnishes used in ceramics and pottery, and color changes to dark brown when manganese oxide is added or green when copper oxide is added. These varnishes continue to be used in many countries today. It can be dissolved with organic acids. Food or acidic liquids (vinegar, wine, lemon juice, etc.) should not be placed in containers painted with this type of varnish. Lead hydrocarbonate is mixed with linen oil and used to obtain white paint. However, these dyes darken with the effect of hydrogen sulfide in the air and also have toxic effects. Although its use is prohibited in many countries, residues in old settlements can cause some poisoning cases. Arsenate lead was formerly used as insecticide in agriculture, including in Turkey, before being banned in many countries because of its toxicity. However, some residues from past use can be found in the soil due to its durability. Lead chromate is also called "chrome yellow" and can be used to obtain yellow dye. Lead stearate is used to reinforce some plastic materials. The basic iodide and acetate of lead -lead iodide and lead acetate (soluble in water)- are used in dermatology to benefit from solubilizing effect in the form of a solution. Basic lead acetate is a compound made by neutralizing lead acetate with lead oxide (PbO). Lead naphthenate is used as an additive in some moisturizing drugs. Alcoholic derivatives of tetraethyl-lead and tetramethyl-lead are organic derivatives of lead that are added to gasoline as an antidetonante (0.5%). Often ethylene dibromide and ethylene dichloride compounds are added to engines to prevent metallic lead build-up. Exhaust fumes from cars operating with leaded gasoline contain lead halides and oxides as well as a small amount of undissolved lead-alcohol (1–10%). In many countries, the proportion of lead in gasoline is legally limited (0.5 g/l) and unleaded gasoline vehicles are now produced. These organic lead derivatives added to gasoline play an important role in air pollution (especially in big cities) [1, 20].

Lead is taken into the organism generally through digestion (in the form of food, liquid drinks, vapor or particles) and through the respiratory tract. It is a highly toxic metal that enters biological systems by mixing with respired air and nutrients from the air, water and soil. Up to 90% of the lead in the atmosphere in solid and gaseous form is absorbed by the lungs. Small lead dust or vapor, especially in the air, cause lead compounds (oxides or salts) to enter the organism by inhalation. Very small particles can reach the alveoli, while large particles are discharged from the upper respiratory tract or can be reabsorbed by entering the digestive tract through sputum and saliva. When organic lead compounds contaminate the skin, they can be absorbed significantly, especially from lesioned areas. When the carcinogenic effect is examined, human studies are insufficient regarding lead exposure and increased cancer risk. In animal studies, kidney tumors were reported in rats and mice orally exposed to lead. Lead is assessed as B2 (possible human carcinogen - based on evidence of adequate carcinogenicity in animals) according to the Guidelines for Carcinogenic Risk Assessment (US EPA, 1986) [1, 21].

2.3 Mercury and mercury compounds

Elemental mercury (Hg) is a silver-white metal with an atomic weight of 200.59 g/mol. Mercury is liquid at room temperature and has a vapor pressure of 0.002 mm Hg at 25°C. Mercury can be found in three oxidation states; elemental-metallic (Hg), precious mercury-mercurous (Hg +) and divalent mercury-mercuric (Hg ++), as well as both inorganic and organic compounds. Inorganic mercury compounds include mercury chloride, and mercury sulfide. Organic mercury compounds include mercuric acetate, methylmercuric chloride, dimethyl mercury, and phenylmercuric acetate. These different forms transform into each other in the natural environment (for example, methylation of inorganic Hg, oxidation of metallic mercury to inorganic compounds, etc.). All types of mercury are highly toxic, and the toxic properties of these different mercury compounds are quite different: inorganic and organic mercury differ from each other in terms of entry and absorption routes [1]. Mercury is one of the main elements involved in the formation of the earth's crust. Mercury, which is generally found in superficial layers, is used in many areas today. However, its usage areas are narrowing due to the risk it creates. Mercury (Hg) compounds are used in industry, paint, explosives, electronic devices, accumulators, thermometers, etc. It is used in construction. The main source of mercury contamination in the world is the production of alkali chlorine to obtain electrolytes from chlorine and sodium hydroxide in industry [1, 22].

Elementary mercury (metallic mercury) is a very dense liquid and volatile at ambient temperature. A major source of exposure for elemental mercury occurs via inhalation in occupational settings. Elemental mercury is used in thermometers, barometers and pressure-sensing devices. It is also used in batteries, lamps, industrial processes, and in oils used for purification and lubrication. Another source of exposure to low levels of elemental mercury in the general population is elemental mercury released from amalgam fillings in teeth. It is generally taken into the body in the form of steam. Mercury vapors readily cross the alveolar barrier. Inorganic mineral mercury salts can be mono or divalent (mono- or divalent). Today, the general population is not exposed to significant inorganic mercury compounds. Most products containing these compounds are now banned. Until 1990, mercury compounds were used in latex paint to prevent bacterial and fungal growth. The EPA banned all interior paints containing mercury in 1990, and exterior use after 1991. Although agricultural and pharmaceutical use of inorganic mercury has ceased in the United States, mercury chloride is still used as a disinfectant and pesticide. Mercury chloride (bivalent), also known as sublimation, has a corrosive (caustic, burning) effect. Inorganic mercury (mono) salts, such as calomel, are less soluble than divalent compounds and show lower toxicity. After exposure to inorganic mercury salts, the main storage locations are kidney and brain tissue. Organic mercury compounds are in the form of alkyl or aryl derivatives. Methyl-Hg (methyl mercury) is the most important of these compounds in terms of toxicity. Methyl mercury has no industrial use; it forms by methylation of inorganic mercury ions in the environment. Methyl mercury exposure mostly occurs through diet, especially fish and fish products. In the past, sources of methyl mercury exposure included fungicide-treated cereals and meats from animals fed such cereals. However, fungicides containing mercury are banned in the USA today and exposure through this route does not occur at present [1, 23].

Mercury is listed as a contaminant of concern by the EPA due to its persistence in the environment, its bioaccumulative potential, and its toxicity to humans and the environment. In today's world where environmental health conditions are decreasing, people are constantly exposed to metal (elements) and organic mercury compounds. Contamination of the environment with mercury as a result of distorted industrialization, use of mercury in fossil sourced petroleum products, fungicides, chemical production, used batteries, use in hospitals and use against mold cause intense exposure of humans and foods (fish, plants, etc.) to mercury. Significant amounts of mercury come from the atmosphere with annual rainwater. Pollution with mercury is more common in closed seas and inland waters than in open seas. For this reason, mercury accumulation occurs more rapidly in creatures living in contaminated waters. Mercury is found in the water system in elemental, inorganic and organic forms. Organic mercury in water is found either in the form of covalently-bonded organic mercury (methyl mercury) and dimethyl mercury (less toxic than methylmercuride) or mercury complexed with organic matter (humic substances). Methyl mercury is a metabolized form of mercury and is highly toxic to human embryos and fetuses. It is shaped by the effect of microorganisms in fish and accumulates to a degree that threatens human and animal health. Especially large fish (predators) that feed on small fish pose a problem in this regard. High levels of methyl mercury can be obtained from fish that are fed or caught in contaminated areas [1, 24]. Between 1953 and 1956, a type of mercury poisoning called "Minamata disease" occurred in Japanese fishermen. It was understood that these fishermen had eaten fish they caught in waters contaminated with organic mercury wastes from a factory producing plastic in the vicinity for a long time [25]. The use of organic mercury compounds in agriculture poses a risk in terms of environmental and food pollution [26, 27].

Various studies were conducted on elemental mercury and cancer in humans. However, the EPA classified elemental mercury as group D, which cannot be classified as a human carcinogenic, based on insufficient human and animal data. There are no studies on the carcinogenic effects of inorganic mercury in humans. The EPA classified inorganic mercury compounds in the possible human carcinogen group C, as there is insufficient data in humans and animals. There are no studies on the carcinogenic effects of methyl mercury in humans. Based on insufficient data in humans and limited evidence of carcinogenicity in animals, the EPA classified it in group C as a possible human carcinogen [28].

2.4 Cadmium and cadmium compounds

Cadmium is a soft silver-white metal often found in combination with other elements. Although cadmium compounds are not normally soluble, their solubility in water is quite good. The chemical symbol of cadmium is Cd and its atomic weight is 112.41 g/mol. Cadmium is found in nature in the form of yellow cadmium sulfide (CdS), together with zinc ore [29]. Cadmium, one of the most dangerous heavy metal pollutants in the ecosystem, is a toxic metal discovered by Friedrich Stromeyer in 1817. The industrial use of cadmium goes back 50 years. Due to its noncorrosive features, it is widely used in the aircraft industry, as a neutron absorber in nuclear power plants, as a stabilizer in insecticide formulations, plastic production, and in the paint and nickel cadmium battery industry. It occurs as a by-product in lead production. This situation is significant in terms of environmental pollution. Cadmium contaminates water and soil by transfer into soil and water from industrial waste and residues. Due to the long-term use of phosphorus fertilizer and treatment sludge, agricultural soils in many parts of the world are exposed to low or moderate cadmium accumulation. Due to its high mobility in water and the soil–plant system, cadmium can easily be included in the food chain; thus, it can be dangerous for plant, animal and human health. The sources of cadmium in the air are the burning of fossil fuels such as coal or oil and waste. Cadmium can also be emitted into the air as a byproduct of the smelting of ores such as zinc, lead or copper. Cadmium in the atmosphere is also increased by the burning of nickel-cadmium batteries and the use of cadmium dyes.

Various salts of cadmium (acetate, bromide, fluoride, iodide, carbonate, chloride, nitrate, oxide, salicylate, cyanide, tungstate) are free or in the form of alloys with zinc, nickel, silver and lead. They are used in cells, accumulators, porcelain, nuclear material, amalgam in dentistry, plastic, metal coating, vacuum tubes, photographic materials etc. Cadmium oxide and anthranilate are used as anthelmintic agents in pigs and chickens in production and veterinary medicine [1, 30–32].

Poisoning with cadmium occurs by inhalation of dust or vapors at low concentrations over a long period of time and ingestion of dissolved cadmium and its salts mixed with food. Smoking is an important source of cadmium exposure. Smokers have twice as much cadmium in their bodies than non-smokers. For non-smokers, food is the largest source of cadmium exposure. Cadmium levels in some foods can be increased by applying phosphate fertilizer or sewage sludge to agricultural areas. The dark brown fumes that occur when cadmium is heated outdoors cause poisoning. The presence of cadmium's main mineral compounds in the environment or workplace air causes occupational diseases [1]. In the most obvious example of Cd poisoning today; in the city of Toyama, Japan, 200 people were shown to have permanent bone and kidney disorders due to severe Cd poisoning due to excessive consumption of Cd-contaminated rice [33].

Cadmium poisoning is on the occupational diseases list in EU countries. The Japanese named cadmium poisoning itai itai (pain-pain) disease. The EPA considers cadmium to be a possible human carcinogen (cancer-causing agent) and classified it as a Group B1 carcinogen [34].

2.5 Chromium and chromium compounds

Chromium (Cr) is a gray solid steel with a high melting point and atomic weight of 51.996 g/mol. It has oxidation steps ranging from chromium (-II) to chromium (+VI). It forms a large number of compounds in both chromium (III) and chromium (VI) forms. Chromium compounds are stable in the trivalent state and the hexavalent form is the second most stable state. While chromium (VI) compounds are easily soluble in water, chromium (III) compounds are slightly soluble in water and the water-soluble part can be found as chromium (III) complexes. Chromium is an element naturally found in rocks, animals, plants, soil, volcanic dust and gases. It is mostly in the form of naturally occurring chromium (III), which is an important nutrient, and chromium (VI), most commonly produced by industrial processes, along with the lesser known metallic chromium (0). Chromium (III) is essential for normal glucose, protein and fat metabolism and is therefore an essential dietary element. Chromium (III) is an essential element that should be consumed daily by adults at 50–200 μg/day. The body has various systems to reduce chromium (VI) to chromium (III). The most important industrial chromium sources in the atmosphere are related to ferrochrome production. Metal chrome is mostly used to make steel and other alloys. Chromium compounds in chromium (III) or chromium (VI) forms are used for chrome plating, paint and pigment production, leather and wood protection, and treatment of cooling tower water. It is used for toner for copiers and in smaller quantities in various industries such as drilling muds and textiles. The general population is exposed to chromium (chromium (III) in general) by food and drinking water and breathing air containing chemicals. Chromium (VI) is much more toxic than chromium (III) for both acute and chronic exposure. The respiratory tract is the primary target organ for chromium (VI) following inhalation exposure in humans [1, 35, 36]. The EPA concluded that only chromium (VI) should be classified as a human carcinogen. Animal studies show that chromium (VI) causes lung tumors with respiratory exposure. The EPA classified chromium (VI) as Group A, a known human carcinogen, for inhalational exposure. There is no data on the carcinogenic potential of chromium (III) compounds. The EPA classified chromium (III) as group D [37].

2.6 Nickel and nickel compounds

The chemical symbol for nickel is Ni and its atomic weight is 58.71 g/mol. It is a silvery-white hard metal found in nature as a component of silicate, sulfide or arsenide ores. Nickel is found in the environment primarily with oxygen or sulfur as oxides or sulfides. The most common form in nature is the Ni + 2 state. Each form of nickel has different physical properties. Soluble nickel salts include nickel chloride, nickel sulfate and nickel nitrate. Nickel carbonyl, an extremely unstable form, does not occur naturally and decomposes rapidly. Since nickel is a natural element in the earth's crust, it is found in small amounts in food, water, soil and air. Foods are the main sources of nickel exposure. Average intake for adults is approximately 100–300 μg/day. In addition, people working in professions involved in the production, processing and use of it, or people who come into contact with everyday products such as jewelry and stainless-steel cooking and dining utensils containing nickel, and who consume cigarette tobacco may be exposed to nickel. The most toxic nickel compound is nickel carbonyl (Ni (CO)4). Nickel carbonyl, used in the metallurgy and refining industry, is obtained by the reaction of metallic nickel with CO2. If air inhaled for 20 minutes contains 30 ppm of nickel carbonyl, it can lead to lethal toxic effects [1, 38, 39]. Nickel is found at very low levels in ambient air as a result of emissions from oil and coal burning, nickel refining, sewage sludge burning, production facilities and other sources. Due to its high instability, nickel carbonyl exposure is extremely rare. In industry, it is used in electroplating, batteries, (cadmium-nickel mixture) coins, industrial installations, spark plugs, machine parts, stainless steel, nickel-chromium resistance wires and catalysts [1, 40].

Nickel salts, nickel sulfate via inhalation and drinking water and nickel acetate are not carcinogenic in rats or mice. The EPA has not evaluated soluble salts of nickel for potential human carcinogenicity. Lung tumors were reported to develop in rats exposed to nickel carbonyl through inhalation. The EPA classified nickel carbonyl as group B2, possible human carcinogen. An increased risk of lung and nasal cancer was reported for nickel refinery powders and nickel subsulfide among nickel refinery workers. Nickel refinery powder is a mixture of many nickel compounds, the main component being nickel subsulfide. Additionally, animal studies reported lung tumors due to nickel refinery dust and nickel subsulfide exposure. The EPA classified nickel refinery powder and nickel subsulfide as group A human carcinogens [41].

3. Mitigation of heavy metals

Heavy metal toxicity caused by various anthropogenic and geological activities poses a global threat to the environment and people [42]. Humans have long used chemical, physical and biological methods to combat the threat of heavy metal toxicity, both to humans and the surrounding environment [8]. For example, in the literature, Spirulina (Arthrospira), a photosynthetic thread-like cyanobacterium commonly known as blue green algae, was shown to attenuate experimentally-induced heavy metal toxicity. A total of 58 preclinical studies demonstrated the mitigating effect of Spirulina against experimental arsenic, cadmium, lead and mercury toxicities [42]. Some important antioxidants such as flavonoids, pectin, and phytic acid are also used to remove heavy metals from the human body [43].

The removal of heavy metals from environmental compartments has become an extremely important issue. Current cleaning (or remediation) technologies to reduce harmful effects in heavy metal-contaminated areas include excavation (physical removal of the contaminated material), stabilization of metals in the site soil, and the use of plants to stop the spread of contamination or remove metals from the soil (bioremediation). Along with a range of phytochemicals, microorganisms are thought to be able to act as heavy metal removal agents in both humans and the surrounding environment [43]. The bioremediation approach is an effective method for removing toxic wastes from the environment using biological agents. It can help remove many contaminants such as metals, pesticides, explosives and oil compounds. It is mostly performed on-site and is widely accepted. Microorganisms and plants use different mechanisms for bioremediation of contaminated water and soils [44–46]. The bioremediation process occurs through a variety of mechanisms like redox, adsorption, complexation, ion exchange, precipitation and electrostatic attraction [47]. Microorganisms may begin metal mobilization/immobilization with redox reactions and hence affect bioremediation processes. Heavy metals like Fe, As, Cr and Hg enter oxidation and reduction cycles. Remediation of heavy metals from water uses microorganisms including algae and bacteria, along with fungi (*Aspergillus sp., Saccharomyces cerevisiae* etc.) and consortium organisms (Acinetobacter sp. & Arthrobacter sp., P.aeruginosa & B.subtilis, S.cerevisiae & P.aeruginosa, S.cerevisiae and B.subtilis). Biosorption by bacteria ((*Bacillus*), cocci (*Streptococcus*), filamentous (*Actinomyces*) and spiral (*Vibrio cholera*)) is an inexpensive and efficient technique to remove pollutants, like heavy metals, from wastewater [48, 49]. Different species of algae are present in large amounts in marine ecosystems [50]. Among the three algal groups, i.e., Phaeophyta, Rhodophyta and Chlorophyta (i.e. brown, green and red, respectively), brown algae were reported to possess better biosorption capacity (phycoremediation) [51, 52].

Phytoremediation is proven to be a promising alternative to traditional approaches as it is cost effective, environmentally friendly and esthetically pleasing. To date, approximately 500 taxa were identified as hyperaccumulators of one or more metals based on their natural extraction ability. In addition, more research is needed that integrates biotechnological approaches with extensive multidisciplinary research to improve plant tolerance and reduce toxic metal deposition in the soil [8, 53]. In the agricultural field, phytoremediation, phyto-extraction, phyto-polymerization, phytostabilization, phytodegradation and rhizophilization are effectively used for the removal of heavy metals [45]. Plants also help prevent wind, rain, and groundwater runoff from carrying pollutants from the site to the surrounding areas or deeper underground [46] (**Figure 1**). Silicon (Si) is the second most abundant element in the soil. Research to date suggests that Si can alleviate heavy metal toxicity in plants at both plant and soil levels, with the use of Si supplementation providing a beneficial role in reducing abiotic factors, including heavy metal stress, as well as stress exerted by biota [54, 55].

Integrated processes are a combination of two different methods to achieve a synergistic and effective method to remove heavy metals. Combining both microorganisms and plants is a bioremediation approach that enables more efficient

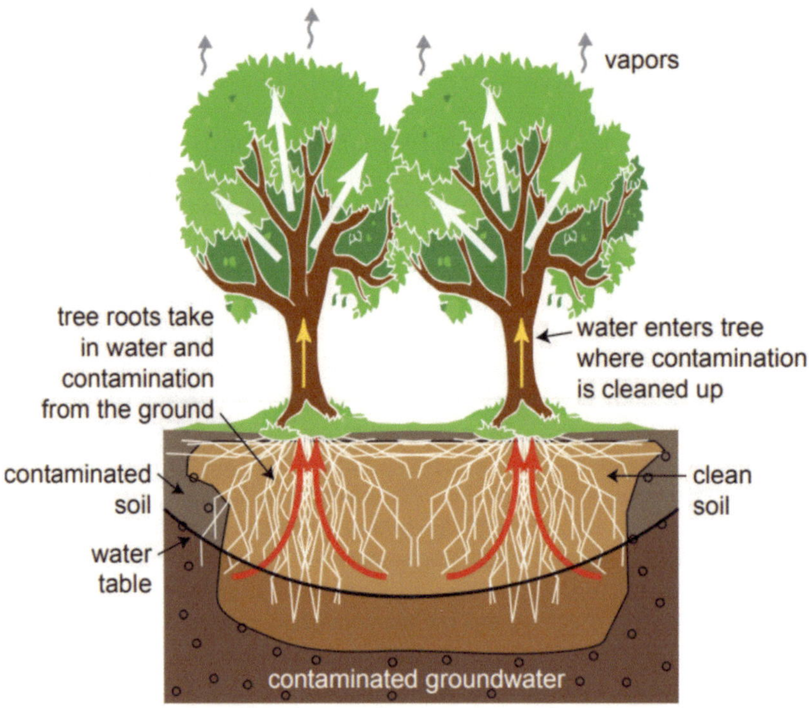

Figure 1.
Mitigation of heavy metal pollution [46].

cleaning of heavy metal contaminated soils. However, the success of this approach largely depends on the organisms involved [44, 56].

4. Conclusion

Heavy metal pollution in water and soil has increased in the last few decades due to anthropogenic activities. For this reason, the issue of dealing with heavy metal pollution in environmental matrices still maintains its importance. Legal regulations are enforced by governments, and monitoring studies, assessment of environmental impacts, use of soluble and non-toxic compounds in environmental compartments (air, water, soil and plants) in industrial processes, heavy metal-free pesticides, appropriate wastewater treatment plants and renewable energy sources instead of fossil fuels are priority measures to reduce the concentrations of heavy metals in the environment. In addition, heavy metals, which cause serious health problems even at very low concentrations due to their long-term persistence, must be removed from receiving environments in order to protect the integrity of the ecosystem and human health. As a bioremediation approach, removing toxic wastes from the environment by using bioaccumulatory organisms such as plants or mussels maintains its importance among studies aimed at recovery. Studies have shown that integrated methods - especially the combination of suitable plants and microorganisms - are very effective in mitigating the effects of heavy metals in the environment.

DOI: http://dx.doi.org/10.5772/intechopen.95103

Author details

Ayşe Handan Dökmeci
Department of Emergency and Disaster Management, Health of School,
Tekirdag Namik Kemal University, Tekirdağ, Turkey

*Address all correspondence to: hdokmeci@nku.edu.tr

References

[1] Dökmeci AH. Toxicological-Environmental and Industrial-Disasters. Nobel Tıp Kitabevi, Istanbul. Turkey 2018; 99-120 (In Turkish)

[2] Tansel B. From electronic consumer products to e-wastes: Global outlook, waste quantities, recycling challenges. Environmental International. 2017; **98**: 35-45. DOI:10.1016/j.envint.2016.10.002

[3] Medina-Estévez F, Zumbado M, Luzardo OP, Rodríguez-Hernández A, Boada LD, Fernández-Fuertes F, Santandreu-Jimenez ME, and Henríquez-Hernández LA. Association between Heavy Metals and Rare Earth Elements with Acute Ischemic Stroke: A Case-Control Study Conducted in the Canary Islands (Spain). Toxics. 2020; **8**(3): 66. DOI:10.3390/toxics8030066

[4] Jaishankar M, Tseten T, Anbalagan N, Mathew BB, Beeregowda KN. Toxicity, mechanism and health effects of some heavy metals. Interdisciplinary Toxicology. 2014; **7**(2): 60-72.

[5] Cucu AK, Topkaya M, Erdogan G, Aboul-Enein HY. Quantitative Determination of Heavy Metal Contamination in Horse Mackerel and Whiting Caught in the Sea of Marmara. Bulletin of Environmental Contamination and Toxicology. 2019; **102**, 498-503 DOI:doi:10.1007/s00128-019-02574-5

[6] Krishnamoorthy R, Venkatramanan V, Senthilkumar R, Anandham R, Kumutha K, Sa T. Management of Heavy Metal Polluted Soils: Perspective of Arbuscular Mycorrhizal Fungi. In: Shah S., Venkatramanan V., Prasad R. (eds). Sustainable Green Technologies for Environmental Management. Springer, Singapore . 2019; 67-85

[7] Duffus JH. “Heavy metals” a meaningless term? (IUPAC Technical Report) Pure Appl. Chem. 2002;**74**, 793-807. DOI: 10.1351/pac200274050793

[8] Li C, Zhou K, Qin W, Tian C, Qi M, Yan X and Han W.A Review on Heavy Metals Contamination in Soil: Effects, Sources, and Remediation Techniques, Soil and Sediment Contamination: An International Journal, 2019; **28**(4), 380-394, DOI: 10.1080/15320383.2019.1592108

[9] Patniak P. Handbook of Environmental Analysis. Chemical Pollutants in Air, Water, Soil and Solid Wastes. Second Adition, CRC Press Taylor and Francis Group, Boca Raton. 2010; 770

[10] United States Agency for Toxic Substances and Disease Registry (ATSDR). Toxicological Profile for Lead (Draft for Public Comment). Department of Health and Human Services Atlanta, GA, USA. 2019;582

[11] Misra S., Misra KG. Phytoremediation: An Alternative Tool Towards Clean and Green Environment. In: Shah S., Venkatramanan V., Prasad R. (eds) Sustainable Green Technologies for Environmental Management. Springer, Singapore; 2019 https://doi.org/10.1007/978-981-13-2772-8_5

[12] Dökmeci AH, Yıldız T, Öngen A, and Sivri N. Heavy metal concentration in deepwater rose shrimp species *Parapenaeus longirostris* Lucas 1846 collected from the Marmara Sea Coast in Tekirdağ, Environmental Monitoring Assessment. 2014;**186**,2449-2454

[13] International Agency for Research on Cancer (IARC). Monographs on the Evaluation of Carcinogenic Risks to Humans. Arsenic and Arsenic Compounds. Vol. 100c. Lyon, France: IARC; 2011

[14] California Environmental Protection Agency (CalEPA).

Technical Support Document for the Determination of Noncancer Chronic Reference Exposure Levels. . Office of Environmental Health Hazard Assessment, Berkeley, CA. 2008.

[15] Seid MA, Kapri A, Goel R. Heavy Metal Pollution: Source, Impact, and Remedies. Biomanagement of Metal-Contaminated Soils. Editors: Khan MS, Zaidi A, Goel R, Musarrat J. (Eds.). Springer Science+Business Media B.V.2011; 1-28

[16] U.S. Environmental Protection Agency (USEPA). Toxics Release Inventory Guidance for Reporting the Lead and Lead Compounds Category. USA, 2020;20-34

[17] Masindi V., Muedi K. Environmental contamination by heavy Metals. In: Hosam El Din M., Saleh Refaat Aglan F., editors. Heavy Metals. IntechOpen; Rijeka, Croatia: 2018

[18] Centers for Disease Control and Prevention (CDC). Lead in the environment. https://www.cdc.gov/nceh/lead/infographic.htm [Date of access: 01.11.2020]

[19] Kumar D, Malik DS, Patel S, Land Gupta V. In: Contaminants in Agriculture and Environment: Health Risks and Remediation Chapter 6. Human health risk assessment and mitigation of heavy metal pollution in agriculture and environment. DOI: 10.26832/AESA-2019-CAE-0157-06

[20] Wanı Al, Ara A, Usmanı JA. Lead toxicity: a review. Interdisciplinary Toxicology. 2015;**8**(2): 55-64. DOI: 10.1515/intox-2015-0009

[21] U.S. Environmental Protection Agency (USEPA). Integrated Risk Information System (IRIS). Chemical Assessment Summary. Lead and compounds (inorganic); CASRN 7439-92-1

[22] U.S. Environmental Protection Agency (USEPA). National Elemental Mercury Response Guidebook March 2019; 5-6

[23] Rani L, Basnet B, and Kumar A. Encyclopedia of Environmental Health. 2011; 705-712

[24] Dökmeci AH. An Assessment of the Heavy Metals in Wild Mussels Mytilus Galloproincialis from the Marmara Sea Coast of Tekirdag, Turkey, Fresenius Environmental Bulletin 2017; **26**, 7608-7612

[25] Dökmeci AH, Şabudak T, Dalmış V. Bioaccumulation of essential and toxic metals in four different species of bottomfish in the Marmara Sea, Tekirdag, Turkey: risk assessment to human health, Desalination and Water Treatment 2019; **148**, 213-221

[26] Türkdoğan MK, Kilicel F, Kara K, Tuncer I and Uygan I. Heavy Metals in Soil, Vegetables and Fruits in The Endemic Upper Gastrointestinal Cancer Region of Turkey", Environmetal Toxicology and Pharmacology. 2002; **13**, 175-79

[27] Kueh CSW and Lam JYC. Monitoring of Toxic Substances in The Hong Kong Marine Environment. Marine Pollution Bulletin. 2008; 57, 744-57

[28] U.S. Environmental Protection Agency. National Elemental Mercury Response Guidebook March 2019;8-10

[29] Marković JP, Jović MD , Smičiklas ID, Šljivić-Ivanović MZ, Smiljanić SN, Onjia AE and Popović AR. Estimation of Cadmium uptake by tobacco plants from laboratory leaching tests. Journal of Environmental Science and Health, Part A. 2018; 53:4, 352-361, DOI: 10.1080/10934529.2017.1401396

[30] Elements and Their Compounds in the Environment: Occurrence, Analysis and Biological Relevance, Second Edition Editor(s): E. Merian M. Anke M. Ihnat M. Stoeppler. 2004 ISBN:9783527304592 doi:10.1002/9783527619634 Copyright © 2004 WILEY-VCH Verlag GmbH & Co. KGaA

[31] Bernhoft RA. Cadmium toxicity and treatment. Scientific World Journal. 3; 2013:394652. DOI: 10.1155/2013/394652. PMID: 23844395; PMCID: PMC3686085.

[32] Rafati Rahimzadeh M, Rafati Rahimzadeh M, Kazemi S, Moghadamnia AA. Cadmium toxicity and treatment: An update. Caspian J Intern Med 2017; **8**(3): 135-145.

[33] Dökmeci AH, Öngen A, Dağdeviren S. Environmental Toxicity Of Cadmium and Health Effect, Journal of Environmental Protection and Ecology. 2009;**10**,84-93

[34] IARC Working Group on the Evaluation of Carcinogenic Risks to Humans. Arsenic, Metals, Fibres and Dusts. Lyon (FR): International Agency for Research on Cancer; 2012. (IARC Monographs on the Evaluation of Carcinogenic Risks to Humans, No. 100C) Cadmıum and Cadmıum Compounds. Available from: https://www.ncbi.nlm.nih.gov/books/NBK304372/

[35] Agency for Toxic Substances and Disease Registry (ATSDR). Toxicological Profile for Chromium. U.S. Public Health Service, U.S. Department of Health and Human Services, Atlanta, GA. 1998

[36] IARC Working Group on the Evaluation of Carcinogenic Risks to Humans. Chromium, Nickel and Welding. Lyon (FR): International Agency for Research on Cancer; 1990. (IARC Monographs on the Evaluation of Carcinogenic Risks to Humans, No. 49.) Chromıum and Chromıum Compounds. Available from: https://www.ncbi.nlm.nih.gov/books/NBK519246/

[37] U.S. Environmental Protection Agency. Integrated Risk Information System (IRIS) on Chromium VI. National Center for Environmental Assessment, Office of Research and Development, Washington, DC. 1999

[38] Tchounwou PB, Yedjou CG, Patlolla AK, Sutton DJ. Heavy Metals Toxicity and the Environment. EXS. 2012; **101**: 133-164. DOI: 10.1007/978-3-7643-8340-4_6

[39] IARC Working Group on the Evaluation of Carcinogenic Risks to Humans. Arsenic, Metals, Fibres and Dusts. Lyon (FR): International Agency for Research on Cancer; 2012. (IARC Monographs on the Evaluation of Carcinogenic Risks to Humans, No. 100C.) Nıckel And Nıckel Compounds. Available from: https://www.ncbi.nlm.nih.gov/books/NBK304378/

[40] National Research Council (US) Committee on Acute Exposure Guideline Levels. Acute Exposure Guideline Levels for Selected Airborne Chemicals: Volume 6. Washington (DC): National Academies Press (US); 2008. 9, Nickel Carbonyl Acute Exposure Guideline Levels. Available from: https://www.ncbi.nlm.nih.gov/books/NBK207867/

[41] U.S. Environmental Protection Agency. Health Assessment Document for Nickel. EPA/600/8-83/012F. National Center for Environmental Assessment, Office of Research and Development, Washington, DC. 1986.

[42] Sanjib Bhattacharya The Role of Spirulina (Arthrospira) in the Mitigation of Heavy-Metal Toxicity: An Appraisal. Journal of Environmental

Pathology, Toxicology and Oncology. 2020;**39**;2:149-157, DOI: 10.1615/jenvironpatholtoxicoloncol.2020034375

[43] Sharma S, Rana S, Thakkar A, Baldi A, Murthy RSR and Sharma RK. Physical, Chemical and Phytoremediation Technique for Removal of Heavy Metals. Journal Heavy Metal Toxicity Dis. 2016, 1:2

[44] Chibuike GU and Obiora SC. Heavy Metal Polluted Soils: Effect on Plants and Bioremediation Methods", Applied and Environmental Soil Science, 2014; 2 DOI:https://doi.org/10.1155/2014/752708

[45] Kumar, D., Malik, D.S., Patel, S.L. and Gupta, V. Human health risk assessment and mitigation of heavy metal pollution in agriculture and environment. In: Kumar, V., Kumar, R., Singh, J. and Kumar, P. (eds) Contaminants in Agriculture and Environment: Health Risks and Remediation. 2019;**1**, 66-75 Agro Environ Media, Haridwar, India DOI: https://doi.org/10.26832/AESA-2019-CAE-0157-06

[46] United State Environmental Protection Agency (US EPA), 2012, A Citizen's Guide to Phytoremediation, Office of Solid Waste and Emergency Response (5102G) EPA 542-F-12-016 September 2012. www.epa.gov/superfund/sites

[47] Kumar V, Chopra AK, Kumar S, Singh J and Thakur RK. Effects of pulp and paper mill effluent disposal on soil characteristics in the vicinity of Uttaranchal Pulp and Paper Mill, Haridwar (Uttarakhand), India. International Journal of Agricultural Science Research. 2015; **4**:117-125

[48] Kapahi., M and Sachdeva., S. Bioremediation Options for Heavy Metal Pollution. J Health Pollut. 2019 Nov 27;9(24):191203. doi: 10.5696/2156-9614-9.24.191203

[49] Mustapha MU, Halimoon N. Microorganisms and biosorption of heavy metals in the environment: a review paper. J Microb Biochem Technol. 2015;7(5):253-6.

[50] Coelho., L, Rezende., H.C, Coelho., L.M, Sousa., P.A.R., Melo D.F.O and Coelho., N.M.M. (2015). Chapter: Bioremediation of Polluted Waters Using Microorganisms. Book:Advances in Bioremediation of Wastewater and Polluted Soil. IntechOpen. DOI: 10.5772/60770

[51] Brinza L, Dring M, Gavrilescu M. Marine micro and macro algal species as biosorbents for heavy metals. Environ Eng Manag J. 2007 May/Jun;6(3):237-51.

[52] Oyedepo TA. Biosorption of lead (II) and copper (II) metal ions on *Calotropis procera* (Ait.). Sci J Pure Appl Chem [Internet]. 2011 [cited 2019 Aug 22]: [7 p.]. Available from: https://www.sjpub.org/ sjpac.html

[53] Dökmeci AH, Adiloğlu S. The Phytoremediation of Chromium from Soil UsingCirsium Vulgare and the Health Effects, Biosciences Biotechnology Research Asia. 2020;17(3), 535-541

[54] Bhat AA, Shivaraj SM, Singh P, Navadagi DB, Tripathi DK, Dash PK, Solanke AU, Sonah H, Deshmukh R. Role of Silicon in Mitigation of Heavy Metal Stresses in Crop Plants Plants (Basel) 2019; **8**(3):71. DOI: 10.3390/plants8030071

[55] Adrees M, Ali S, Rizwan M, Rehman MZ, Ibrahim M, AbbasF, FaridM, QayyumMF, IrshadMK. Mechanisms of silicon-mediated alleviation of heavy metal toxicity in plants: A review. Ecotoxicology

and Environmental Safety, 2015; 119, 186-197.

[56] Selvi A, Rajasekar A, Theerthagiri J, Ananthaselvam A, Sathishkumar K, Madhavan J and Rahman PKSM. Integrated Remediation Processes Toward Heavy Metal Removal/ Recovery From Various Environments-A Review. Frontiers in Environmental Science. 2019; 7, 66. DOI: 10.3389/ fenvs.2019.00066

Chapter 9

Metallothioneins: Diverse Protein Family to Bind Metallic Ions

Ettiyagounder Parameswari, Tamilselvan Ilakiya, Veeraswamy Davamani, Periasami Kalaiselvi and Selvaraj Paul Sebastian

Abstract

Metallothionein's (MTs) are the lower molecular weight (6-7 kDa) proteins that are found to be present in almost all organism types ranging from prokaryotes to eukaryotes species. MT are the metal detecting proteins that can mitigate the effect caused by the excess metal ions. They are also found to be involved in cellular process such as cell growth regulation, ROS (Reactive Oxygen Species) and DNA repair. The protein was termed as Metallothionein due to the unusual higher metal (metallo) and the sulfur (thiol) content. They are further grouped into 3 classes *viz.,* class I, II and III. The Class I and II MTs are polypeptides that were obtained from direct gene products, the class III MTs are from the cysteine-rich non-translational molecules that are termed as phytochelatins. The metal ions are been sequestered through the MTs with Cys rich motifs. All the cysteines are present in the reduced form and are been co-ordinated through the mercaptide bonds. The cysteines present in the MTs are preserved across the species, it is supposed that, cysteines are essential for the function and the MTs are required for the life. Metallothionins structure, conservation in evolution, their ubiquitous nature of occurrence, the genes redundancy and the programmed MTs synthesis in development, regeneration and reproduction of living organisms are some of the weighty arguments in suspecting MTs to also serve others and perhaps the high particular metal-related cellular roles. In this chapter, there is a detailed discussion about Metallothionein its structure, occurrence and function.

Keywords: Metallothioneins, proteins, heavy metals, tolerance

1. Introduction

Many of the elements are play a significant role in all kind of organisms and pose some functional outcomes in living organisms. Based on this outcome, elements are classified as essential and non-essential elements. Recent trace element research showed many of the metals also play a major significant role in metabolisms but they required very trace quantities. The metal ions of iron (Fe), zinc (Zn), manganese (Mn), copper (Cu), cobalt (Co) molybdenum (Mo) and the non-metals selenium (Se) and iodine (I) bromine (Br), rubidium (Rb), aluminum (Al), nickel (Ni), titanium (Ti) and barium (Ba) etc. exceeded certain limits, it become toxic to organisms. Predicting the role of metals in living organisms is difficult since its intrinsic characteristics, redox properties, charge, and, above all, size. Toxicity of the metals mainly through altering the cell metabolism and thereby blocking synthesis of fundamental enzymes. To mitigate this, organisms have developed different detoxifying mechanisms such as blocking of toxic metals through oligopeptides or proteins and chelating metals encoding by genes, which are called metallothioneins (MTs) which possess residues of cysteine (Cys) having thiol or sulfhydryl groups (R-SH) and thereby immobilize the metallic ions. The prime function of MTs is the storage, transportation and binding of metals. Margoshes and Vallee's discovery of Cd-binding, cystein-rich protein in the horse kidney marked a birth field for the research that focused to study a novel polypeptide super family, the metallothioneins (MTs). The MTs are family consisting of low molecular weight (6-7 kDa), cysteine (Cys)-rich, non-enzymatic proteins that are found ubiquitously in animals, in higher plants, in eukaryotic microorganisms, and many prokaryotes [1, 2]. The metal ions is sequestered through MTs by binding with Cys rich metal binding – motifs Cys-Cys, Cys-X-Cys, or Cys-X-X-Cys) that [3, 4]. Based on the Cys residues arrangement, MTs grouped into different classes [5, 6]. All the cysteines present in the reduced form are been coordinated to the metal ions that gives rise to the spectroscopic appearance of the metal-thiolate groups. Cysteines in the MTs are preserved across the species, it is suspected that, cysteines are essential for the function and the MTs is required for the life. MTs have some composition of unusual amino acid as they do not contain any aromatic amino acids and the most important is that, one third of its residues contacysteines. Moreover, MTs show spectroscopic features of metal thoilates (mercaptides) [7]. MTs can bind Cd, Ni, Zn, Cu, and Pb with the affinities that depend up on the bioavailability and concentration of heavy metals. Three major scientist groups have contributed in the field of MTs include:

1. Toxicologists and physiologists for analyzing MTs role in the heavy metal metabolism and the de-toxification.

2. Protein chemists and spectroscopies intrigued by its usual structural features.

3. Molecular biologist interested in the gene regulation and use of MTs (promoter) sequences for the genetic engineering experiments.

MTs play a critical role in protecting the cell from harmful factors i.e. metals, free radicals, etc. through controlling cell growth, differentiation, proliferation, explicating their nuclear vs.cytoplasmic localization [8]. MTs is cytoplasmic protein, accumulated in lysosomes, trandpoted to the nucleus and intermediated space in mitochondria [9, 10]. MTs transported the metals from one cell to other cell by receptor- mediated mechanism and metal is transported to the cytosol but protein remains in an endocytotic compartment [11, 12]. According to Wang et al. [13], MTs have exhibits four main functions and it is illustrated in the **Figure 1**.

Nevertheless, although MTs were discovered over four decades back, their physiological functions are not fully understood still [14]. Based on the this prelude this review is being taken up to refresh the MTs structure, classification and role in metal detoxification.

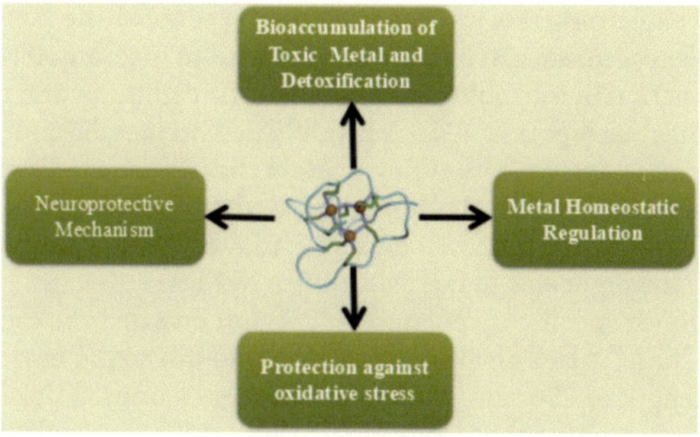

Figure 1.
MTs - four main functions. (adopted from [13]).

2. Various scientific developments in metallothionein research

The scientific research history of MTs begins in 1950s almost seven decades back. The developments in MTs research includes the contributions from scientists in fields of environmental hygiene/medicine, biochemistry, medicine, pathology, nutrition and toxicology and are best illustrated in number of articles that are been published. During this time, in protein research methods, there is a considerable development in isolation and analytical methods for identification. In beginning the interest in MTs was focused on two main topics. The major interest were in the protein chemistry and in the toxicological, *i.e.* the metabolic pathways for the metal toxicology specifically in relation to the cadmium kinetics. In the recent years, the key contributions are brought in the molecular biology. Consensus and the conclusions reached at various scientific meetings on the metallothionein are based up on the published data as depicted in the **Table 1**.

In 1957, the data related to cadmium binding protein in the equine tissue were published. However, in the beginning this study was first initiated through a report that was in a form of abstract that deals with cadmium in the human organs. In most of the animal species, cadmium is present in small amounts in tissues and the

Year	Scientific developments
1960,1961	Metallothionein
1964	Induction by cadmium
1971	Modification of Cd-toxicity
1972	Amino acid composition
1976	Sequence
1979	1st international meeting - Consesus nomenclature
1979	Radioimmunoassay
1983	International meeting - Metallothionein and nephrotoxicity
1985	2nd International meeting on metallothionein
1992	3rd International meeting on metallothionein
1996	Workshop on metallothionein

Table 1.
Various scientific events held on the metallothionein.

body fluids. For explaining this unexpected finding some of the hypotheses were postulated. Either cadmium can be present as bound to macromolecule or so had some natural function in the biological system or the cadmium also could be just a contaminant. The first report in 1960 was published on metallothionein [15]. For the first time that a cadmium containing protein that are isolated from the equine renal tissue, was termed as 'metallothionein' based on the higher content of sulfur (4.1%) followed by Cadmium (2.9%) and Zinc (0.6%). Later researchers estimated that the molecular weight was found as 10000 ± 260 Da.

The absorption at specific wavelength of 250 nm revealed the cadmium mercaptide bonds. The MTs lack an aromatic amino acid due to the non absorption at precise wavelength of 280 nm. These were further examined through amino acid analyses. Higher sulfur concentration of about 4.1% was explained through the amino acid analyses that revealed high cysteine content. The eactive protein mercapto groups at this time were examined through titration with the p-chloro mercuri benzoate (CMB), N-ethylmaleimide (NEM) and silver ions. The Amino acids were determined using both ion exchange chromatography and two-dimensional paper chromatography. The cysteine residues were found as cysteic acid after metallothionein oxidation with the performic acid and as N-ethylmaleide derivatives. The Sedimentation constant were determined by Schlieren diagram with the use of sedimentation through ultracentrifugation to 1.75 (S_{020w}). Partial specific volume, diffusion constant, and friction ratio were obtained. The estimated protein molecular weight was still varying between 9790 and 10500 Da. These were partly explained by several artifacts during the preparation. Metal analyses for Cd and Zn reported 5.9% of MTs weight and 5.2 g atoms mol^{-1} for cadmium and for zinc as 2.2% by weight and 3.3 g atoms mol^{-1}. Some exchange between Cd and Zn was taking place obviously. It was suggested that the bonding with 3 SH-groups together with an atom of either Cd and Zn was formed.

In the years between 1970s and early 1980s, there were only few research groups performed researches that are related to MTs. A workshop was arranged with nearly 25 invited participants who had submitted their background manuscripts and the tentative report [16] was prepared and were distributed in advance to each of the participant. A consensus report was agreed during the meeting held in the year of 1978.

During this I[st] international meeting about MTs, held in the Zurich consensus was reached about, e.g. nomenclature and methods for protein preparing. This first meeting had been followed by another meeting held at Zurich in 1985 that were more open but still with a workshop consensus. In between, a meeting was arranged at Aberdeen in1981.

3. Structure and occurrence of metallothionein

The structure of MTs was first reported by Kagi and Schafer [15]. MTs have 2 domains consists of a cluster with three and other with 4 metal atoms. The gene are located on the chromosome 16. This protein contains a number of isoforms that are coded with several alleles. The ratio of the mRNA for MT-I and MT-II genes remains constant during the induction with metals, e.g. Cd, Cu and Zn. The MTs amino acid sequences from mammalian sources reveals that all of them contain nearly 61 amino acids that are more similar in their composition. All contain 20 cysteine residues which remain invariant to the amino acid sequence. Every cysteines do participate in the coordination of the seven moles of Zinc or cadmium per mole of the MT [17]. The MTs are the cytoplasmic proteins; but, their occurrence during fetal development in nucleus are given by immuno histochemistry. The Plant MTs were first recognized in roots of the copper-resistant strain of *Agrotis gigantean* and also in roots of the tomato plants that are been exposed to higher concentrations of Cadmium chloride in medium are found in numerous other plants like Cd-treated maize, tobacco, wheat, rice and cabbage and also added found in some cell suspension cultures of plants [18, 19].

Metallothioneins from mammals contain 20 cysteins residues and bind seven cadmium or zinc ions in two discrete clusters. The thiolate side chains of these proteins act as bridging and terminal ligands to form Cd3SCy9 and Cd4 SCy11 clusters.

4. Classes of metallothionein

Based upon their similarities in structure, the MTs are divided into 3 classes:

Class I: This type present in mammalians, oysters, some fishes, crabs and mussels as Cys - Cys proteins.

Class II: Occurs in fungi, plants, yeasts, drosophila, cyanobacteria and nematodes in the form of Cys – x - Cys.

Class III: Presence noticed in phytochelatins that present in all plants, some fungi and also algae as Cys – x – x - Cys.

The MTs belongs to the class I & II are been produced through direct mRNA translation. Whereas, the class III MTs belongs to the metal-binding polypeptides and have peptide bonds which are biologically synthesized as some of the enzymatic products. All vertebrates that are examined contain 2 or more distinct isoforms of MT, that are grouped into 2 classes, designated as MT-I and MT-II, depending upon elution position from the DEAE cellulose. In many cases each of the class actually consists of various different proteins that were designated as MT-IA, MT-IB, MT-IC, etc.

General characteristics of Metallothioneins are,

1. MT I and MT II's molecular weight is 6-7 kDa, 61 amino acids, 20 Cysteine (30%), C-alanine, N-acetylmethionine, no histidine, no aromatics.

2. The have an unique amino acid sequence, metal clusters/tertiary structure

3. Metal content; Cu, Cd, Hg, Zn; 5-10% W/w

4. Light absorption are at 250 nm (Cd), 275 nm (Cu), 225 nm (Zn), 300 nm (Hg)

5. Induced synthesis through Zn and Cd.

6. There are no disulfide bonds, heat stability

7. Isoforms

8. Localization on chromosome and cytoplasm

MT expression in the crops serve as the key index in selecting for the heavy metal tolerance. In attempting for shedding light on their function, the researchers relied on the RNA blot hybridization for studying genes expression of MT during the development and also in response to several other environmental factors. Detailed localization of the MT mRNAs or gene promoter activity is obtained through reporter gene expression studies and *in situ* hybridization. The RT–PCR are evaluated as an alternative for Northern blot analysis for quantifying mRNA levels of MT. But the sensitivities of the two techniques were identical for detecting the induced mRNA levels in MT. Thus the above results suggests that the RT–PCR would be a quantitative and a sensitive method for evaluating gene expression of MT [20]. Most of the MT genes are been expressed at a higher level in the plant tissues, in the terms of the transcript abundance. The direct evidence are from the rice gene profiling experiments using SAGE (serial analysis of gene expression) protocol. The transcripts from four MT genes comprised about 3% of the transcripts in 2 weeks old seedlings. Metallothionin gene belonging to Type II contributes about 1.26% of all transcripts that was found to be most expressed form in that tissue. The MT, 2 forms of transcripts as Type 3 genes counted as extra 1.25% of mRNAs. Numerous MT genes that are identified through differential screening of the cDNA libraries indicates that the RNAs encoding to MTs are higher in most of the other plant species also.

In general, affinity of the metal ions for binding sites follows the typical order of thiolate model complexes, *i.e.,* Zn (II) < Pb (II) < Cd (II) < Cu (I), Ag (I), Hg (II), Bi (III) [15]. The MT are usually detected by its virtue of high content of the metals (detected by labelling with radionuclides or by atomic absorption spectrophotometer) or by cysteine. Some of the procedures include, reversed phase High Pressure Liquid Chromatography (HPLC) gel filtration be elevated as a variety of mammalian metallothionins and utilized for immune-electrophoresis, radioimmunoassay and immunodiffusion. These protocols are mainly helpful for the detection of minimum quantity in MT. Basic structure of MT are examined by several biochemical and bio-physical techniques such as NMR spectroscopy, UV, ESR,CD partial proteolysis and X-ray crystallography.

5. Metallothionein redox cycle and function

The MT redox cycle play an important role in the MT biologic function. It might link to homeostasis of the essential metals, detoxification of the toxic metals and protection against the oxidative stress. Advancement in MT research is shown by demonstrating redox regulation of the Zn-S interaction and zinc coupling. The Zn-MT cluster structure provides some chemical basis through which the cysteine ligand induces the oxido reductive properties. These structure permits for the thermo-dynamic stability of Zn in the MT. The decrease of oxidized MT can restore Zn binding ability, thereby replicates endogenous reservoir of zinc [21]. These process constitutes the redox cycle of MT, as depicted in the **Figure 2**, that provides some new perspective upon MT biologic function.

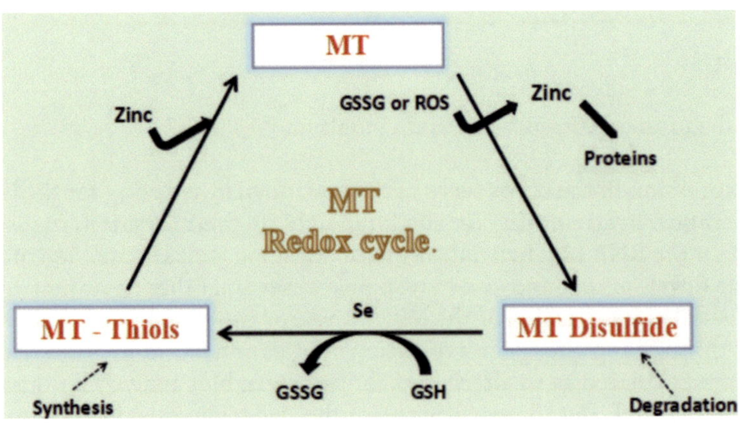

Figure 2.
Redox cycle of MT.

Under the physiologic conditions, Zn that bounds to the MT are been released from the thiolate group oxidation. The MT disulfide and/or thionin formation follows a Zn release. The MT disulfide and/or the thionin undergoes some degradation, however, when the oxidized environment gets reduced. This reduction progress is more in occurrence of a catalyst (selenium). These processes constitute the MT redox cycle, which plays a critical role in biologic function of MT.

6. Proposed mechanism for metallothionein induction by metal cations

Cellular effects of exposure to the divalent metal cations and roles of particular resistance mechanism of the metallothionein expression, and expression of stress protein GroEL is shown in the **Figure 3**. GroEL forms essential component which reduces heavy metal stress [22]. The MTs is a cysteine rich protein which binds with the metals and thereby detoxified through reducing its presense [23]. The heavy metals enter the cells by the transport of the proteins which binds the metals. This metals could further binds SH groups with repressor protein of smtB. These attached metals further alter conforms the repressor protein and then released it. These process permits RNA polymerase to initiate the transcription from smtA, resulted the metallothionein expression.

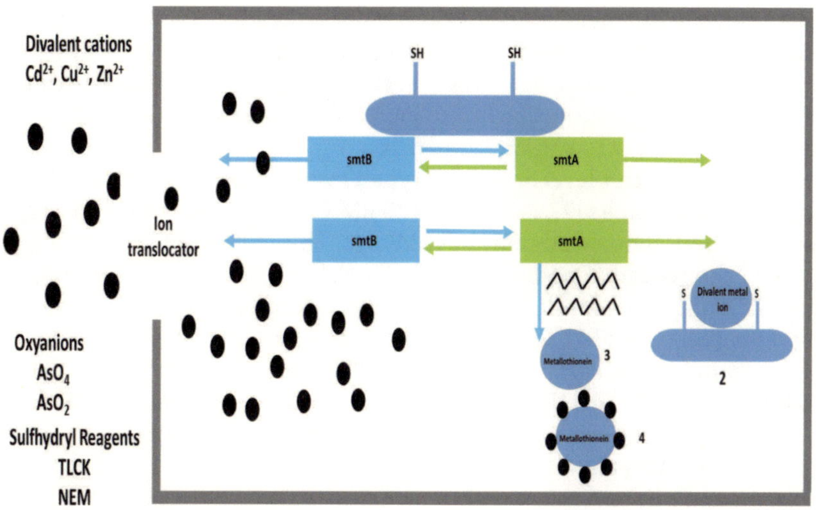

Figure 3.
Proposed mechanism for metallothionein induction by metal cations and arsenic oxyanions. (1) divalent metal cation challenge from the environment, (2) sulfhydryl (-SH) modification by metal or arsenate binding, (3) SmtA expression producing metallothionein proteins, (4) metal sequestration and detoxification by metallothionein proteins.

7. Plant metallothioneins

After the Phytochelatin discovery, as a key metal ligand essential for the plant tolerance to Cadmium, a MT protein was identified in the wheat, and numerous MT genes was isolated from the plants. The plant Class II MT proteins are further classified based upon the amino acid sequence. In the year 1993, Robinson and his co-workers identified the two plant based MT which was based on cysteine residue positions in the predicted proteins. Since then, number of characterized plant MT genes has been increased drastically, as many do not conform to these 2 groups, some additional categories are added [24, 25].

The majority of the plant MT genes are identified in angiosperms. Numerous species, like *Arabidopsis*, sugarcane and rice contain some genes encoding all 4 MTs types. This shows that the evolution of 4 plant MT types predates the separation of the monocots and dicots, and are likely that the major flowering plants also contains 4 different MT types. The presence of 4 MTs types in the plants with a distinct arrangement of the cysteines contrasts with situation in animals. For an example, the 4 mouse MTs all contain the same conserved cysteines, although they do differ in tissue expression [17]. The diversity of plant MT gene family suggests that these might differ not only in the sequence but also in the function. There is only little information about the MT genes in the non-flowering plant species. However, the genes encoding the Type3 MTs are cloned from different gymnosperms. One of the MT-encoding gene are isolated from a brown alga, *Fucus vesiculocus*, [26]. This MT do not fit readily into any of the 4 plant types described above but, primarily relies on basis of cysteine residues, is equally similar to the Arabidopsis MT1a and an oyster MT.

8. Expression pattern under influence of various factors

8.1 Metal ions

As homologs of animal MTs, it would be reasonable to assume that the MTs in plants are related to metal metabolism and detoxification of excess metal ions. Indeed, the expression of both Ec proteins and several class I MT genes vary under metal treatment. The wali l (type 2) in wheat was originally isolated as an aluminum-induced gene. Its expression is also increased by treatment with other toxic metal ions like Cd, Fe, Zn, Cu, Ga, In and La.

In rice, the transcript level of RgMT (type 1) was elevated by 250 μM $CuCl_2$, in suspension-cultured cells [27, 28], while 100 μM $CuSO_4$ increased ricMT (type 4) transcripts in both shoots and roots of seedlings. This expression-enhancing effect of Cu^{2+} are been reported in the type 2 MT genes from *Arabidapsis* and tobacco, and in type 3 MT from *Arabidopsis*. In contrast, copper ion are shown to have a suppressive effect on type 2 MT in soybean and *Vicia faba*; in the copper-tolerant plant *Mimulus guttatus*, transcription of the type l MT was inhibited by 5 μM CuS04, as well as by 5 μM $CdSO_4$ and 15 μM $ZnSO_4$. In addition, the mRNA Ievel of ricMT in the root of rice seedlings was reduced by 100 μM aqueous solutions of $ZnSO_4$, $CdCl_2$, $FeCl_3$, Pb $(Ac)_2$ and $AlC1_3$, respectively.

8.2 Senescence

Coupe *et al.* [29] isolated the type 2 MT gene JET12 from abscission zones of the ethylene-treated leaves of *Sambucus nigra* L. Northern analysis reported that JET12 gene was expressed only in senescent leaves, and in the ethylene-treated abscission zone.

Similarly, the type 3 MT LSC54 in rape (*Brassica napus*) was expressed during leaf senescence, and as senescence progressed, the mRNA level of LSC54 increased. Yu *et al.* [28] showed that the expression of rgMT (type 1) increased as young leaves progressed to senescence. These findings suggest that these MT genes may play a key role in senescence-associated metabolism of plants.

8.3 Other stresses

Considerable effort are made to clarify the relationships between plant MTs and environmental factors other than metal ions. The expression of type 2 MT seems to be easily affected by stresses- For example; the MT in tobacco was strongly induced by tobacco mosaic virus and wounding [30] and expression of MT in *Vicia faba* was reduced when plants were treated with salt, cold, salicylic acid or UV-light. Furthermore, transcripts of OsMT-2 in rice can be induced by sucrose sta, heat shock and ABA and the expression of wali 1 in wheat was increased by low-nutrient conditions and wounding.

Regarding type I MT, the mRNA level of rgMT in rice increased under treatment with heat shock, a toxicant (DMSO) and starvation, and decreased in presence of abscisic acid.

9. Fungal metallothioneins

Fungal MTs, study are more sporadic and are described usually as copper binding proteins. Still, few are characterized functionally in the positions of metal organization in chemical properties or complementation tests in mutants of metal sensitive yeasts. Reputed MTs are mostly designated within collections of ESTs (Expressed Sequence Tag) based on sequence resemblance. These are occasion for the MT like arrangements that are present in the mycorrhizea, specifically in *Pisolithus tinctorius* (ectomycorrhizal fungi), the ericoid fungus (*Oidiodendron maius)* and the arbuscular mycorrhizal fungi (*Gigaspora rosea*, *Gigaspora margarita* and *Glomus intraradices)* [31, 32].

The *P. involutus* gene (*Pimt1*) codes for short MT (34 amino acids) and contains domain bearing classical C–X–C motifs, compared with longer and canonical MTs that usually have 2 Cys-rich domains. These are found in the other fungal MTs too; it appears to be that length is not critical required for binding of metal, as even some shorter protein of MT has proved to efficiently chelate the metal ions [33].

Heterologous expression systems for demonstrating unambiguously that *Pimt1* gene product can sequester metal ions, therefore conferring *in vivo* protection against the metals, in particular with cadmium and copper [34]. The functional complementation assays was done using 3 distinct mutants of metal-hypersensitive yeast. Similar approach was done for the characterization of 2 MT- encoding genes from the endomycorrhizal fungi [31]. It explains in what way Genome Omission Project for *Saccharomyces* offered materials towards the efficient genomics researches in area of the metal forbearance. Availability of abundant metal sensitive strains lets *in vivo* segmentation of the metal protection mechanism and ultimately lead for identifying specific molecular roles that are played by the DNA sequence of interest.

There are more evidence for MTs role or MT-like proteins in the mutualistic and the pathogenic interactions in between fungi and plants. The genes of fungal MT are mostly regulated transcriptionally in the life stages mainly, during the colonization of plant. The first report was in 1995, when the 2 genes similar to MTs were expressed uniquely during appressorium formation by of conidia in *Colletotrichum gloeosporioides* which are induced by the host surface wax. The 2 putative MT genes from biotrophic pathogen (*Uromyces fabae)* are upregulated strongly in the parasitic mycelium that are colonizing tissues of leaves [35]. So far as the mycorrhizal fungi are concerned, the *Pimt1* was found in macroarray experiments as upregulated in ectomycorrhizal tissues compared with the saprotroph growth condition. In contrast, MTs from arbuscular mycorrhizal fungi appear to be downregulated when the fungus colonizes the root tissues. It is worth noting that plant MTs are also upregulated in ectomycorrhizal associations of *Betula pendula* with *Eucalyptus globulus* and *P. involutus* with *P. tinctorius*, respectively. To support further MTs role in the plant–fungus interaction, Tucker *and* his co-workers described one unusual MT-like protein with 22 amino acids long with 6 cysteines in *Magnaporthae grisea* (fungal pathogen), that showed high affinity to zinc. Gene has no effect in the metal tolerance and it is shown to confer the pathogenicity, through playing role in the biochemical differentiation of appressorium cell wall [33].

10. Cyanobacterial metallothionein

Numerous mechanisms of metal resistance exist in cyanobacteria. The first mechanism involves extracellular binding. Cells might synthesize and release some organic materials that could chelate the metals and reduce their bioavailability or metal ions might bound to outer cell surface [36]. These complex forms are not readily transported into cell due to its complexity and structure. Secondly, cells could increase excretion rate of some metal ions using the energy-driven efflux pumps [37]. The sequestration of internal metal, a 3rd resistance mechanism, is a key mechanisms through which the bacteria combats the exposure of the heavy metal and their subsequent accumulation. In cyanobacteria, metal ion sequestration inside the cell are performed by the Class II MTs.

The Class II MTs are a cysteine-rich, thiol-containing and are metal-binding proteins that can sequester the metals, thereby stopping the buildup of potentially some deadly free metals in cell. Metal binding happens by interactions of the metals with thiol clusters of the cysteine remainders. MTs gene was organized as operon termed as smt locus, that contains together the metallothionein protein (smtA) and the regulatory protein, repressor (smtB) genes. The SmtBs, are the managing repressor of the countenance from smtA promoter- operator province. The MTs countenance, from gene to the efficient protein, was encouraged through metals and regulation of the transcript to mRNA is based on the interface among the metals and repressor protein that regulates transcript, over by the interface by a thiol clusters that are contemporary on the repressor type of protein [38, 39].

11. Mammalian metallothionein

The MT genes were readily induced by several toxicologic and physiologic stimuli. Since the MT cysteines are conserved across the species, it is suspected that the cysteines are essential for the function and the MTs are required for the life. In attempts for determining the MT function(s), researches were carried out with 4 different experimental paradigms: (i) the animals were injected with chemicals that induce MT; (ii) the cells adapted to grow and survive in higher MT concentrations - that induces toxicants; (iii) the cells are transfected with the MT gene; and (iv) the MT-null mice and the MT transgenic. Very often, the reports from the researches with the first 3 approaches had indicated several purposes of MT in the cell biology is as follows: the MT (a) is "warehouse" for the "Zn" (b) the free radical hunter (c) protect from cadmium harmfulness. However, researches with MT transgenic with the null rats did not toughly reinforced initial 2 proposed purposes then they sturdily sustenance its purpose in the protection in contradiction of the cadmium harmfulness. Recurrent cadmium management to the MT null rats resulted in the nephron harmfulness at the 1/10 of the quantity that products the toxicity in the control rats. Researches in human being showed as 7% of overall populations had some renal malfunction by cadmium exposures. Consequently, if the human being did not consume the MT, "standard" exposure of cadmium will be toxic to human beings [17]. Thus, it appears that during evolution, the MT ability to protect against the Cd toxicity would had taken a chief role for the life processes maintenance, as compared with its other proposed functions (*i.e.* storehouse for zinc and free radical scavenger) [40, 41].

12. Studies conducted at Tamil Nadu agricultural university

Expression of metallothionein 3-like protein mRNA in the sorghum cultivars under the chromium (VI) stress was estimated [20]. The *MT3* gene expression in roots after imposition of 100 μM Cr (VI) for five days in fifteen days old seedlings of 2 sorghum cultivars, *viz.* K 10 and CO - 27 (susceptible and tolerant respectively, based on earlier studies on growth and antioxidant polymerase chain reaction (RT–PCR), as described by Kaplana *et al.* [42]. After the reports attained it is shown as the sorghum variety CO - 27 pretreated with the Chromium (VI) represented some great strength band which matches the genetic factor of attention. These suggestions can be increased MT transcript degrees under the chromium strain, predominantly in the forbearing variety. Now it is probable that Reactive Oxygen Species (ROS) and hydrogen peroxide that are produced under the chromium stress acts as signal for inducing MT mRNA transcription. The phytochelatin functions in regulation of essential metals and in the detoxification of several toxic metals. Distinct absence of the phytochelatin are shown in the plants under the Cr (VI) stress. These suggested that there can be enhanced role for the MTs in plants under the chromium stress.

13. Limitations

Study on the MT structure of plant and their function are impaired by difficulties that are encountered in the purification of this proteins since the innate bases, mostly because of intrinsic variability of the cysteine rich polypeptides with the occurrence of oxygen. Very few MTs are recuperated from the plant cells, and are uniform some as integral, noncleaved collections of metal [6]. The heterologous countenance approaches stunned this complications, but the early results exhibited as the pea MTs produced in *Escherichia coli* as innate protein were likewise vulnerable to proteolysis. Alternative strategies for producing the plant MTs as the fusion proteins generated some intact proteins that retained their metal binding and their antigenic features [43]. Inappropriately, the cleavage and the retrieval of MTs portion were not made and so the description of the consistent collections of metals were not shown. But, GST founded expression system are functional to various inborn MT forms allows amalgamation of huge quantity of the identical metal MT provisions. The investigation of this developments by the spectrometric and the spectroscopic systems have given some unparalleled data about the MT collective stoichiometric and the manageable performance of MTs [44].

14. Conclusion

MTs are stress related protein and might be involved in essential metal homeostasis and detoxification of excess metals during organism's developmental process. The MT genes are readily induced by several toxicologic and physiologic stimuli. Due to the cysteines in MT are absolutely conserved across the species, it is suspected that cysteines are essential for the function and MT is required for the life. A clear role for MTs is yet to be established, although they are certainly thought to play a key role in the metal metabolism. MTs may function as the antioxidants and another possibility is the role in repair of the plasma membrane. Although MTs are expressed ubiquitously and conserved in living organisms, determining their function remains a future challenge. So further research work is necessary to better understand MT-metal interaction, the mechanism of MT expression and molecular transformation of novel MT genes to plant and microbial system for increasing heavy metal detoxification in the environment.

Author details

Ettiyagounder Parameswari*, Tamilselvan Ilakiya, Veeraswamy Davamani, Periasami Kalaiselvi and Selvaraj Paul Sebastian
Tamil Nadu Agricultural University, Coimbatore, India

*Address all correspondence to: parameswariphd@gmail.com

References

[1] Bourdineaud JP, Baudrimont M, Gonzalez P, Moreau JL. Challenging the model for induction of metallothionein gene expression. Biochimie. 2006 Nov 1;88(11):1787-1792.

[2] Xu X, Duan L, Yu J, Su C, Li J, Chen D, Zhang X, Song H, Pan Y. Characterization analysis and heavy metal-binding properties of Cs MTL 3 in Escherichia coli. FEBS open bio. 2018 Nov;8(11):1820-1829.

[3] Freisinger E. Structural features specific to plant metallothioneins. JBIC Journal of Biological Inorganic Chemistry. 2011 Oct 1;16(7):1035-1045.

[4] Sutherland DE, Stillman MJ. The "magic numbers" of metallothionein. Metallomics. 2011;3(5):444-463.

[5] Chaturvedi AK, Mishra A, Tiwari V, Jha B. Cloning and transcript analysis of type 2 metallothionein gene (SbMT-2) from extreme halophyte Salicornia brachiata and its heterologous expression in E. coli. Gene. 2012 May 15;499(2):280-287.

[6] Cobbett C, Goldsbrough P. Phytochelatins and metallothioneins: Roles in heavy metal detoxification and homeostasis. Annual review of plant biology. 2002 Jun;53(1):159-182.

[7] Messerle BA, Schäffer A, Vašák M, Kägi JH, Wüthrich K. Three-dimensional structure of human [113Cd7] metallothionein-2 in solution determined by nuclear magnetic resonance spectroscopy. Journal of molecular biology. 1990 Aug 5;214(3):765-779.

[8] Dziegiel P. Expression of metallothioneins in tumor cells. Pol J Pathol. 2004 Jan 1;55(1):3-12.

[9] Tsujikawa K, Imai T, Kakutani M, Kayamori Y, Mimura T, Otaki N, Kimura M, Fukuyama R, Shimizu N. Localization of metallothionein in nuclei of growing primary cultured adult rat hepatocytes. FEBS letters. 1991 Jun 3;283(2):239-242.

[10] Ye B, Maret W, Vallee BL. Zinc metallothionein imported into liver mitochondria modulates respiration. Proceedings of the National Academy of Sciences. 2001 Feb 27;98(5):2317-2322.

[11] Moltedo O, Verde C, Capasso A, Parisi E, Remondelli P, Bonatti S, Alvarez-Hernandez X, Glass J, Alvino CG, Leone A. Zinc transport and metallothionein secretion in the intestinal human cell line Caco-2. Journal of Biological Chemistry. 2000 Oct 13;275(41):31819-31825.

[12] Wolff NA, Abouhamed M, Verroust PJ, Thevenod F. Megalin-dependent internalization of cadmium-metallothionein and cytotoxicity in cultured renal proximal tubule cells. Journal of Pharmacology and Experimental Therapeutics. 2006 Aug 1;318(2):782-791

[13] Wang, W.C., Mao, H., Ma, D.D and Yang ,W.X. Characteristics, functions, and applications of metallothionein in aquatic vertebrates. Front. Mar. Sci. 2014. 1:34

[14] Liu J, Shi X, Qian M, Zheng L, Lian C, Xia Y, Shen Z. Copper-induced hydrogen peroxide upregulation of a metallothionein gene, OsMT2c, from Oryza sativa L. confers copper tolerance in Arabidopsis thaliana. Journal of Hazardous Materials. 2015 Aug 30; 294:99-108.

[15] Kaegi JH, Schaeffer A. Biochemistry of metallothionein. Biochemistry. 1988 Nov 1;27(23):8509-8515.

[16] Nordberg M, Kojima Y. Metallothionein and other low

molecular weight metal-binding proteins. InMetallothionein 1979 (pp. 41-124). Birkhäuser, Basel.

[17] Klaassen CD, Choudhuri S, McKim Jr JM, Lehman-McKeeman LD, Kershaw WC. In vitro and in vivo studies on the degradation of metallothionein. Environmental health perspectives. 1994 Sep;102(suppl 3):141-146.

[18] Hasan M, Cheng Y, Kanwar MK, Chu XY, Ahammed GJ, Qi ZY. Responses of plant proteins to heavy metal stress—A review. Frontiers in plant science. 2017 Sep 5;8:1492.

[19] Joshi R, Pareek A, Singla-Pareek SL. Plant metallothioneins: Classification, distribution, function, and regulation. InPlant Metal Interaction 2016 Jan 1 (pp. 239-261). Elsevier.

[20] Shanker AK, Djanaguiraman M, Sudhagar R, Jayaram K, Pathmanabhan G. Expression of metallothionein 3-like protein mRNA in sorghum cultivars under chromium (VI) stress. Current Science. 2004 Apr 10;86(7):901-902.

[21] Kang YJ. Metallothionein redox cycle and function. Experimental biology and Medicine. 2006 Oct;231(9):1459-1467.

[22] Weissman JS, Rye HS, Fenton WA, Beechem JM, Horwich AL. Characterization of the active intermediate of a GroEL–GroES-mediated protein folding reaction. Cell. 1996 Feb 9;84(3):481-490.

[23] Zhou J, Goldsbrough PB. Functional homologs of fungal metallothionein genes from Arabidopsis. The Plant Cell. 1994 Jun 1;6(6):875-884.

[24] Freisinger E. Plant MTs—Long neglected members of the metallothionein superfamily. Dalton Transactions. 2008(47):6663-6675.

[25] Hassinen VH, Tervahauta AI, Schat H, Kärenlampi SO. Plant metallothioneins–metal chelators with ROS scavenging activity?. Plant Biology. 2011 Mar;13(2):225-232.

[26] Morris CA, Nicolaus B, Sampson V, Harwood JL, Kille P. Identification and characterization of a recombinant metallothionein protein from a marine alga, *Fucus vesiculosus*. Biochemical Journal. 1999 Mar 1;338(2):553-560.

[27] Akashi K, Nishimura N, Ishida Y, Yokota A. Potent hydroxyl radical-scavenging activity of drought-induced type-2 metallothionein in wild watermelon. Biochemical and Biophysical Research Communications. 2004 Oct 8;323(1):72-78.

[28] Yu LH, Liu JY, Umeda M, UCHIMIYA H. Molecular characterization of metallothionein genes in plants. Plant biotechnology. 1998 Dec 1;15(4):167-172.

[29] Coupe SA, Taylor JE, Roberts JA. Characterisation of an mRNA encoding a metallothionein-like protein that accumulates during ethylene-promoted abscission of Sambucus nigra L. leaflets. Planta. 1995 Oct 1;197(3):442-447.

[30] Robinson NJ, Wilson JR, Turner JS. Expression of the type 2 metallothionein-like gene MT2 from Arabidopsis thaliana in Zn^{2+}–metallothionein-deficient Synechococcus PCC 7942: Putative role for MT2 in Zn^{2+} metabolism. Plant Molecular Biology. 1996 Mar 1;30(6): 1169-1179.

[31] González-Guerrero M, Azcón-Aguilar C, Ferrol N. GintABC1 and GintMT1 Are Involved in Cu and Cd Homeostasis in Glomus Intraradices. In5th International Conference on Mycorrhiza, Granada, Spain 2006 Jul.

[32] Nguyen H, Rineau F, Vangronsveld J, Cuypers A, Colpaert JV,

Ruytinx J. A novel, highly conserved metallothionein family in basidiomycete fungi and characterization of two representative SlMTa and SlMTb genes in the ectomycorrhizal fungus Suillus luteus. Environmental microbiology. 2017 Jul;19(7):2577-2587.

[33] Tucker SL, Thornton CR, Tasker K, Jacob C, Giles G, Egan M, Talbot NJ. A fungal metallothionein is required for pathogenicity of Magnaporthe grisea. The Plant Cell. 2004 Jun 1;16(6):1575-1588.

[34] Bellion M, Courbot M, Jacob C, Blaudez D, Chalot M. Extracellular and cellular mechanisms sustaining metal tolerance in ectomycorrhizal fungi. FEMS microbiology letters. 2006 Jan 1;254(2):173-181.

[35] Jakupoviae, M, Heintz M, Reichmann P, Mendgen K and Hahn M. Microarray analysis of expressed sequence tags from haustoria of the rust fungus Uromyces fabae. Fungal Genetics Biol., 2006. 43: 8-19.

[36] Liu T, Nakashima S, Hirose K, Shibasaka M, Katsuhara M, Ezaki B, Giedroc DP, Kasamo K. A novel cyanobacterial SmtB/ArsR family repressor regulates the expression of a CPx-ATPase and a metallothionein in response to both Cu (I)/Ag (I) and Zn (II)/Cd (II). Journal of Biological Chemistry. 2004 Apr 23;279(17): 17810-17818.

[37] Gupta A, Morby AP, Turner JS, Whitton BA, Robinson NJ. Deletion within the metallothionein locus of cadmium-tolerant Synechococcus PCC 6301 involving a highly iterated palindrome (HIP1). Molecular microbiology. 1993 Jan;7(2):189-195.

[38] Acharya C, Blindauer CA. Unexpected interactions of the cyanobacterial metallothionein smta with uranium. Inorganic Chemistry. 2016 Feb 15;55(4):1505-1515.

[39] Erbe JL, Taylor KB, Hall LM. Metalloregulation of the cyanobacterial smt locus: Indentification of SmtB binding sites and direct interaction with metals. Nucleic acids research. 1995 Jul 11;23(13):2472-2478.

[40] Babula P, Masarik M, Adam V, Eckschlager T, Stiborova M, Trnkova L, Skutkova H, Provaznik I, Hubalek J, Kizek R. Mammalian metallothioneins: Properties and functions. Metallomics. 2012;4(8):739-750.

[41] Vasak M, Meloni G. Mammalian metallothionein-3: New functional and structural insights. International journal of molecular sciences. 2017 Jun;18(6):1117.

[42] Kaplana, LAE., Kathleen VC, Virgina I. and Joseph FC. 2000. Mar. Environ. Res.,2000. 39: 137-141.

[43] Murphy A, Zhou J, Goldsbrough PB, Taiz L. Purification and immunological identification of metallothioneins 1 and 2 from Arabidopsis thaliana. Plant Physiology. 1997 Apr 1;113(4):1293-1301.

[44] Capdevila M, Cols N, Romero-Isart N, Gonzalez-Duarte R, Atrian S, Gonzalez-Duarte P. Recombinant synthesis of mouse Zn3-β and Zn4-α metallothionein 1 domains and characterization of their cadmium (II) binding capacity. Cellular and Molecular Life Sciences CMLS. 1997 Sep 1;53(8):681-688.

Chapter 10

Removal of Heavy Metals from Wastewater by Adsorption

Athar Hussain, Sangeeta Madan and Richa Madan

Abstract

Adsorption processes are extensively used in wastewater treatment for heavy metal removal. The most widely used adsorbent is activated carbon giving the best of results but it's high cost limits its use. It has a high cost of production and regeneration. As the world today faces a shortage of freshwater resources, it is inevitable to look for alternatives that lessen the burden on existing resources. Also, heavy metals are toxic even in trace concentrations, so an environmentally safe method of their removal necessitated the requirement of low cost adsorbents. Adsorption is a cost-effective technique and gained recognition due to its minimum waste disposal advantage. This chapter focuses on the process of adsorption and the types of adsorbent available today. It also encompasses the low-cost adsorbents ranging from agricultural waste to industrial waste explaining the adsorption reaction condition. The cost-effectiveness, technical applicability and easy availability of raw material with low negative impact on the system are the precursors in selecting the adsorbents. The novelty of the chapter lies in covering a wide range of adsorbents with their efficiency in removal of heavy metals from wastewater.

Keywords: adsorption, low-cost adsorbent, isotherm, wastewater, heavy metals

1. Introduction

Heavy metals are toxic elements having specific gravity greater than 5g/cm^3 e.g. Zn, Fe, Cu, Cr, Hg, Pb, Ni, Co, etc. [1]. The main natural sources of heavy metals include volcanic processes, weathering of rocks and soil erosion. While the anthropogenic sources include mineral processing, fuel combustion and industrial activities like mining, metal processing, chemical fertilizers and dye manufacturing etc. Heavy metals are non-biodegradable, recalcitrant and have high mobility in aqueous media, so they tend to accumulate in soils and living organisms leading to environmental repercussions. Heavy metals are taken up by plants which biomagnify through food chains in animals and humans causing serious negative health effects due to their carcinogenicity [2–5]. **Table 1** gives the maximum contaminant level (MCL) in drinking water given by USEPA [7] along with their harmful effects.

Heavy metals have a high tendency to form complexes, are highly reactive and have increased biochemical activity which makes them very persistent in the environment. They are transported through aqueous medium and can concentrate in soil and water resources. This makes them extremely dangerous to all kinds of life forms and the environment. Hence, it is necessary to remove these toxic metals

S. No.	Heavy Metal	MCL (mg/L)	Harmful effects
1.	Zn	0.80	Skin irritation, nausea, depression, anemia, neurological symptoms
2.	Hg	$3.0 X 10^{-5}$	Neurotoxin, Kidney dysfunction, Circulatory & Neurological Disorder
3.	Pb	$6.0 X 10^{-3}$	Central Nervous System Damage, Cerebral Disorders, Kidney, Liver Reproductive System Dysfunction
4.	Ni	0.20	Carcinogen, Dermatitis, Gastrointestinal Disorder, Lung, Kidney Damage
5.	Cu	0.25	Liver Damage, Convulsions, Insomnia
6.	Cr	0.05	Carcinogen, Nausea, Diarrhea
7.	Cd	0.01	Carcinogen, Kidney Dysfunction
8.	As	0.05	Skin Problems, Visceral Cancer

Table 1.
Harmful Effects of Heavy Metals [6]

from wastewater before discharge to prevent further detrimental consequences. Conventional methods like membrane filtration [8–12], chemical precipitation [13–17], ion exchange [17–22], etc. have been used to remove of heavy metals from wastewater. However, these methods suffer from some disadvantages like low efficiency, high energy requirement, precipitation of toxic substances, cost ineffectiveness, etc. [3, 23]. To get past these demerits, processes like adsorption are investigated, since it greatly impacts the bioavailability and transport of toxic metals. It is low-cost and efficient technique for remediation of heavy metals from wastewater. Adsorption process is often reversible in many cases, so the adsorbent can be regenerated back adding another advantage to this process [24, 25]. Many factors such as temperature, pH, initial concentration, contact time and rotation speed affect the efficacy of adsorbents [23, 25].

1.1 Overview of adsorption process

Adsorption is a surface phenomenon in which a solution containing the adsorbate gets adsorbed on the surface of an adsorbent. Adsorption phenomenon can be of two types; one is physiosorption, in which the adsorbate binds to adsorbent due to van der Waals forces, and other is chemisorption, which occurs due to chemical reactions between adsorbate and adsorbent. Physiosorption is reversible, weak and is usually endothermic, while chemisorption is irreversible, selective and exothermic [26–28].

1.2 Adsorption isotherm and models

Adsorption isotherms are representations that estimate the amount the solute that is adsorbed on the surface of the adsorbent per unit weight as a function of equilibrium concentration at a constant temperature. The most commonly used are Langmuir and Freundlich isotherms that describe the adsorption process [29]. Some other models are also used such as Redlich and Peterson [30], Radke and Prausnitz [31], Sips [32], Toth [33] and Koble and Corrigan [34].

1.3 Types of adsorbents

Adsorbents are typically classified on the basis of their origin i.e. natural and synthetic. Natural adsorbents include clays, minerals, charcoal, ores and zeolites. While the synthetic adsorbents are prepared from industrial wastes, agricultural wastes, waste sludge etc.

2. Removal of heavy metals from wastewater by adsorption

Adsorption is presumed to be an efficient and cost-effective method as compared to other wastewater treatment technologies for heavy metal removal. The main advantage this method provides is the production of a high-quality effluent. The process of adsorption has an edge over other processes since it is an economic method for heavy metal remediation. In most cases, the adsorbent can be regenerated back and can be used further [35]. Adsorption is easy to use and does not generate any toxic pollutants, hence it is an environment friendly technique [36]. The prominent criteria of selection of adsorbents include their cost effectiveness, high surface area and porosity, distribution of functional groups and their polarity [37, 38]. Conventional and commercial adsorbents comprise of activated carbon [39–42], zeolites [43–46], graphenes and fullerenes [47–51] and carbon nanotubes [52–56]. Carbons and their derivatives are the most prominently used adsorbents due their great adsorption efficiency. Their exceptional ability comes from their structural characteristics giving them a large surface area with easy chemical modifications which makes them universally acceptable to a wide spectrum of pollutants [57]. The activated carbons suffer from a few flaws which makes their use quite limited. They are expensive to manufacture; the spent activated carbon is difficult to dispose and their regeneration is cumbersome and not economical. Thus, there was extensive research in the area of low-cost adsorbents. The non-conventional adsorbents are cheap, abundantly available and have great complexing capacity due to their varied structure which binds the pollutant ions. They range from agricultural waste to industrial waste sludge and spent slurry [58, 59].

2.1 Activated carbon adsorbents

Activated carbon (AC) is one of the most widely used adsorbents due to its high efficiency, porosity and high surface area. It is commercially manufactured from the carbonization of like coal and wood, so it is expensive and its use is limited [24, 60–62]. They are mainly produced by pyrolysis of carbonaceous material at temperatures lower than 1000°C. The preparation of activated carbon involves two steps, one is the carbonization of raw material at temperature less than 800°C in inert atmosphere, second is activation of the produced product at temperature between 950°C and 1000°C [63]. Hence, most of the carbonaceous material can be used as raw material for activated carbon production, though the characteristics of the final product will rely on the raw material used and activated conditions [63]. Carbon is the main component of activated carbon adsorbent, other elements such as hydrogen, oxygen sulfur and nitrogen are also present. They are produced in both powdered and granular forms. The powdered one has large pores and smaller internal surface area; while the granular one has large internal area and small pores. The adsorptive capacity of an activated carbon is determined by its high porosity and surface are along with its chemical structure. Hence, other low cost raw materials such as agricultural wastes are looked upon for increasing the cost effectiveness of activated carbon.

Kobya studied adsorptive removal of Cr^{4+} from aqueous solutions by AC prepared from hazelnut shell and reached a maximum removal of 170mg/g at pH 1.0 [64]. This was found to be higher than other adsorbents like coconut shell and wood AC [65] which had a removal of 58.5 and 87.6mg/g respectively. Karthikeyan et al. studied removal of Cr^{6+} from wastewater using activated carbon prepared from wood saw dust. The adsorption capacity of Cr^{6+} reached a maximum at 44mg/g at an optimum pH 2.0 [66]. This was significantly higher than other adsorbents for instance coconut shell carbon [67], treated saw dust derived from Indian rose wood [68], coconut tree saw dust [69] and sugarcane bagasse [70]. In these studies, the maximum adsorption was found to be 10.88, 10, 3.60 and 13.40 mg/g respectively. Kongsuwan et al. used eucalyptus bark for preparation of AC in the adsorption of Cu^{2+} and Pb^{2+} from low strength wastewater. The adsorption capacity for Cu^{2+} and Pb^{2+} was maximum at was 0.45 and 0.53 mmol/g, respectively [71]. El-Ashtoukhy et al. studied Cu^{2+} and Pb^{2+} removal from aqueous solutions by AC prepared from pomegranate peel. Batch adsorption experiments were conducted as a function of adsorbent dosage, contact time and pH. The removal of both the metals reached a saturation at 120 min with optimum pH 5.8, 5.6 for Cu^{2+} and Pb^{2+} [72]. Kavand et al. studied adsorptive removal of Pb^{2+}, Cd^{2+} and Ni^{2+} from aqueous solution using granulated activated carbon. The removal was in the order $Pb^{2+} > Cd^{2+} > Ni^{2+}$ at an optimum pH of 2, adsorbent dose of 2g/L and contact time of 80 minutes [73]. Kim et al. conducted a study on the removal of Zn^{2+}, Ni^{2+} and Cr^{2+} from electroplating wastewater using powdered AC and modified powdered AC. A removal efficiency of around 90% was achieved for both the adsorbents at neutral pH [74].

2.2 Zeolites

They are alumino silicates with a crystalline structure that occur naturally or are manufactured industrially. They are one of the best adsorbents for heavy metal removal as they consist of hydrated aluminosilicate minerals comprising of interlinked alumina and silica. Zeolites possess appreciable ion exchange capacities, hydrophilic properties and high specific surface area which makes then exceedingly good adsorbents for heavy metal remediation [75]. Zeolites can also be modified which attain a better adsorption capacity as compared to unmodified ones. NaX zeolite is one of the most widely used nanosized zeolite for removal of heavy metals from wastewater [76–79]. Rad et al. prepared NaX nanozeolite followed by polyvinylacetate polymer/NaX nanocomposite nanofibers to study removal of Cd^{2+}. The maximum adsorption capacity was reported to 838.7mg/g at pH 5.0 [79]. Javadian et al. used fly ash for preparation of amorphous zeolite and obtained a maximum adsorption capacity of 26.246mg/g for Cd^{2+} at 5 optimum pH [80]. Similar studies were conducted by Visa who reported that zeolites have a high surface area and porosity which aid in adsorption of heavy metals [81]. Kobayashi et al. studied removal of Hg^{2+} and Pb^{2+} from aqueous solutions using zeolites prepared from fly ash. The maximum amount of Hg^{2+} and Pb^{2+} adsorbed were 22.4 mg/g and 30.7mg/g respectively at optimum pH of 5 [82].

2.3 Clay minerals

Bentonite, a clay mineral holds the highest cation exchange capacity, is regenerable and around 20 times cheaper than activated carbon [83, 84]. Clay minerals have less removal capacity of heavy metals when compared to zeolites. But they are still used owing to their advantages such as brilliant physical, chemical and surface properties [84–87]. Jiang et al. studied removal of Ni^{2+}, Pb^{2+}, Cu^{2+} and Cd^{2+} from wastewater using kaolinite clay and it was found that concentration of Pb^{2+} decreased from 160.00 to 8.00 mg/L [88]. Bertagnolli et al. conducted a study on bentonite clay for removal of Cu^{2+} and achieved a maximum adsorption capacity of 11.89mg/g [89]. Chai et al. conducted a study using raw kaolinite and acid activated kaolinite for the removal of Ni^{2+} and Cu^{2+} from aqueous solutions and cemetery wastewater. The raw kaolinite adsorbed 69.23% Cu^{2+} and 63.37% Ni^{2+} whereas acid activated kaolinite adsorbed 77.47% and 68.32% at optimum pH of 7, contact time 60 min and temperature 25°C [90].

2.4 Nanostructured materials

In the last decade, carbon nanotubes [91], fullerenes [92] and graphene [93] have occupied an important place in the area of adsorption of heavy metals from effluents. They possess exceptional mechanical and chemical properties, strength, exchange capacity, electrical conductivity and thermal stability. A high surface area along with numerous intermolecular interactions gives them an edge over other adsorbents in remediation of heavy metals.

2.4.1 Carbon nanotubes, fullerenes and graphene

Iijima discovered carbon nanotubes (CNTs) in 1991 [94]. They exist as long carbon cylindrical in shape with a continuous hexagonal graphite sheets. They are of two types: single walled CNT, which have a single graphite sheet and multi walled CNTs which have multiple sheets. They have portrayed excellent potential for heavy metal from wastewater for copper [95, 96] lead, [97, 98], chromium [99, 100], nickel [100, 101] and cadmium [100, 102]. CNTs prove to be excellent adsorbents owing to the advantages such as mechanical and surface properties electrical and semiconductor properties [102, 103]. They also provide a high specific surface area (150-1500m^2/g) and the presence of mesopores increases their adsorption efficiency [104–107]. The presence of different functional groups containing elements such as oxygen, nitrogen and sulfur directly and indirectly affect the adsorption mechanisms that enhance the adsorption of heavy metals [108–111].

Oxidized CNTs also portray exceedingly high adsorption capacity for the removal of Cr^{6+}, Pb^{2+} and Cd^{2+} from wastewater [112–115]. Wang et al. (2007a) carried out a study using MWCNTs activated with conc. HNO_3 which escalated the adsorption capacity due to creation of more oxygen functional groups. The equilibrium time for Pb^{2+} adsorption was found to be 120 min at an optimum pH 2.0 [116]. Nanocomposites are also prepared using CNTs with ferrous, zirconium, aluminium oxides by coprecipitation method for removal of Pb, As, Cu, Ni and Cr ions [117–122]. Luo et al. synthesized Fe_2O_3/MnO_2/acid oxidized MWCNT nanocomposites for removal of Cr^{6+}. At an optimum pH of 2.0 a maximum removal capacity of 85% was achieved by the nanocomposite [123]. Ge et al. prepared magnetic Fe@MgO nanocomposites for the removal of Pb^{2+} from water. A maximum adsorption efficiency of 14746.4 mg/g was achieved for Pb^{2+} at 120 min contact time [124]. Stafiej and Pyrzynska stated some facts related to adsorption capacity of CNTs and reached a conclusion that pH and concentration of heavy metals significantly affect the CNTs efficiency [125]. CNTs portray excellent adsorption efficiency due to their surface morphology, electrochemical potential and ion exchange capacity [126, 127]. The ability of CNTs to be easily modified makes them selective adsorbents with the merit of enhanced adsorption efficiency [113, 127–130]. They are instituted as great adsorbents in the field of wastewater treatment due to their appreciable mechanical and surface characteristics, mechanical and magnetic properties and high stability [131]. But the use is restricted due to the accumulation of the active sites by the adsorbate. Hence, activation of CNTs offers the advantage of increasing the sites with functional groups which in turn increases their adsorption efficiency for heavy metal removal from water and wastewater [132–138].

The discovery of fullerenes in 1985 led to another breakthrough in adsorption science [139, 140]. They have a closed-cage structure containing pentagonal and hexagonal carbon rings with the formula C_{20+m}, m being an integer. Their adsorption efficiency can also be attributed to their surface morphology and presence of mesopores which gives then higher ion affinity and higher specific surface area for remediation of heavy metal ions from water and wastewater [141, 142]. Alekseeva et al. conducted a study using fullerenes for the removal of Cu^{2+} and explained the mechanism through Langmuir model [143]. The maximum adsorption efficiency was found to be 14.6 mmol/g. Spherical fullerene containing 60 carbon atoms is the most explored one. Its striking features comprise of hydroxyl and epoxy functional groups on surface, large surface to volume ratio, hydrophobicity, high electron affinity and low aggregation capacity which make it beneficial for heavy metal removal [144–146]. But their use is often restricted due to their high price. So, research on incorporation of other conventional adsorbents with fullerenes has come up. It was revealed that fullerenes enhance the porous structure of adsorbent leading to increase in the removal efficiency of heavy metals. It was found that adsorption capacity of ACs escalated by 1.5-2.5 times after introduction of fullerenes into their structure [147, 148].

Graphene came into the scene in 2004 and is a 2-D hexagonal lattice of carbon atoms. It also possesses structural, chemical and mechanical properties which aid its use in wastewater treatment. It has a high surface area, active functional groups and sites on its surface which enhance its adsorption capacity [149–151]. Graphene can also be activated by oxidation to increase functional groups which surge the adsorption capacity for removal of heavy metals [114, 152–155]. Deng et al. 2010 conducted a study using functionalized graphene for removal of Pb^{2+} from aqueous solution. At an optimum pH of 5.0 the maximum adsorption capacity reached was 406.6mn/g within 40 min [156]. Several studies were conducted to study the properties of graphene oxides for adsorption [157–161]. It was revealed that graphene oxides can also be magnetically modified which increases their adsorption capacity [162, 163]. A study by Zhao et al. used layered graphene oxide for removal of Pb^{2+} from aqueous solution. The adsorbent layers had oxygen functional groups which greatly enhanced the adsorption capacity reaching a maximum of 1850 mg/g [164]. Jian et al. synthesized a bio-adsorbent polyacrylamide/graphene oxide hydrogel grafted with sodium alginate and studied the removal of Cu^{2+} and Pb^{2+} from aqueous solution. The maximum adsorption capacity of Cu^{2+} was 68.76 mg/g at pH 5 and 240.69 mg/g for Pb^{2+} at 5.5 pH [165].

2.5 Low cost adsorbents

Although, ACs are the most widely used adsorbents, their use is limited due to their high cost and low regeneration. Same is with other developed adsorbents such as carbon nanotubes, fullerenes and nanocomposites. To make the process of wastewater treatment speed up and effective, it is vital to look for adsorbents that are cost effective as well administer a high adsorption efficiency. Thus, the need for low cost adsorbents came to be realized. Low cost adsorbents comprise of those non-conventional materials that are easily available and cost effective mainly agricultural and industrial waste.

S. No.	Type of wastewater	Type of Adsorbent	Adsorbent Dosage (g/L)	Metal Ion	Amount Adsorbed (mg/g)	Contact Time (min)	Temperature (°C)	pH	References
1.	Hospital Wastewater	Cassava peels	10.0	Pb^{2+} Cu^{2+}	5.80 8.00	20-120	39.85	8.0	[168]
2.	Aqueous Solution	Ash Gourd Peel Powder	6.0	Cr^{6+}	18.7 0	40-60	28.0	1.0	[169]
3.	Aqueous Solution	Barley Straw	1.0	Cu^{2+}	4.64	120	25.0	6.0-7.0	[170]
4.	Aqueous Solution	Cashew Nut	3.0	Ni^{2+}	18.86	30	30.0	5.0	[171]
5.	Electroplating Wastewater	Chemically Modified Orange peel	2.0	Cu^{2+}	289.0	180	30.0	5.0	[172]
6.	Aqueous Solution	Modified Lawny Grass	0.5	Pb^{2+}	137.12	400	29.85	6.0	[173]
7.	Aqueous Solution	Grapefruit Peel	2.0	U^{6+}	140.79	60-80	24.85	4.0-6.0	[174]
8.	Aqueous Solution	Peanut Shell	1.0	Cr^{6+}	4.32	360	30.0	2.0	[175]
9.	Aqueous Solution	Sugar cane and orange peel biochar	1.0	Pb^{2+}	86.96 and 27.86	30	25.0	5.0	[176]
10.	Electroplating Wastewater	Mango Peel	5.0 5.0 5.0	Ni^{2+} Cu^{2+} Zn^{2+}	39.75 46.09 28.21	120	25.0	6.0	[177]
11.	Aqueous Solution	Wheat Shell	10.0	Cu^{2+}	17.42	60	25.0	7.0	[178]
12.	Aqueous Solution	Sulfonated Biochar	2.0	Pb^{2+} Cd^{2+}	191.07 85.76	5	180.0	4.5	[179]

Table 2.
Agricultural wastes for heavy metal removal.

S. No.	Type of wastewater	Type of Adsorbent	Adsorbent Dosage (g/L)	Metal Ion	Amount Adsorbed (mg/g)	Contact Time (min)	Temperature (°C)	pH	References
1.	Dye and heavy metals Wastewater	Fly Ash	2.0	Cd^{2+} Cu^{2+} Ni^{2+}	6.36 12.78 1.66	60	25.0	4.8-5.3	[191]
2.	Aqueous Solution	Coal Fly Ash	2.5	Cu^{2+} Ni^{2+}	158.0 99.0	120 60	20.0	8.0	[192]
3.	Aqueous Solution	Fly Ash Geopolymer	2.0	Cu^{2+}	152.0	120	45.0	6.0	[193]
4.	Aqueous Solution	Fly Ash Bottom Ash	2.0	Cd^{2+}	142.9 23.3	240	25.0	5.0-6.0	[196]
5.	Aqueous Solution	Red Mud	1.0	Cu^{2+}	5.3	60	30.0	5.5	[197]
6.	Aqueous Solution	Activated Red Mud	4.0	Zn^{2+}	14.9	480	25.0	6.0	[198]
7.	Aqueous Solution	Red Mud	1.0	Co^{2+}	18.0	15	30.0	5.0	[200]
8.	Aqueous Solution	Activated Red Mud	4.0	Pb^{2+}	6.0	30	30.0	4.0	[202]
9.	Aqueous Solution	Basic Oxygen Furnace Slag	0.5	Cu^{2+}	245.2	60	20.0	3.5	[204]
10.	Synthetic Wastewater	LD Slag Geopolymer Raw LD Slag	2.0	Ni^{2+}	84.8 14.3	1440	45	10.0 10.0	[205]

Table 3.
Industrial wastes for heavy metal removal.

2.5.1 Agriculture waste

Agricultural wastes have the constitution of lignin, cellulose, hydrocarbons, sugars, water and starch along with other functional groups which enhances the adsorption capacity of these agricultural wastes. These wastes can range from rice husk to wheat shells, egg shells, coconut husk, palm fruit, bagasse, groundnut shell, fruit peels, biochar etc. These wastes can be used directly in which they are washed and grounded first. Then they are sieved to get the desirable particle size which are used for adsorption tests. They can also be modified into chars and further activated to increase the adsorption sites [166, 167]. **Table 2** shows the different agricultural wastes used for heavy metal ions removal.

2.5.2 Biochar

Biochar is the charred solid material obtained from the carbonization of biomass. The most common method of production of biomass is by pyrolysis which is the thermal decomposition of biomass in absence or limited oxygen. Biochars are less carbonized than AC so more carbon, hydrogen and oxygen remain in their structure. Biochar has shown remarkable potential for remediation of heavy metals from wastewater than other conventional and low-cost adsorbents. They have a mesoporous structure leading to a high surface area and presence of different functional groups and a low ash content makes them excellent and effective adsorbents. Feedstock such as rice husk [180–185], corn husk [180], tea waste [181, 184, 186, 187] and digested sludge have [185, 188] been employed for removal of heavy metals from aqueous solutions as well as wastewater.

2.6 Industrial waste

Industrial activities generate huge quantities of waste that are usually sent to landfill sites for disposal. These wastes possess a good adsorption capacity and are solve the problem of waste treatment. Waste products like fly ash [189–196], red mud [197–203] and slag [204, 205] have been effectively used owing to their appreciable capacity for removal of heavy metals from wastewater. Many industrial waste adsorbents have been employed for the remediation of Zn^{2+} from effluents. The maximum adsorption capacity for lignin was 73.2mg/g, 168mg/g for waste sludge and 55.82mg/g for cassava waste [206, 207]. **Table 3** gives various industrial wastes used for removal of heavy metals from wastewater and aqueous solutions.

3. Comparison of conventional and non-conventional adsorbents

For the adsorption process to be efficient, selection of the most appropriate adsorbent is a crucial step. The main basis of selection of an adsorbent include low cost, high adsorption capacity, effective for a broad spectrum of pollutants and having a low footprint [208, 209]. There has been extensive research in the field of conventional and non-conventional adsorption performances and mechanisms. Different adsorbents follow varied mechanisms because of difference in raw material and adsorbent production conditions. Mainly four mechanisms have been identified for efficient adsorption of pollutants; chemisorption, physisorption, ion-exchange and precipitation [210, 211]. Davis et al. stated that ion-exchange does not necessarily describe the mechanism of adsorption but a lot of other factors and mechanisms co-aid to make the process successful [210]. Some other researchers also explained the adsorption mechanisms [212–215]. Literature evidently points out that activated carbons have proven themselves as brilliant adsorbents due to their high specific surface area, mechanical and structural surface morphology and presence of functional groups which can also be modified. However, non-conventional adsorbents are increasingly employed as low-cost and effectual adsorbents. Their commercialization remains a task but there as upcoming as their available in abundant. More focused research into their engineering and modification can bring them at par some commercial solid adsorbents.

4. Conclusion

Heavy metal pollution is one the most dangerous situations being faced today. They harmful even in trace concentrations. Many of them are carcinogenic, cause birth defects and are extremely fatal. Hence, it is necessary to remove these toxic metals from wastewater before it is discharged into open waters. Adsorption is one such technique that caters not only to the remediation of heavy metal from wastewater, it is also eco-friendly with a low footprint. Adsorbents like activated are widely used, but it is restricted due to its high cost. So, it is necessary to look for options that are sustainable and aim at remediating the larger prospect of the problem. Low-cost adsorbents like agricultural wastes, industrial wastes and biochar aid not only in removal of heavy metals but are also cheap methods. Their raw material is easily available and these adsorbents can be easily manufactured. So, it is a green technology that greatly enhances the process of wastewater treatment. Further research into developing more low-cost adsorbents can help in further remedial of heavy metals.

Author details

Athar Hussain[1], Sangeeta Madan[2] and Richa Madan[2*]

1 Department of Civil Engineering, Netaji Subhas Institute of Technology, West Campus, Jaffarpur, Delhi, India

2 Department of Environment Science, Gurukula Kangri (Deemed to be University), Haridwar, Uttarakhand, India

*Address all correspondence to: richa.madan.92@gmail.com

References

[1] Srivastava NK, Majumder CB. Novel biofiltration methods for the treatment of heavy metals from industrial wastewater. Journal of hazardous materials. 2008 Feb 28;151(1):1-8.

[2] Baldwin DR, Marshall WJ. Heavy metal poisoning and its laboratory investigation. Annals of clinical biochemistry. 1999 May;36(3):267-300.

[3] Barakat MA. New trends in removing heavy metals from industrial wastewater. Arabian journal of chemistry. 2011 Oct 1;4(4):361-77.

[4] Akpor OB, Ohiobor GO, Olaolu DT. Heavy metal pollutants in wastewater effluents: sources, effects and remediation. Advances in Bioscience and Bioengineering. 2014 Nov 10;2(4):37-43.

[5] Harvey PJ, Handley HK, Taylor MP. Identification of the sources of metal (lead) contamination in drinking waters in north-eastern Tasmania using lead isotopic compositions. Environmental Science and Pollution Research. 2015 Aug 1;22(16):12276-88.

[6] Molinari R, Poerio T, Argurio P. Selective separation of copper (II) and nickel (II) from aqueous media using the complexation–ultrafiltration process. Chemosphere. 2008 Jan 1;70(3):341-8.

[7] US Environmental Protection Agency. Framework for metals risk assessment. US Environmental Protection Agency, Office of the Science Advisor: Washington, DC. EPA 120/R-07/001. 2007.

[8] Landaburu-Aguirre J, García V, Pongrácz E, Keiski RL. The removal of zinc from synthetic wastewaters by micellar-enhanced ultrafiltration: statistical design of experiments. Desalination. 2009 May 15;240(1-3):262-9.

[9] Zhao M, Xu Y, Zhang C, Rong H, Zeng G. New trends in removing heavy metals from wastewater. Applied microbiology and biotechnology. 2016 Aug 1;100(15):6509-18.

[10] Abdullah N, Yusof N, Lau WJ, Jaafar J, Ismail AF. Recent trends of heavy metal removal from water/wastewater by membrane technologies. Journal of Industrial and Engineering Chemistry. 2019 Aug 25;76:17-38.

[11] Crini G, Lichtfouse E. Advantages and disadvantages of techniques used for wastewater treatment. Environmental Chemistry Letters. 2019 Mar 1;17(1):145-55.

[12] Kongsricharoern N, Polprasert C. Electrochemical precipitation of chromium (Cr^{6+}) from an electroplating wastewater. Water Science and Technology. 1995 May 1;31(9):109.

[13] Charerntanyarak L. Heavy metals removal by chemical coagulation and precipitation. Water Science and Technology. 1999 Jan 1;39(10-11):135-8.

[14] Mirbagheri SA, Hosseini SN. Pilot plant investigation on petrochemical wastewater treatmentfor the removal of copper and chromium with the objective of reuse. Desalination. 2005 Jan 1;171(1):85-93.

[15] TanongK, TranLH, Mercier, G, Blais, JF. Recovery of Zn (II), Mn (II) Cd (II) and Ni (II) from the unsorted spent batteries using solvent extraction, electrodeposition and precipitation methods, J. Clean. Prod. 148 (2017) 233-244.

[16] Peng H, Guo J. Removal of chromium from wastewater by membrane filtration, chemical precipitation, ion exchange, adsorption electrocoagulation, electrochemical

reduction, electrodialysis, electrodeionization, photocatalysis and nanotechnology: a review. Environmental Chemistry Letters. 2020 Jul 23:1-4.

[17] Abo-Farha, SA, Abdel-Aal, AY, Ashour, IA, Garamon, SE. Removal of some heavy metal cations by synthetic resin purolite C100, J. Hazard. Mater. 169 (2009)190-194.

[18] Alyuz, B, Veli, S. Kinetics and equilibrium studies for the removal of nickel and zinc from aqueous solutions by ion exchange resins, J. Hazard. Mater. 167 (2009) 482-488.

[19] Zanin E, Scapinello J, de Oliveira M, Rambo CL, Franscescon F, Freitas L, de Mello JM, Fiori MA, Oliveira JV, Dal Magro J. Adsorption of heavy metals from wastewater graphic industry using clinoptilolite zeolite as adsorbent. Process Safety and Environmental Protection. 2017 Jan 1;105:194-200.

[20] Huang Y, Zeng X, Guo L, Lan J, Zhang L, Cao D. Heavy metal ion removal of wastewater by zeolite-imidazolate frameworks. Separation and Purification Technology. 2018 Apr 3;194:462-9.

[21] Hong M, Yu L, Wang Y, Zhang J, Chen Z, Dong L, Zan Q, Li R. Heavy metal adsorption with zeolites: The role of hierarchical pore architecture. Chemical Engineering Journal. 2019 Mar 1;359:363-72.

[22] Eccles H. Treatment of metal-contaminated wastes: why select a biological process? Trends in biotechnology. 1999 Dec 1;17(12):462-5.

[23] Fu F, Wang Q. Removal of heavy metal ions from wastewaters: a review. Journal of environmental management. 2011 Mar 1;92(3):407-18.

[24] Agarwal M, Singh K. Heavy metal removal from wastewater using various adsorbents: a review. Journal of Water Reuse and Desalination. 2017 Dec;7(4):387-419.

[25] Tyagi I, Gupta VK, Sadegh H, Ghoshekandi RS, Makhlouf AS. Nanoparticles as adsorbent; a positive approach for removal of noxious metal ions: a review. Science Technology and Development. 2017;34(3):195-214.

[26] Tripathi A, Ranjan MR. Heavy metal removal from wastewater using low cost adsorbents. J BioremedBiodeg. 2015 Jan;6(6):315.

[27] Singh N, Gupta SK. Adsorption of heavy metals: A review. Int. J. Innov. Res. Sci. Eng. Technol. 2016 Feb;5(2):2267-81.

[28] Freundlich HM. Over the adsorption in solution. J. Phys. Chem. 1906 Mar;57(385471):1100-7.

[29] Langmuir I. The constitution and fundamental properties of solids and liquids. Part I. Solids. Journal of the American chemical society. 1916 Nov;38(11):2221-95.

[30] Redlich OJ, Peterson DL. A useful adsorption isotherm. Journal of physical chemistry. 1959 Jun 1;63(6):1024-28.

[31] Radke CJ, Prausnitz JM. Adsorption of organic solutes from dilute aqueous solution of activated carbon. Industrial & Engineering Chemistry Fundamentals. 1972 Nov;11(4):445-51.

[32] Sips R. Combined form of Langmuir and Freundlich equations. J. Chem. Phys. 1948;16(429):490-5.

[33] Toth J. State equation of the solid-gas interface layers. Acta chim. hung. 1971;69:311-28.

[34] Koble RA, Corrigan TE. Adsorption isotherms for pure hydrocarbons. Industrial & Engineering Chemistry. 1952 Feb;44(2):383-7.

[35] Ojedokun AT, Bello OS. Sequestering heavy metals from wastewater using cow dung. Water Resources and Industry. 2016 Mar 1;13:7-13.

[36] Demirbas A. Heavy metal adsorption onto agro-based waste materials: a review. Journal of hazardous materials. 2008 Sep 15;157(2-3):220-9.

[37] Vunain E, Mishra AK, Mamba BB. Dendrimers, mesoporous silicas and chitosan-based nanosorbents for the removal of heavy-metal ions: a review. International journal of biological macromolecules. 2016 May 1;86:570-86.

[38] Ewecharoen A, Thiravetyan P, Wendel E, Bertagnolli H. Nickel adsorption by sodium polyacrylate-grafted activated carbon. Journal of hazardous materials. 2009 Nov 15;171(1-3):335-9.

[39] Rivera-Utrilla J, Sánchez-Polo M, Gómez-Serrano V, Alvarez PM, Alvim-Ferraz MC, Dias JM. Activated carbon modifications to enhance its water treatment applications. An overview. Journal of hazardous materials. 2011 Mar 15;187(1-3):1-23.

[40] Abbaszadeh S, Alwi SR, Webb C, Ghasemi N, Muhamad II. Treatment of lead-contaminated water using activated carbon adsorbent from locally available papaya peel biowaste. Journal of Cleaner Production. 2016 Apr 1;118:210-22.

[41] Wong S, Ngadi N, Inuwa IM, Hassan O. Recent advances in applications of activated carbon from biowaste for wastewater treatment: a short review. Journal of Cleaner Production. 2018 Feb 20;175:361-75.

[42] Duan C, Ma T, Wang J, Zhou Y. Removal of heavy metals from aqueous solution using carbon-based adsorbents: A review. Journal of Water Process Engineering. 2020 Oct 1;37:101339.

[43] Meng Q, Chen H, Lin J, Lin Z, Sun J. Zeolite A synthesized from alkaline assisted pre-activated halloysite for efficient heavy metal removal in polluted river water and industrial wastewater. Journal of environmental sciences. 2017 Jun 1;56:254-62.

[44] Taamneh Y, Sharadqah S. The removal of heavy metals from aqueous solution using natural Jordanian zeolite. Applied Water Science. 2017 Jul 1;7(4):2021-8.

[45] Huang Y, Zeng X, Guo L, Lan J, Zhang L, Cao D. Heavy metal ion removal of wastewater by zeolite-imidazolate frameworks. Separation and Purification Technology. 2018 Apr 3;194:462-9.

[46] Hong M, Yu L, Wang Y, Zhang J, Chen Z, Dong L, Zan Q, Li R. Heavy metal adsorption with zeolites: The role of hierarchical pore architecture. Chemical Engineering Journal. 2019 Mar 1;359:363-72.

[47] Carpio IE, Mangadlao JD, Nguyen HN, Advincula RC, Rodrigues DF. Graphene oxide functionalized with ethylenediamine triacetic acid for heavy metal adsorption and anti-microbial applications. Carbon. 2014 Oct 1;77:289-301.

[48] Yu JG, Yu LY, Yang H, Liu Q, Chen XH, Jiang XY, Chen XQ, Jiao FP. Graphene nanosheets as novel adsorbents in adsorption, preconcentration and removal of gases, organic compounds and metal ions. Science of the total environment. 2015 Jan 1;502:70-9.

[49] Sadegh H, Ali GA, Gupta VK, Makhlouf AS, Shahryari-ghoshekandi R, Nadagouda MN, Sillanpää M, Megiel E. The role of nanomaterials as effective adsorbents and their applications in wastewater treatment. Journal of Nanostructure in Chemistry. 2017 Mar 1;7(1):1-4.

[50] Nimibofa A, Newton EA, Cyprain AY, Donbebe W. Fullerenes: synthesis and applications. J. Mater. Sci. 2018 Jun 30;7:22-33.

[51] Baby R, Saifullah B, Hussein MZ. Carbon nanomaterials for the treatment of heavy metal-contaminated water and environmental remediation. Nanoscale Research Letters. 2019 Dec 1;14(1):341.

[52] Mubarak NM, Sahu JN, Abdullah EC, Jayakumar NS. Removal of heavy metals from wastewater using carbon nanotubes. Separation & Purification Reviews. 2014 Oct 2;43(4):311-38.

[53] Yadav DK, Srivastava S. Carbon nanotubes as adsorbent to remove heavy metal ion (Mn+ 7) in wastewater treatment. Materials Today: Proceedings. 2017 Jan 1;4(2):4089-94.

[54] Yu G, Lu Y, Guo J, Patel M, Bafana A, Wang X, Qiu B, Jeffryes C, Wei S, Guo Z, Wujcik EK. Carbon nanotubes, graphene, and their derivatives for heavy metal removal. Advanced Composites and Hybrid Materials. 2018 Mar 1;1(1):56-78.

[55] Ouni L, Ramazani A, Fardood ST. An overview of carbon nanotubes role in heavy metals removal from wastewater. Frontiers of Chemical Science and Engineering. 2019 May 22:1-22.

[56] Fiyadh SS, AlSaadi MA, Jaafar WZ, AlOmar MK, Fayaed SS, Mohd NS, Hin LS, El-Shafie A. Review on heavy metal adsorption processes by carbon nanotubes. Journal of Cleaner Production. 2019 Sep 1;230:783-93.

[57] Crini G, Lichtfouse E, Wilson LD, Morin-Crini N. Conventional and non-conventional adsorbents for wastewater treatment. Environmental Chemistry Letters. 2019 Mar 1;17(1):195-213.

[58] Crini G (2010) Wastewater treatment by sorption. In: Crini G, Badot PM (eds) Sorption processes and pollution, chap 2. PUFC, Besançon, pp 39-78.

[59] Crini G, Lichtfouse É, Wilson LD, Morin-Crini N (2018) Adsorption-oriented using conventional and non-conventional adsorbents for wastewater treatment. In: Crini G, Lichtfouse É (eds) Environmental chemistry for a sustainable world, green adsorbents for pollutant removal—fundamentals and design, chapter 2, vol 1. Springer, Berlin, pp 23-71. ISBN 978-3-319-92111-2.

[60] Jusoh A, Shiung LS, Noor MJ. A simulation study of the removal efficiency of granular activated carbon on cadmium and lead. Desalination. 2007 Feb 5;206(1-3):9-16.

[61] Kang KC, Kim SS, Choi JW, Kwon SH. Sorption of Cu2+ and Cd2+ onto acid-and base-pretreated granular activated carbon and activated carbon fiber samples. Journal of Industrial and Engineering Chemistry. 2008 Jan 1;14(1):131-5.

[62] Deliyanni EA, Kyzas GZ, Triantafyllidis KS, Matis KA. Activated carbons for the removal of heavy metal ions: A systematic review of recent literature focused on lead and arsenic ions. Open Chemistry. 2015 Jan 13;1(open-issue).

[63] Bansal RC, Goyal M. Activated carbon adsorption. CRC press; 2005 May 24.

[64] Kobya M. Removal of Cr (VI) from aqueous solutions by adsorption onto hazelnut shell activated carbon: kinetic and equilibrium studies. Bioresource technology. 2004 Feb 1;91(3):317-21.

[65] Selomulya C, Meeyoo V, Amal R. Mechanisms of Cr (VI) removal from water by various types of activated carbons. Journal of Chemical Technology & Biotechnology: International Research in Process,

Environmental & Clean Technology. 1999 Feb;74(2):111-22.

[66] Karthikeyan T, Rajgopal S, Miranda LR. Chromium (VI) adsorption from aqueous solution by HeveaBrasilinesis sawdust activated carbon. Journal of hazardous materials. 2005 Sep 30;124(1-3):192-9.

[67] Babel S, Kurniawan TA. Cr (VI) removal from synthetic wastewater using coconut shell charcoal and commercial activated carbon modified with oxidizing agents and/or chitosan. Chemosphere. 2004 Feb 1;54(7):951-67.

[68] Garg VK, Gupta R, Kumar R, Gupta RK. Adsorption of chromium from aqueous solution on treated sawdust. Bioresource technology. 2004 Mar 1;92(1):79-81.

[69] Sharma DC, Forster CF. A preliminary examination into the adsorption of hexavalent chromium using low-cost adsorbents. Bioresource Technology. 1994 Jan 1;47(3):257-64.

[70] Selvi K, Pattabhi S, Kadirvelu K. Removal of Cr (VI) from aqueous solution by adsorption onto activated carbon. Bioresource technology. 2001 Oct 1;80(1):87-9.

[71] Kongsuwan A, Patnukao P, Pavasant P. Binary component sorption of Cu (II) and Pb (II) with activated carbon from Eucalyptus camaldulensisDehn bark. Journal of Industrial and Engineering Chemistry. 2009 Jul 25;15(4):465-70.

[72] El-Ashtoukhy ES, Amin NK, Abdelwahab O. Removal of lead (II) and copper (II) from aqueous solution using pomegranate peel as a new adsorbent. Desalination. 2008 Mar 1;223(1-3):162-73.

[73] Kavand M, Kaghazchi T, Soleimani M. Optimization of parameters for competitive adsorption of heavy metal ions (Pb+ 2, Ni+ 2, Cd+ 2) onto activated carbon. Korean journal of chemical engineering. 2014 Apr 1;31(4):692-700.

[74] Kim TK, Kim T, Choe WS, Kim MK, Jung YJ, Zoh KD. Removal of heavy metals in electroplating wastewater by powdered activated carbon (PAC) and sodium diethyldithiocarbamate-modified PAC. Environmental Engineering Research. 2018;23(3):301-8.

[75] Choi HJ, Yu SW, Kim KH. Efficient use of Mg-modified zeolite in the treatment of aqueous solution contaminated with heavy metal toxic ions. Journal of the Taiwan Institute of Chemical Engineers. 2016 Jun 1;63:482-9.

[76] Erdem E, Karapinar N, Donat R. The removal of heavy metal cations by natural zeolites. Journal of colloid and interface science. 2004 Dec 15;280(2):309-14.

[77] Ibrahim HS, Jamil TS, Hegazy EZ. Application of zeolite prepared from Egyptian kaolin for the removal of heavy metals: II. Isotherm models. Journal of Hazardous materials. 2010 Oct 15;182(1-3):842-847.

[78] Aliabadi M, Irani M, Ismaeili J, Piri H, Parnian MJ. Electrospun nanofiber membrane of PEO/Chitosan for the adsorption of nickel, cadmium, lead and copper ions from aqueous solution. Chemical engineering journal. 2013 Mar 15;220:237-43.

[79] Rad LR, Momeni A, Ghazani BF, Irani M, Mahmoudi M, Noghreh B. Removal of Ni2+ and Cd2+ ions from aqueous solutions using electrospun PVA/zeolite nanofibrous adsorbent. Chemical Engineering Journal. 2014 Nov 15;256:119-27.

[80] Javadian H, Sorkhrodi FZ, KoutenaeiBB.Experimentalinvestigation on enhancing aqueous cadmium

removal via nanostructure composite of modified hexagonal type mesoporous silica with polyaniline/polypyrrole nanoparticles. Journal of Industrial and Engineering Chemistry. 2014 Sep 25;20(5):3678-88.

[81] Visa M. Synthesis and characterization of new zeolite materials obtained from fly ash for heavy metals removal in advanced wastewater treatment. Powder Technology. 2016 Jun 1;294:338-47.

[82] Kobayashi Y, Ogata F, Nakamura T, Kawasaki N. Synthesis of novel zeolites produced from fly ash by hydrothermal treatment in alkaline solution and its evaluation as an adsorbent for heavy metal removal. Journal of Environmental Chemical Engineering. 2020 Apr 1;8(2):103687.

[83] Tripathi A, Ranjan MR. Heavy metal removal from wastewater using low cost adsorbents. J Bioremed Biodeg. 2015 Jan;6(6):315.

[84] Agarwal M, Singh K. Heavy metal removal from wastewater using various adsorbents: a review. Journal of Water Reuse and Desalination. 2017 Dec;7(4):387-419.

[85] Singh SP, Ma LQ, Harris WG. Heavy metal interactions with phosphatic clay: sorption and desorption behavior. Journal of Environmental Quality. 2001 Nov;30(6):1961-8.

[86] Krikorian N, Martin DF. Extraction of selected heavy metals using modified clays. Journal of Environmental Science and Health. 2005 Jan 1;40(3):601-8.

[87] Aşçı Y, Nurbaş M, Açıkel YS. Sorption of Cd (II) onto kaolin as a soil component and desorption of Cd (II) from kaolin using rhamnolipid biosurfactant. Journal of Hazardous Materials. 2007 Jan 2;139(1):50-6.

[88] Jiang MQ, Jin XY, Lu XQ, Chen ZL. Adsorption of Pb (II), Cd (II), Ni (II) and Cu (II) onto natural kaolinite clay. Desalination. 2010 Mar 1;252(1-3):33-9.

[89] Bertagnolli C, Kleinübing SJ, Da Silva MG. Preparation and characterization of a Brazilian bentonite clay for removal of copper in porous beds. Applied Clay Science. 2011 Jul 1;53(1):73-9.

[90] Chai JB, Au PI, Mubarak NM, Khalid M, Ng WP, Jagadish P, Walvekar R, Abdullah EC. Adsorption of heavy metal from industrial wastewater onto low-cost Malaysian kaolin clay–based adsorbent. Environmental Science and Pollution Research. 2020 Feb 8:1-4.

[91] Burakov A, Romantsova I, Kucherova A, Tkachev A. Removal of heavy-metal ions from aqueous solutions using activated carbons: effect of adsorbent surface modification with carbon nanotubes. Adsorption Science & Technology. 2014 Sep;32(9):737-47.

[92] Darwish AD. Fullerenes. Annual Reports Section" A"(Inorganic Chemistry). 2013;109:436-52.

[93] Melezhyk AV, Kotov VA, Tkachev AG. Optical properties and aggregation of graphene nanoplatelets. Journal of Nanoscience and Nanotechnology. 2016 Jan 1;16(1):1067-75.

[94] Iijima S, Helical microtubes of graphitic carbon. Nature. 1991; 354:56-58.

[95] Li Y, Liu F, Xia B, Du Q, Zhang P, Wang D, Wang Z, Xia Y. Removal of copper from aqueous solution by carbon nanotube/calcium alginate composites. Journal of Hazardous Materials. 2010 May 15;177(1-3):876-80.

[96] Rodríguez C, Leiva E. Enhanced heavy metal removal from acid mine drainage wastewater using double-oxidized multiwalled carbon nanotubes. Molecules. 2020 Jan;25(1):111.

[97] Wang H, Zhou A, Peng F, Yu H, Yang J. Mechanism study on adsorption of acidified multiwalled carbon nanotubes to Pb (II). Journal of Colloid and Interface Science. 2007 Dec 15;316(2):277-83.

[98] Kabbashi NA, Atieh MA, Al-Mamun A, Mirghami ME, Alam MD, Yahya N. Kinetic adsorption of application of carbon nanotubes for Pb (II) removal from aqueous solution. Journal of Environmental Sciences. 2009 Apr 1;21(4):539-44.

[99] Pillay K, Cukrowska EM, Coville NJ. Multi-walled carbon nanotubes as adsorbents for the removal of parts per billion levels of hexavalent chromium from aqueous solution. Journal of hazardous materials. 2009 Jul 30;166(2-3):1067-75.

[100] Temitope BM, Saka AA, Alhassan MI, Ochigbo SS, Oladejo TJ, Kamaldeen AO, Roos WD. Selected Heavy Metals Removal from Electroplating Wastewater by Purified and Polyhydroxylbutyrate Functionalized Carbon Nanotubes Adsorbents. Scientific Reports (Nature Publisher Group). 2019 Dec 1;9(1).

[101] Kandah MI, Meunier JL. Removal of nickel ions from water by multi-walled carbon nanotubes. Journal of hazardous materials. 2007 Jul 19;146(1-2):283-8.

[102] Kuo CY, Lin HY. Adsorption of aqueous cadmium (II) onto modified multi-walled carbon nanotubes following microwave/chemical treatment. Desalination. 2009 Dec 15;249(2):792-6.

[103] Haddon RC, Carbon nanotubes. Acc. Chem. Res. 2002; 35: 977-1113.

[104] Fiyadh SS, AlSaadi MA, Jaafar WZ, AlOmar MK, Fayaed SS, Mohd NS, Hin LS, El-Shafie A. Review on heavy metal adsorption processes by carbon nanotubes. Journal of Cleaner Production. 2019 Sep 1;230:783-93.

[105] Duan C, Ma T, Wang J, Zhou Y. Removal of heavy metals from aqueous solution using carbon-based adsorbents: A review. Journal of Water Process Engineering. 2020 Oct 1;37:101339.

[106] Gupta VK, Agarwal S, Saleh TA. Chromium removal by combining the magnetic properties of iron oxide with adsorption properties of carbon nanotubes. Water research. 2011 Mar 1;45(6):2207-12.

[107] Wang X, Guo Y, Yang L, Han M, Zhao J, Cheng X. Nanomaterials as sorbents to remove heavy metal ions in wastewater treatment. J. Environ. Anal. Toxicol. 2012 Nov;2(7):154-8.

[108] Sharma M, Singh J, Hazra S, Basu S. Adsorption of heavy metal ions by mesoporous ZnO and TiO2@ZnO monoliths: adsorption and kinetic studies. Microchemical journal. 2019 Mar 1;145:105-12.

[109] Yang X, Wan Y, Zheng Y, He F, Yu Z, Huang J, Wang H, Ok YS, Jiang Y, Gao B. Surface functional groups of carbon-based adsorbents and their roles in the removal of heavy metals from aqueous solutions: a critical review. Chemical Engineering Journal. 2019 Jun 15;366:608-21.

[110] Verma B, Balomajumder C. Surface modification of one-dimensional Carbon Nanotubes: A review for the management of heavy metals in wastewater. Environmental Technology & Innovation. 2020 Feb 1;17:100596.

[111] Li YH, Wang S, Wei J, Zhang X, Xu C, Luan Z, Wu D, Wei B. Lead adsorption on carbon nanotubes. Chemical Physics Letters. 2002 May 10;357(3-4):263-6.

[112] Li YH, Wang S, Luan Z, Ding J, Xu C, Wu D. Adsorption of cadmium

(II) from aqueous solution by surface oxidized carbon nanotubes. Carbon. 2003 Jan 1;41(5):1057-62.

[113] Robati D. Pseudo-second-order kinetic equations for modeling adsorption systems for removal of lead ions using multi-walled carbon nanotube. Journal of nanostructure in Chemistry. 2013 Dec 1;3(1):55.

[114] Xu J, Cao Z, Zhang Y, Yuan Z, Lou Z, Xu X, Wang X. A review of functionalized carbon nanotubes and graphene for heavy metal adsorption from water: Preparation, application, and mechanism. Chemosphere. 2018 Mar 1;195:351-64.

[115] Rodríguez C, Leiva E. Enhanced heavy metal removal from acid mine drainage wastewater using double-oxidized multiwalled carbon nanotubes. Molecules. 2020 Jan;25(1):111.

[116] Wang HJ, Zhou AL, Peng F, Yu H, Chen LF. Adsorption characteristic of acidified carbon nanotubes for heavy metal Pb (II) in aqueous solution. Materials Science and Engineering: A. 2007 Sep 25;466(1-2):201-6.

[117] Yang S, Li J, Shao D, Hu J, Wang X. Adsorption of Ni (II) on oxidized multi-walled carbon nanotubes: effect of contact time, pH, foreign ions and PAA. Journal of hazardous materials. 2009 Jul 15;166(1):109-16.

[118] Gupta VK, Agarwal S, Saleh TA. Synthesis and characterization of alumina-coated carbon nanotubes and their application for lead removal. Journal of hazardous materials. 2011 Jan 15;185(1):17-23.

[119] Ntim SA, Mitra S. Adsorption of arsenic on multiwall carbon nanotube–zirconia nanohybrid for potential drinking water purification. Journal of colloid and interface science. 2012 Jun 1;375(1):154-9.

[120] Tang WW, Zeng,GM, Gong, JL, Liu, Y, Wang, XY, Liu, YY, Liu, ZF, Chen, L, Xhang, XR, Tu, DZ. Simultaneous adsorption of atrazine and Cu(II) from wastewater by magnetic multi-walled carbon nanotubes. Chem. Eng. J. 2012; 21-212: 470-478.

[121] Wang B, Wu T, Angaiah S, Murugadoss V, Ryu JE, Wujcik EK, Lu N, Young DP, Gao Q, Guo Z. Development of nanocomposite adsorbents for heavy metal removal from wastewater. ES Materials & Manufacturing. 2018 Dec 4;2(2):35-44.

[122] Sarma GK, Gupta SS, Bhattacharyya KG. Nanomaterials as versatile adsorbents for heavy metal ions in water: a review. Environmental Science and Pollution Research. 2019 Mar 8;26(7):6245-78.

[123] Luo C, Tian Z, Yang B, Zhang L, Yan S. Manganese dioxide/iron oxide/acid oxidized multi-walled carbon nanotube magnetic nanocomposite for enhanced hexavalent chromium removal. Chemical Engineering Journal. 2013 Dec 1;234:256-65.

[124] Ge L, Wang W, Peng Z, Tan F, Wang X, Chen J, Qiao X. Facile fabrication of Fe@ MgO magnetic nanocomposites for efficient removal of heavy metal ion and dye from water. Powder Technology. 2018 Feb 15;326:393-401.

[125] Stafiej A, Pyrzynska K. Adsorption of heavy metal ions with carbon nanotubes. Separation and purification technology. 2007 Dec 1;58(1):49-52.

[126] Gao Z, Bandosz TJ, Zhao Z, Han M, Qiu J. Investigation of factors affecting adsorption of transition metals on oxidized carbon nanotubes. Journal of Hazardous Materials. 2009 Aug 15;167(1-3):357-65.

[127] Musielak M, Gagor A, Zawisza B, Talik E, Sitko R. Graphene oxide/carbon

nanotube membranes for highly efficient removal of metal ions from water. ACS applied materials & interfaces. 2019 Jul 18;11(31):28582-90.

[128] Rao GP, Lu C, Su F. Sorption of divalent metal ions from aqueous solution by carbon nanotubes: a review. Separation and purification technology. 2007 Dec 1;58(1):224-31.

[129] Chen GC, Shan XQ, Zhou YQ, Shen XE, Huang HL, Khan SU. Adsorption kinetics, isotherms and thermodynamics of atrazine on surface oxidized multiwalled carbon nanotubes. Journal of Hazardous Materials. 2009 Sep 30;169(1-3):912-8.

[130] Verma B, Sewani H, Balomajumder C. Synthesis of carbon nanotubes via chemical vapor deposition: an advanced application in the Management of Electroplating Effluent. Environmental Science and Pollution Research. 2020 Feb 8:1-2.

[131] Egbosiuba TC, Abdulkareem AS, Kovo AS, Afolabi EA, Tijani JO, Roos WD. Enhanced adsorption of As (V) and Mn (VII) from industrial wastewater using multi-walled carbon nanotubes and carboxylated multi-walled carbon nanotubes. Chemosphere. 2020 Apr 21:126780.

[132] Ihsanullah, Al-Khaldi FA, Abu-Sharkh B, Abulkibash AM, Qureshi MI, Laoui T, Atieh MA. Effect of acid modification on adsorption of hexavalent chromium (Cr (VI)) from aqueous solution by activated carbon and carbon nanotubes. Desalination and Water Treatment. 2016 Apr 2;57(16):7232-44.

[133] Tawabini BS, Al-Khaldi SF, Khaled MM, Atieh MA. Removal of arsenic from water by iron oxide nanoparticles impregnated on carbon nanotubes. Journal of Environmental Science and Health, Part A. 2011 Jan 31;46(3):215-23.

[134] Chen H, Li J, Shao D, Ren X, Wang X. Poly (acrylic acid) grafted multiwall carbon nanotubes by plasma techniques for Co (II) removal from aqueous solution. Chemical engineering journal. 2012 Nov 1;210:475-81.

[135] Hayati B, Maleki A, Najafi F, Daraei H, Gharibi F, McKay G. Synthesis and characterization of PAMAM/CNT nanocomposite as a super-capacity adsorbent for heavy metal (Ni2+, Zn2+, As3+, Co2+) removal from wastewater. Journal of Molecular Liquids. 2016 Dec 1;224:1032-40.

[136] Baby R, Saifullah B, Hussein MZ. Carbon nanomaterials for the treatment of heavy metal-contaminated water and environmental remediation. Nanoscale Research Letters. 2019 Dec 1;14(1):341.

[137] Ouni L, Ramazani A, Fardood ST. An overview of carbon nanotubes role in heavy metals removal from wastewater. Frontiers of Chemical Science and Engineering. 2019 May 22:1-22.

[138] Verma B, Balomajumder C. Surface modification of one-dimensional Carbon Nanotubes: A review for the management of heavy metals in wastewater. Environmental Technology & Innovation. 2020 Feb 1;17:100596.

[139] Zhang BT, Zheng X, Li HF, Lin JM. Application of carbon-based nanomaterials in sample preparation: A review. Analytica chimica acta. 2013 Jun 19;784:1-7.

[140] Malik S. Structural and electronic properties of nano-carbon materials such as graphene, nanotubes and fullerenes. Nature. 1985;318:162-3.

[141] Valcárcel M, Cárdenas S, Simonet BM, Moliner-Martínez Y, Lucena R. Carbon nanostructures as sorbent materials in analytical processes. TrAC Trends in Analytical Chemistry. 2008 Jan 1;27(1):34-43.

[142] Zhang M, Gao B, Cao X, Yang L. Synthesis of a multifunctional graphene–carbon nanotube aerogel and its strong adsorption of lead from aqueous solution. Rsc Advances. 2013;3(43):21099-105.

[143] Alekseeva OV, Bagrovskaya NA, Noskov AV. Sorption of heavy metal ions by fullerene and polystyrene/fullerene film compositions. Protection of Metals and Physical Chemistry of Surfaces. 2016 May 1;52(3):443-7.

[144] Lalia, B.S.; Kochkodan, V.; Hashaikeh, R.; Hilal, N. A review on membrane fabrication: Structure, properties and performance relationship. Desalination 2013, 326, 77-95.

[145] Djordjevic, A.; Srdjenovic, B.; Seke, M.; Petrovic, D.; Injac, R.; Mrdjanovic, J. Review of Synthesis and Antioxidant Potential of Fullerenol Nanoparticles. J. Nanomater. 2015, 2015, 1-15.

[146] Jani M, Arcos-Pareja JA, Ni M. Engineered Zero-Dimensional Fullerene/Carbon Dots-Polymer Based Nanocomposite Membranes for Wastewater Treatment. Molecules. 2020 Jan;25(21):4934.

[147] Samonin VV, Nikonova VY, Podvyaznikov ML. Sorption properties of fullerene-modified activated carbon with respect to metal ions. Protection of Metals. 2008 Mar 1;44(2):190-2.

[148] Samonin VV, Nikonova VY, Podvyaznikov ML. Carbon adsorbents on the basis of the hydrolytic lignin modified with fullerenes in producing. Russian Journal of Applied Chemistry. 2014 Feb 1;87(2):190-3.

[149] Geim AK. Graphene: status and prospects. science. 2009 Jun 19;324(5934):1530-4.

[150] Novoselov KS, Fal VI, Colombo L, Gellert PR, Schwab MG, Kim K. A roadmap for graphene. nature. 2012 Oct;490(7419):192-200.

[151] Gopalakrishnan A, Krishnan R, Thangavel S, Venugopal G, Kim SJ. Removal of heavy metal ions from pharma-effluents using graphene-oxide nanosorbents and study of their adsorption kinetics. Journal of Industrial and Engineering Chemistry. 2015 Oct 25;30:14-9.

[152] Huang ZH, Zheng X, Lv W, Wang M, Yang QH, Kang F. Adsorption of lead (II) ions from aqueous solution on low-temperature exfoliated graphene nanosheets. Langmuir. 2011 Jun 21;27(12):7558-62.

[153] Zhao G, Li J, Ren X, Chen C, Wang X. Few-layered graphene oxide nanosheets as superior sorbents for heavy metal ion pollution management. Environmental science & technology. 2011 Dec 15;45(24):10454-62.

[154] Cong HP, Ren XC, Wang P, Yu SH. Macroscopic multifunctional graphene-based hydrogels and aerogels by a metal ion induced self-assembly process. ACS nano. 2012 Mar 27;6(3):2693-703.

[155] Yu G, Lu Y, Guo J, Patel M, Bafana A, Wang X, Qiu B, Jeffryes C, Wei S, Guo Z, Wujcik EK. Carbon nanotubes, graphene, and their derivatives for heavy metal removal. Advanced Composites and Hybrid Materials. 2018 Mar 1;1(1):56-78.

[156] Deng X, Lü L, Li H, Luo F. The adsorption properties of Pb (II) and Cd (II) on functionalized graphene prepared by electrolysis method. Journal of Hazardous Materials. 2010 Nov 15;183(1-3):923-30.

[157] Ren X, Li J, Tan X, Wang X. Comparative study of graphene oxide, activated carbon and carbon nanotubes as adsorbents for copper decontamination. Dalton Transactions. 2013;42(15):5266-74.

[158] Zhang M, Gao B, Cao X, Yang L. Synthesis of a multifunctional graphene–carbon nanotube aerogel and its strong adsorption of lead from aqueous solution. Rsc Advances. 2013;3(43):21099-105.

[159] Arshad F, Selvaraj M, Zain J, Banat F, Haija MA. Polyethylenimine modified graphene oxide hydrogel composite as an efficient adsorbent for heavy metal ions. Separation and Purification Technology. 2019 Jan 31;209:870-80.

[160] Wu Z, Deng W, Zhou W, Luo J. Novel magnetic polysaccharide/ graphene oxide@ Fe3O4 gel beads for adsorbing heavy metal ions. Carbohydrate polymers. 2019 Jul 15;216:119-28.

[161] Liu X, Ma R, Wang X, Ma Y, Yang Y, Zhuang L, Zhang S, Jehan R, Chen J, Wang X. Graphene oxide-based materials for efficient removal of heavy metal ions from aqueous solution: A review. Environmental pollution. 2019 Sep 1;252:62-73.

[162] Farooq MU, Jalees MI. Application of Magnetic Graphene Oxide for Water Purification: Heavy Metals Removal and Disinfection. Journal of Water Process Engineering. 2020 Feb 1;33:101044.

[163] Li L, Zhao L, Ma J, Tian Y. Preparation of graphene oxide/chitosan complex and its adsorption properties for heavy metal ions. Green Processing and Synthesis. 2020 Jun 6;9(1):294-303.

[164] Zhao G, Ren X, Gao X, Tan X, Li J, Chen C, Huang Y, Wang X. Removal of Pb (II) ions from aqueous solutions on few-layered graphene oxide nanosheets. Dalton transactions. 2011;40(41):10945-52.

[165] Jiang H, Yang Y, Lin Z, Zhao B, Wang J, Xie J, Zhang A. Preparation of a novel bio-adsorbent of sodium alginate grafted polyacrylamide/graphene oxide hydrogel for the adsorption of heavy metal ion. Science of The Total Environment. 2020 Nov 20;744:140653.

[166] Bhatnagar A, Sillanpää M. Utilization of agro-industrial and municipal waste materials as potential adsorbents for water treatment—a review. Chemical engineering journal. 2010 Mar 1;157(2-3):277-96

[167] Bhatnagar A, Sillanpää M, Witek-Krowiak A. Agricultural waste peels as versatile biomass for water purification–A review. Chemical Engineering Journal. 2015 Jun 15;270:244-71.

[168] Owamah HI. Biosorptive removal of Pb (II) and Cu (II) from wastewater using activated carbon from cassava peels. Journal of Material Cycles and Waste Management. 2014 Apr 1;16(2):347-58.

[169] Sreenivas KM, Inarkar MB, Gokhale SV, Lele SS. Re-utilization of ash gourd (Benincasahispida) peel waste for chromium (VI) biosorption: Equilibrium and column studies. Journal of Environmental Chemical Engineering. 2014 Mar 1;2(1):455-62.

[170] Pehlivan E, Altun T, Parlayici Ş. Modified barley straw as a potential biosorbent for removal of copper ions from aqueous solution. Food chemistry. 2012 Dec 15;135(4):2229-34.

[171] Kumar PS, Ramalingam S, Kirupha SD, Murugesan A, Vidhyadevi T, Sivanesan S. Adsorption behavior of nickel (II) onto cashew nut shell: Equilibrium, thermodynamics, kinetics, mechanism and process design. Chemical Engineering Journal. 2011 Feb 15;167(1):122-31.

[172] Feng N, Guo X, Liang S. Adsorption study of copper (II) by chemically modified orange peel. Journal of Hazardous Materials. 2009 May 30;164(2-3):1286-92.

[173] Lu D, Cao Q, Cao X, Luo F. Removal of Pb (II) using the modified lawny grass: Mechanism, kinetics, equilibrium and thermodynamic studies. Journal of Hazardous Materials. 2009 Jul 15;166(1):239-47.

[174] Zou W, Zhao L, Zhu L. Efficient uranium (VI) biosorption on grapefruit peel: kinetic study and thermodynamic parameters. Journal of Radioanalytical and Nuclear Chemistry. 2012 Jun 1;292(3):1303-15.

[175] Al-Othman ZA, Ali R, Naushad M. Hexavalent chromium removal from aqueous medium by activated carbon prepared from peanut shell: adsorption kinetics, equilibrium and thermodynamic studies. Chemical Engineering Journal. 2012 Mar 1;184:238-47.

[176] Abdelhafez AA, Li J. Removal of Pb (II) from aqueous solution by using biochars derived from sugar cane bagasse and orange peel. Journal of the Taiwan Institute of Chemical Engineers. 2016 Apr 1;61:367-75.

[177] Iqbal M, Saeed A, Kalim I. Characterization of adsorptive capacity and investigation of mechanism of Cu2+, Ni2+ and Zn2+ adsorption on mango peel waste from constituted metal solution and genuine electroplating effluent. Separation Science and Technology. 2009 Oct 30;44(15):3770-91.

[178] Basci N, Kocadagistan E, Kocadagistan B. Biosorption of copper (II) from aqueous solutions by wheat shell. Desalination. 2004 Apr 1;164(2):135-40.

[179] Yu W, Hu J, Yu Y, Ma D, Gong W, Qiu H, Hu Z, Gao HW. Facile preparation of sulfonated biochar for highly efficient removal of toxic Pb (II) and Cd (II) from wastewater. Science of The Total Environment. 2020 Aug 6;750:141545.

[180] Sanka PM, Rwiza MJ, Mtei KM. Removal of Selected Heavy Metal Ions from Industrial Wastewater Using Rice and Corn Husk Biochar. Water, Air, & Soil Pollution. 2020 May;231:1-3.

[181] Khalil U, Shakoor MB, Ali S, Rizwan M, Alyemeni MN, Wijaya L. Adsorption-reduction performance of tea waste and rice husk biochars for Cr (VI) elimination from wastewater. Journal of Saudi Chemical Society. 2020 Nov 1;24(11):799-810.

[182] Shen Z, Zhang Y, McMillan O, Jin F, Al-Tabbaa A. Characteristics and mechanisms of nickel adsorption on biochars produced from wheat straw pellets and rice husk. Environmental Science and Pollution Research. 2017 May 1;24(14):12809-19.

[183] Shi J, Fan X, Tsang DC, Wang F, Shen Z, Hou D, Alessi DS. Removal of lead by rice husk biochars produced at different temperatures and implications for their environmental utilizations. Chemosphere. 2019 Nov 1;235:825-31.

[184] Khalil U, Shakoor MB, Ali S, Rizwan M, Alyemeni MN, Wijaya L. Adsorption-reduction performance of tea waste and rice husk biochars for Cr (VI) elimination from wastewater. Journal of Saudi Chemical Society. 2020 Nov 1;24(11):799-810.

[185] Gao LY, Deng JH, Huang GF, Li K, Cai KZ, Liu Y, Huang F. Relative distribution of Cd2+ adsorption mechanisms on biochars derived from rice straw and sewage sludge. Bioresource Technology. 2019 Jan 1;272:114-22.

[186] Fan S, Zhang L. Production and characterization of tea waste–based biochar and its application in treatment of Cd-containing wastewater. Biomass Conversion and Biorefinery. 2019 Nov 26:1-4.

[187] Akgül G, Maden TB, Diaz E, Jiménez EM. Modification of tea biochar

with Mg, Fe, Mn and Al salts for efficient sorption of $PO4^{3-}$ and Cd2+ from aqueous solutions. Journal of Water Reuse and Desalination. 2019 Mar 1;9(1):57-66.

[188] Ho SH, Yang ZK, Nagarajan D, Chang JS, Ren NQ. High-efficiency removal of lead from wastewater by biochar derived from anaerobic digestion sludge. Bioresource Technology. 2017 Dec 1;246:142-9.

[189] Bayat B. Combined removal of zinc (II) and cadmium (II) from aqueous solutions by adsorption onto high-calcium Turkish fly ash. Water, Air, and Soil Pollution. 2002 May 1;136(1-4):69-92.

[190] Wang S, Li L, Zhu ZH. Solid-state conversion of fly ash to effective adsorbents for Cu removal from wastewater. Journal of Hazardous Materials. 2007 Jan 10;139(2):254-9.

[191] Visa M, Bogatu C, Duta A. Simultaneous adsorption of dyes and heavy metals from multicomponent solutions using fly ash. Applied Surface Science. 2010 Jun 15;256(17):5486-91.

[192] Sočo E, Kalembkiewicz J. Adsorption of nickel (II) and copper (II) ions from aqueous solution by coal fly ash. Journal of Environmental Chemical Engineering. 2013 Sep 1;1(3):581-8.

[193] Al-Harahsheh MS, Al Zboon K, Al-Makhadmeh L, Hararah M, Mahasneh M. Fly ash based geopolymer for heavy metal removal: A case study on copper removal. Journal of Environmental Chemical Engineering. 2015 Sep 1;3(3):1669-77.

[194] Ma J, Qin G, Zhang Y, Sun J, Wang S, Jiang L. Heavy metal removal from aqueous solutions by calcium silicate powder from waste coal fly-ash. Journal of Cleaner Production. 2018 May 1;182:776-82.

[195] Nguyen TC, Loganathan P, Nguyen TV, Kandasamy J, Naidu R, Vigneswaran S. Adsorptive removal of five heavy metals from water using blast furnace slag and fly ash. Environmental Science and Pollution Research. 2018 Jul 1;25(21):20430-8.

[196] Park JH, Eom JH, Lee SL, Hwang SW, Kim SH, Kang SW, Yun JJ, Cho JS, Lee YH, Seo DC. Exploration of the potential capacity of fly ash and bottom ash derived from wood pellet-based thermal power plant for heavy metal removal. Science of The Total Environment. 2020 Oct 20;740:140205.

[197] Nadaroglu H, Kalkan E, Demir N. Removal of copper from aqueous solution using red mud. Desalination. 2010 Feb 1;251(1-3):90-5.

[198] Sahu RC, Patel R, Ray BC. Adsorption of Zn (II) on activated red mud: Neutralized by CO2. Desalination. 2011 Jan 31;266(1-3):93-7

[199] Bhatnagar A, Vilar VJ, Botelho CM, Boaventura RA. A review of the use of red mud as adsorbent for the removal of toxic pollutants from water and wastewater. Environmental Technology. 2011 Feb 1;32(3):231-49.

[200] Nadaroglu H, Kalkan E. Removal of cobalt (II) ions from aqueous solution by using alternative adsorbent industrial red mud waste material. International Journal of Physical Sciences. 2012 Feb 23;7(9):1386-94.

[201] Pulford ID, Hargreaves JS, Ďurišová J, Kramulova B, Girard C, Balakrishnan M, Batra VS, Rico JL. Carbonised red mud–a new water treatment product made from a waste material. Journal of Environmental Management. 2012 Jun 15;100:59-64.

[202] Sahu MK, Mandal S, Dash SS, Badhai P, Patel RK. Removal of Pb (II) from aqueous solution by acid activated red mud. Journal of Environmental

Chemical Engineering. 2013 Dec 1;1(4):1315-24.

[203] Chen X, Guo Y, Ding S, Zhang H, Xia F, Wang J, Zhou M. Utilization of red mud in geopolymer-based pervious concrete with function of adsorption of heavy metal ions. Journal of Cleaner Production. 2019 Jan 10;207:789-800.

[204] Xue Y, Wu S, Zhou M. Adsorption characterization of Cu (II) from aqueous solution onto basic oxygen furnace slag. Chemical Engineering Journal. 2013 Sep 1;231:355-64.

[205] Sarkar C, Basu JK, Samanta AN. Removal of Ni2+ ion from waste water by Geopolymeric Adsorbent derived from LD Slag. Journal of Water Process Engineering. 2017 Jun 1;17:237-44.

[206] Srivastava SK, Singh AK, Sharma A. Studies on the uptake of lead and zinc by lignin obtained from black liquor–a paper industry waste material. Environmental Technology. 1994 Apr 1;15(4):353-61.

[207] Zwain HM, Vakili M, Dahlan I. Waste material adsorbents for zinc removal from wastewater: a comprehensive review. International Journal of Chemical Engineering. 2014 Jan 1;2014.

[208] Crini G. Recent developments in polysaccharide-based materials used as adsorbents in wastewater treatment. Progress in polymer science. 2005 Jan 1;30(1):38-70.

[209] Crini G (2010) Wastewater treatment by sorption. In: Crini G, Badot PM (eds) Sorption processes and pollution, chap 2. PUFC, Besançon, pp 39-78.

[210] Davis TA, Volesky B, Mucci A. A review of the biochemistry of heavy metal biosorption by brown algae. Water research. 2003 Nov 1;37(18):4311-30.

[211] Veglio F, Beolchini F. Removal of metals by biosorption: a review. Hydrometallurgy. 1997 Mar 1;44(3):301-16.

[212] Srivastava S, Goyal P. Novel biomaterials: decontamination of toxic metals from wastewater. Springer Science & Business Media; 2010 Apr 26.

[213] Naja G, Volesky B. The mechanism of metal cation and anion biosorption. InMicrobial biosorption of metals 2011 (pp. 19-58). Springer, Dordrecht.

[214] Michalak I, Chojnacka K, Witek-Krowiak A. State of the art for the biosorption process—a review. Applied biochemistry and biotechnology. 2013 Jul 1;170(6):1389-416.

[215] Robalds A, Naja GM, Klavins M. Highlighting inconsistencies regarding metal biosorption. Journal of hazardous materials. 2016 Mar 5;304:553-6.

Chapter 11

Biotechnological Approaches to Facilitate Gold Recovery from Double Refractory Gold Ores

Keiko Sasaki and Kojo T. Konadu

Abstract

Double refractory gold ore (DRGO) not only include ppt levels of gold grains locked in sulfide minerals but also a problematic amount of carbonaceous matter. This causes a significant recovery loss of gold during cyanidation because of the strong affinity of the $Au(CN)_2^-$ with the carbonaceous matter. Combustion decreases the carbonaceous matter content, but also emits pollutant gases like CO_2, SO_2 and As_2O_3. Therefore, environmentally-friendly solutions have been explored by using biotechnology. Due to the very small amount of the above targets in the ore, it is challenging to show evidential changes in solid-phase before and after the biomineral processing of DRGO. This chapter introduces the mineralogical and chemical changes in the various solid residues produced during a sequential biotreatment, consisting of the liberation of gold from sulfides by an iron-oxidizer and decomposition of carbonaceous matter by lignin-degrading enzymes (lignin peroxidase, manganese peroxidase, laccase) secreted from a white rot-fungus, which successfully improved of gold recovery to over 90%. In addition, further development of biotechnology in the recovery of gold from DRGO is addressed.

Keywords: double refractory gold ore, carbonaceous matter, enzyme reaction, QEMSCAN map, biooxidation of sulfides

1. Introduction

Carbonaceous refractory gold ores are classified as double refractory gold ores (DRGO) due to containing sulfide minerals and carbonaceous matter. In DRGO, the sulfide minerals (pyrite and arsenopyrite) tend to have a significant amount of gold enclosed in it compared to other minerals in the ore and the gold confined in sulfides is difficult to recovery due to the minerals' stability during cyanidation, leading to poor recovery from sulfides. Furthermore, the gold that is successfully dissolved as $Au(CN)_2^-$ in the cyanide leaching step, can be adsorbed by the organic carbon (preg-robbing), therefore, using cyanidation without DRGO pretreatment can lead to 30–70% gold recovery losses [1–9]. DRGO is produced in various parts of the world such as Ghana, Brazil, USA, Canada, Kazakhstan, Russia, Malaysia, Indonesia and China [9], and the gold production from DRGO attains to about 1/100 of the total gold production in the world. DRGO also accounts for about one-third of available gold deposits [8].

 IntechOpen

DRGO is generally subjected to flotation to recovery a sulfide concentrate that is then sent for pretreatment prior to cyanidation. However, due to the poor separation of carbonaceous matter and sulfide minerals, the carbonaceous matter also reports in the flotation concentrate [10–13]. Although, some recent studies have looked at pre-flotation as a means of improving the separation of carbonaceous matter and sulfides from DRGO [14–17]. But, for most current industrial operations, the flotation concentrate contains both the carbonaceous matter and the sulfides and therefore, it is extremely difficult to treat. It is no exaggeration to say that there is a global need for technological development to improve the gold recovery from DRGO.

The pretreatment of carbonaceous matter and sulfides in DRGO to minimize preg-robbing and liberate gold has changed with time to become more environmentally friendly (**Figure 1**). Oxidation is one of the most prominent means to pre-treat DRGO and improves gold recovery [4, 10–13, 18, 19]. In the past, thermal oxidation was used to decompose carbonaceous matter and convert sulfide minerals to iron oxides. However, the control of the roasting temperature is very important but difficult because at ≤500°C, there is incomplete sulfide oxidation, and also the removal of volatile matter and decarboxylation reaction in the carbonaceous matter at 100–300°C and 400–500°C, respectively [18, 20]. In such a case, there is an incomplete liberation of gold from sulfides, and the carbonaceous matter becomes a more activate adsorbent, resulting in a higher preg-robbing ability during cyanidation. Furthermore, harmful gases such as SOx and As_2O_3 are generated, making this process no longer as attractive, although recent advances in using gas-scrubbing techniques, the environmental concerns have been reduced [21]. In fact, some works have tried to encourage the gas-scrubbing process by proposing methods like chlorination roasting in which, the solid gold is converted to $AuCl_3$ gas and leaves the furnace along with other harmful gases [22, 23]. In this case, the recovery of the valuable gold necessitates the recovery of all the gases produced from the furnace and significantly decreasing the environmental impact of roasting while also increasing the cost of furnaces operation.

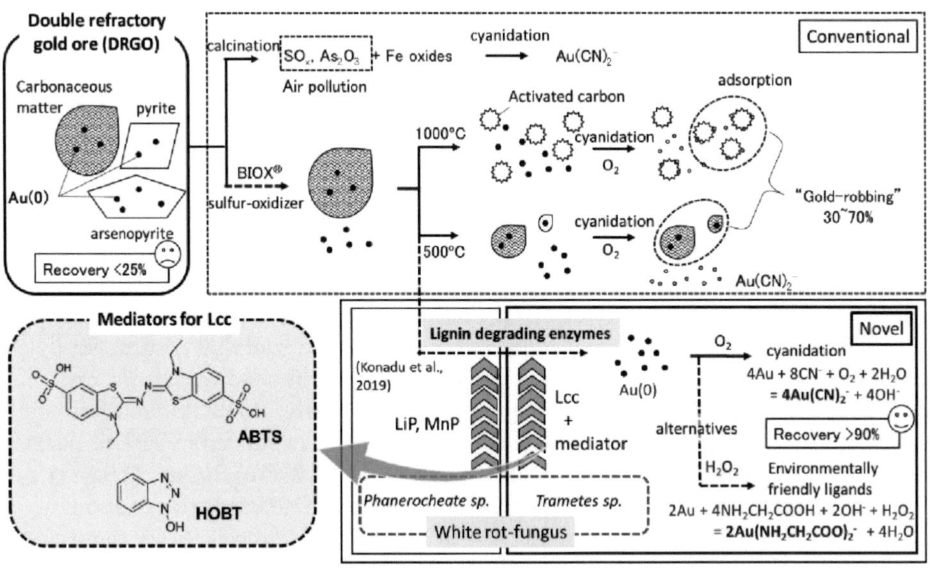

Figure 1.
Conventional and novel methods in (bio) mineral processing of graphitic carbonaceous gold ores.

In place of thermal oxidation, biological (BIOX) and pressure oxidation have been applied to remove sulfides from DRGO, but these processes have a minimal effect on the carbonaceous matter [11–13, 19] (**Figure 1**). After the sulfide oxidation, the carbonaceous matter-containing residue can either undergo a treatment like roasting or blinding to minimize preg-robbing before gold recovery [24]. The roasting of the BIOX or pressure oxidation residue come with the same problems as alluded to before while blinding of the carbonaceous matter leads to the transfer of blinding reagents like kerosene and diesel oil on to the activated carbon used for $Au(CN)_2^-$ recovery from the carbon-in-leach (CIL) or carbon-in-pulp (CIP) process.

A newly alternative carbonaceous matter treatment is the use of lignin-degrading enzyme released by fungi and bacteria [10–13, 25–27]. Although they are more prevalent in white-rot and brown-rot fungi, these enzymes have been isolated from some of the fungi responsible for the oxidation of lignin and a very complex polyaromatic polymer [28]. These lignin-degrading enzymes include lignin peroxidase (LiP), manganese peroxidase (MnP), laccase (Lcc) and versatile peroxidase (VP) [29–31]. These enzymes accelerate the oxidative degradation of C=C and C=O bonds. These enzymes were selected for the present purpose because lignin is a precursor for the carbonaceous matter in DRGO, and it was expected that the lignin-degrading enzymes could successfully attack and oxidize this substance. The lignin-degrading enzyme treatment proceeds under very mild temperature and pH conditions and therefore has a low environmental impact [29, 30]. Additionally, the enzymes' effectiveness of these enzymes can be improved by the inclusion of mediators like veratryl alcohol and ABST [32, 33]. Although, the environmental impact assessment is subject to change once a fuller understanding of the bio-products of the process is known. Additionally, the oxidizing condition generated by these fungi also aids the dissolution of sulfides to liberate gold grains [34–36]. Several studies have shown that using lignin-degrading enzymes produced by fungi like *Phanerochaete chrysosporium* and *Trametes versicolor* can increase gold recovery by 10–20% [10–13]. Although these enzymes are produced in some quantities from white-rot fungi and are beginning to be used for decomposing harmful polyaromatic compounds such as dioxins and for producing biofuels, there is no application research to the mining industry [37].

So that the present study reviews DRGO treatment, with particular focus on the carbonaceous matter treatment by lignin-degrading enzymes. It covers the application of these enzymes to surrogates for the carbonaceous matter to understand the enzyme-substrate interactions and finally move on to using these enzymes on the DRGO to improve gold recovery.

2. Utilization of lignin-degrading enzymes on a surrogate for carbonaceous matter in DRGO

White-rot fungus and brown-rot fungi are a class of microbes known for producing very valuable enzymes. In wood chemistry, a lot of research has focused on determining the properties, reactions, and enzymes at the molecular level [28–33]. A very popular white-rot fungus is *P. chrysosporium*, which grows at a flexible pH range, relatively moderate temperature and produces a variety of lignin-degrading enzymes with low substrate specificity. These include Lip, MnP and Lcc, which have been used been to oxidize substrates from several industries including pulp, agricultural waste, dye treatment but not in the mining industry [37]. There are very few studies that have looked at the performance of these enzymes against substrates with high crystallinity as can be found in DRGO.

Invariably, authors considering the application of the lignin-degrading enzymes to carbonaceous matter in DRGO have started by working with surrogates in a simplified setup to get a better understanding of the oxidation mechanism and bioproduct. Some authors have used activated carbon [9], coal [38] and elemental carbon extracted from DRGO [39] as a model of the graphitic carbon in DRGO. The interaction between the lignin-degrading enzymes and the carbonaceous matter surrogate were facilitated by either growing the fungi in the presence of the carbonaceous matter or harvesting the spent culture liquid and using it for the treatment. It has generally been observed that the lignin-degrading enzymes attack the aromatic C=C bonds in the graphitic carbon and converted it to aliphatic C-C and oxygen-containing functional groups like carbonyl C=O and alcohol C-O. This was confirmed by FTIR and solid-state ^{13}C-NMR spectra, showing the relative intensity of aromatic carbon C=C decreasing after the application of the lignin-degrading enzymes and the relative increase in aromatic carbon C-H and aliphatic carbon C-H (**Figure 2**).

Some consequences of the oxidation of the aromatic C=C, which serve as the backbone for these surrogate materials, is an increase in the surface roughness and a reduction in the specific surface area as shown in **Figure 3**. Several works have shown a reduction in the specific surface area by 76% for anthracite, 34.5% for carbonaceous matter extracted from DRGO and 38% for activated carbon [9, 38, 39]. All of these chemical and physical changes in the graphitic carbon resulting from the interactions with the lignin-degrading enzymes lead to a significant reduction in the $Au(CN)_2^-$ uptake ability.

There is still some amount of work required to improve our understanding of the impact of the lignin-degrading enzymes on the carbonaceous matter. These include establishing a relationship between the amount of enzyme consumed to decompose the carbonaceous matter and a better characterization of the bioproducts of the treatment.

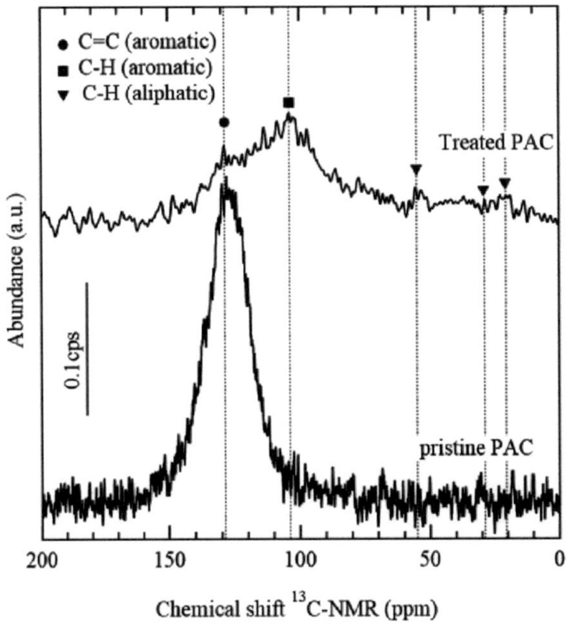

Figure 2.
^{13}C-NMR spectra for powdered activated carbon (PAC) before and after treatment by spent medium of P. chrysosporium *(modified [9]).*

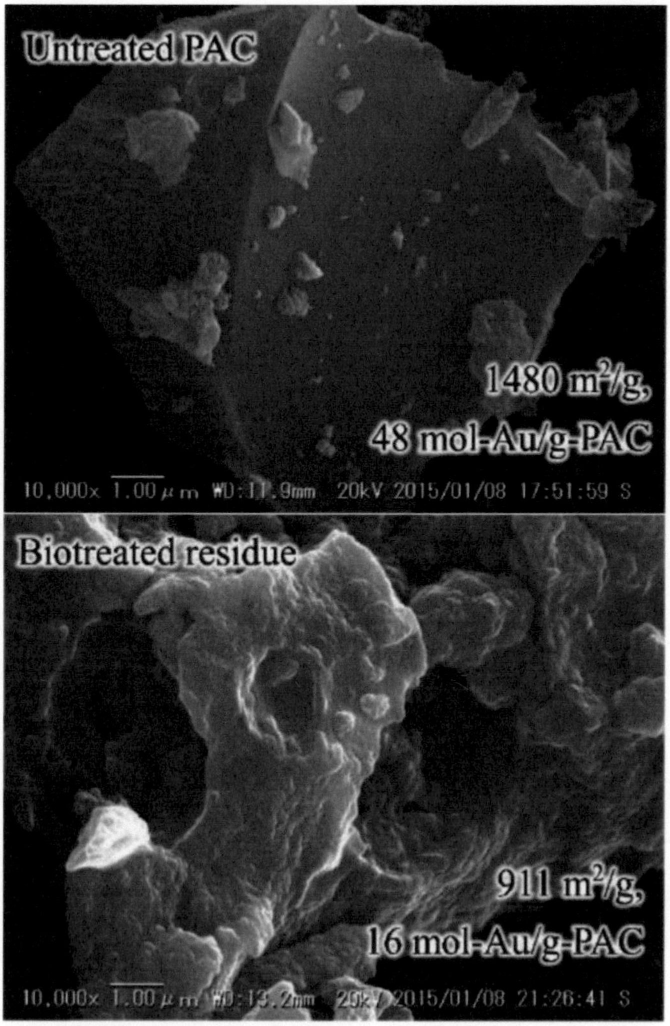

Figure 3.
SEM images of powdery activated carbon before (top) and after (bottom) treatment by spent medium of P. chrysosporium. *Horizontal bars indicate 1.00 um.*

3. Utilization of lignin-degrading enzymes on DRGO

3.1 Decomposition of sulfide minerals

In DRGO, gold (Au(0)) is mainly confined in sulfide minerals, especially arsenopyrite (FeAsS) and pyrite (FeS_2), which must be decomposed to liberate gold. Sulfide minerals in DRGO typically make up 10 ~ 20% after flotation [10–13]. Most of the other major mineral components in are quartz and clay minerals, which do not participate in the oxidative decomposition of sulfide minerals. The standard BIOX treatment of sulfides utilizes chemoautotrophic bacteria *Acidithiobacillus ferrooxidans*, *Acidithiobacillus thiooxidans*, *Leptospirillum ferrooxidans*, heterotrophic bacteria *Sulfolobus acidocaldarius*, *Sulfolobus* sp., *Sulfobacillus* sp., etc., to oxidize and dissolve, or decompose, sulfides by using Fe^{3+} ions as oxidizing agents by oxidizing Fe^{2+} ions in strongly acidic solutions to Fe^{3+} ions [11, 40].

Some white-rot fungi have been shown to improve sulfide oxidation. *P. chrysosporium* produced a mixture of enzymes like LiP and MnP, which needed to be activated by H_2O_2 [18]. Some studies have shown that the H_2O_2 produced by the fungi might be involved in a Fenton reaction with pyrite and arsenopyrite in the DRGO [34–36]. The oxidizing conditions lead to the generation of Fe^{3+} ions as an additional oxidizing agent. However, the competition for H_2O_2 between the lignin-degrading enzymes and the Fenton reaction leads to the inadequate oxidation of both the carbonaceous matter and sulfides [11]. Also, the pH 4–5 used in this treatment results in precipitation of the Fe and reduces the efficacy of Fe^{3+} as an oxidizing agent. Therefore, most studies that have achieved very high gold recoveries have used a sequential treatment.

3.2 Sequential decomposition sulfides minerals and carbonaceous matter

DRGO often contains less than 7% of carbon content, and it is considered to have a more non-uniform structure compared to activated carbon, making it relatively easier to decompose [8]. However, the susceptibility to carbonaceous matter to decomposition by lignin-degrading enzymes should vary depending on the ore. Raman spectroscopy was used to generate **Figure 4**, which shows the difference in graphiticity of carbonaceous matter from various gold ores. The C=C peak for continuous graphitic carbon near 1580 cm^{-1} (G-band) and the C=C peak for vibration adjacent to the graphite defect

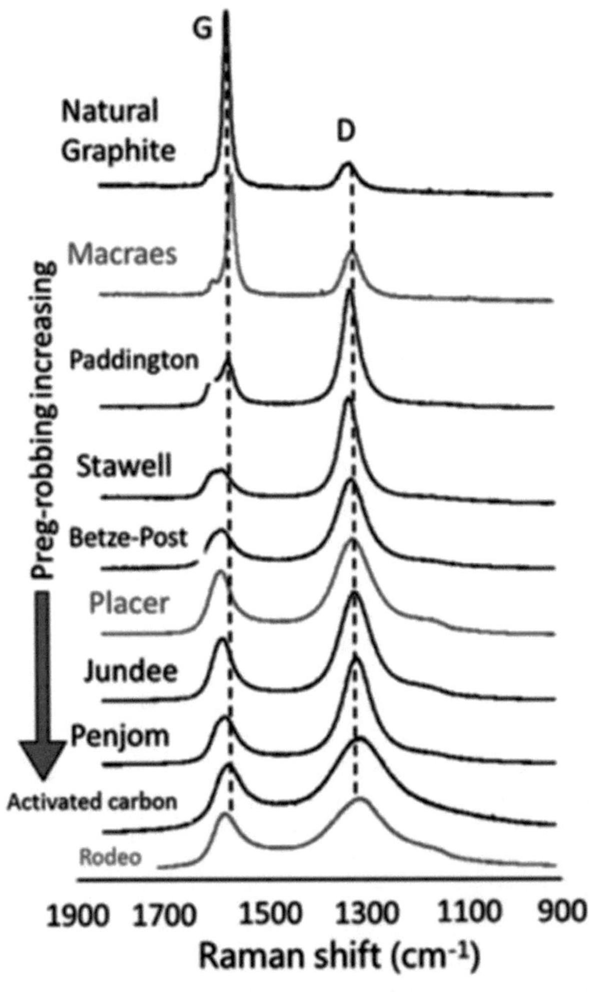

Figure 4.
Raman spectra for DRGOs from different mines (modified [41]).

near 1320 cm^{-1} (D-band) was distinguished, and the relative intensity I_D/I_G and can be used as an index of the degree of the defect [42]. Relative intensity I_D/I_G varies with ore and is an indicator of the preg-robbing ability of the carbonaceous matter. The gold ore from the Paddington mine in Australia has a very high defect compared to carbonaceous matter from the Macraes mine in New Zealand, which has a very aromatic structure. This indicates that the carbonaceous matter from Paddington mine might be more susceptible to lignin-degrading enzymes than the ore from Macraes mine.

The sequential treatment of DRGO has shown significant results, as seen in **Table 1**. Using lignin-degrading enzymes to treat DRGO began with work by Yen et al., [10] who applied the fungi *T. versicolor*, but information about this work is limited because it is a patent. This was followed by Ofori-Sarpong et al. [11] who built open their previous work using *P. chrysosporium* to decompose coals of various ranks [38]. They found that applying the lignin-degrading enzymes directly to the DRGO to decompose both the sulfides and carbonaceous matter improved the gold recovery from 41% - 78% which was slightly less than 81% recovery obtained when the sample was only subjected to sulfide oxidation by chemoautotrophic bacteria. This indicates that the fungal treatment was less effective at oxidizing both refractory materials. Therefore, sequential treatment was used, with the sulfide oxidation by bacteria preceding the carbonaceous matter treatment by *P. chrysosporium*, and they reported a final gold recovery of 94%.

They attempted to further understand the effect of the *P. chrysosporium* on the DRGO by using only the spent medium to treat the DRGO [12]. This was done to check the ability of the lignin-degrading enzymes to oxidize the carbonaceous matter while also avoiding complications like the fungal biomass. This work did not yield as high a gold recovery compared to when the *P. chrysosporium* was cultured with the DRGO, but it did show that it could be viable for DRGO treatment.

Authors	Gold grade (g/t)	Carbon content (%)	Fungus	Treatment conditions	Au recovery (%)	
					Before treatment	After treatment
Yen et al. [10]	—	—	*Trametes versicolor*	Flotation concentrate	54.1–64.5	95.2
Ofori-Sarpong et al. [11]	30.2	3.6	*Phanerochaete chrysosporium*	Flotation concentrate, 30% pulp density, 21 days, pH 4	41 (78)*1	94
Ofori-Sarpong et al. [12]	30.2	3.6	*Phanerochaete chrysosporium* (spent medium)	Flotation concentrate, 30% pulp density, 3 days, pH 4	38	66
Liu et al. [36]	2.18	0.2	*Phanerochaete chrysosporium*	Artificial DRGO	44	62
Konadu et al. [13]	40.4	5.8	*Phanerochaete chrysosporium* (spent medium)	Flotation concentrate, 5% pulp density, 15 days, pH 4	24 (77)*1	92

*1*Numbers in brackets mean the gold recovery after 1st step of microbiological oxidation and before 2nd step of enzyme treatment.*

Table 1.
Summary of the previous works on bio-treatment of carbonaceous matter in DRGO by lignin-degrading enzymes.

To this end, Konadu et al. [13] used the spent medium of *P. chrysosporium* to treat a DRGO, which had a Raman spectrum similar to the Rodeo gold ore in **Figure 4**. It was observed that starting with an iron oxidizer, followed by the spent medium of *P. chrysosporium*, help to improve the gold recovery from 24–92%. Although, an alkaline washing step had to be incorporated after the spent medium treatment to remove some of the byproducts of the carbonaceous matter decomposition to allow for the final 15% recovery to attain overall the 92% recovery. On the other hand, using the spent medium before the iron oxidizer appeared to inhibit sulfide oxidation and lead to a final gold recovery of 45% for that sequence.

The changes in the mineral phases were observed by Quantitative Evaluation of Minerals using Scanning Electron Microscopy (QEMSCAN) [3, 43]. The sulfides (yellow color) were observed to be relatively liberated while the carbonaceous matter was associated with illite as seen in by the dark green color (**Figure 5a**). After the best sequential treatment condition (sulfide oxidation followed by carbonaceous matter treatment by spent medium, DAC) was applied, most of the sulfide minerals were decomposed and disappeared (**Figure 5b**). Additionally, a new mineral phase (C-Si-Al) containing currently unknown amounts of carbon, aluminum and silicon was observed. A backscatter image of two organic carbon-containing particles is

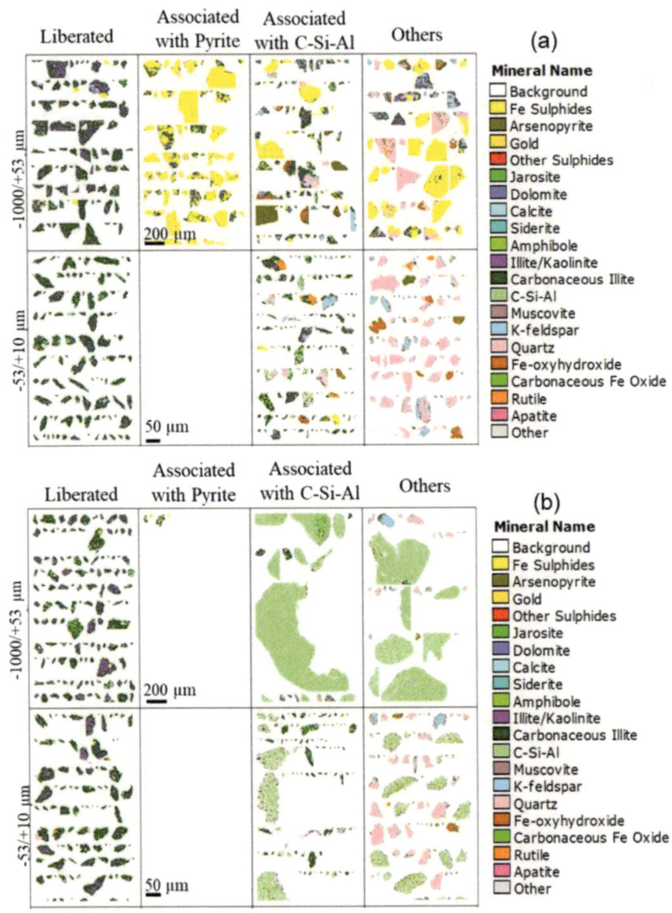

Figure 5.
QEMSCAN maps for DRGO (a) and DAC (b): Particles are classified into liberated carbonaceous illite, associated with illite, associated with quartz, and others according to particle sizes. Notable changes from DRGO to DCA are the disappearance of pyrite and formation of larger particles than 100 µm which are categorized to C-Si-Al, and formation of larger particles than 100 µm which are rich in C, Si and Al (modified [43]).

DOI: http://dx.doi.org/10.5772/intechopen.94334

shown in **Figure 6**. **Figure 6a** shows a particle in which the carbonaceous matter is associated with the illite in the original DRGO while **Figure 6b** shows a residue produced after the sequential treatment was applied (DAC). The sediment-like morphology of particle in **Figure 6b** indicated that it was most likely a product of the carbonaceous matter decomposition.

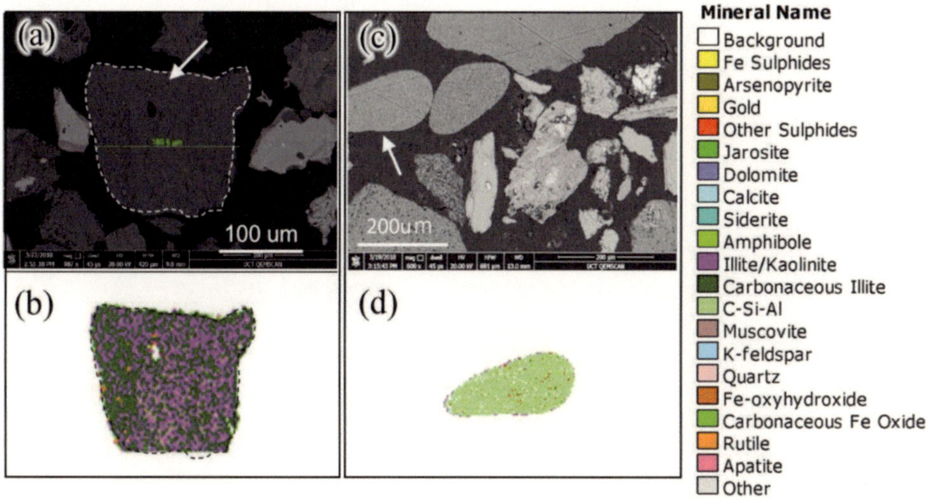

Figure 6.
Backscattering images of (a) DRGO and (c) residue after sequential biotreatment with an iron-oxidizer and spent medium of P. chrysosporium (DAC). Images (b) and (d) are false color renditions of the arrowed particles in (a) and (c), respectively.

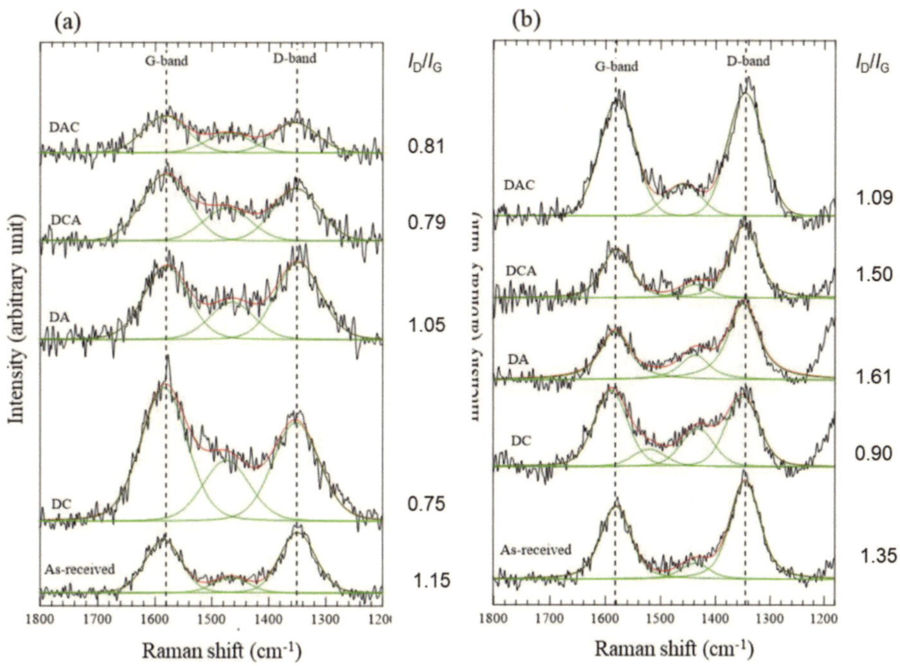

Figure 7.
Raman spectra (a) before and (b) after 1 M NaOH washing of the as-received ore, and the solid residues after treated by CFSM (DC), A. brierleyi (DA), CFSM followed by A. brierleyi (DCA), and A. brierleyi followed by CFSM. Numbers beside figures indicate the intensity ratio (I_D/I_G) for the relative quantity of the defect in all samples with graphitic structures (modified [43]).

Raman spectra of the samples before and after each treatment showed a relative decrease in the D-band compared to the G-band (**Figure 7a**). This change was most significant after the spent medium treatment indicating that the lignin-degrading enzymes might have preferentially oxidized aromatic C=C carbon with some type of physical or chemical defects [43].

After the sequential treatment, the gold recovered before and after the spent medium treatment was unchanged at 77%, indicating that some of the by-products of this treatment were interfering with the recovery (DAC, **Figure 8**). Therefore, 1 M NaOH washing was incorporated into the sequential treatment to remove what

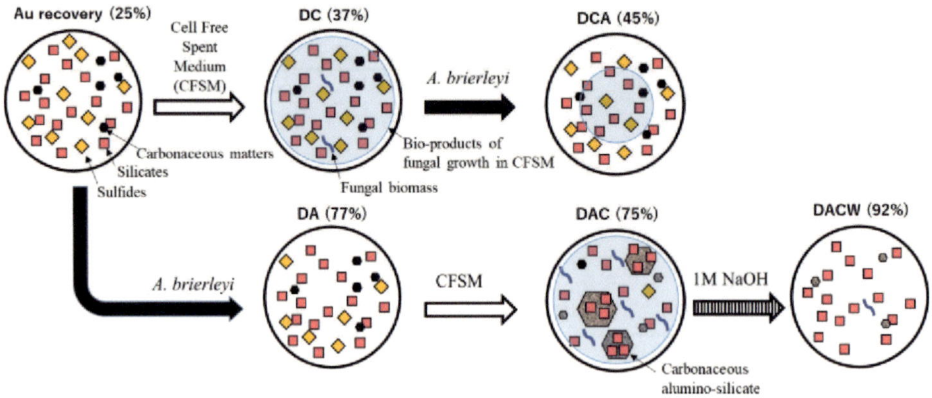

Figure 8.
Gold recovery in each step of sequential biotreatment of DRGO (modified [13]).

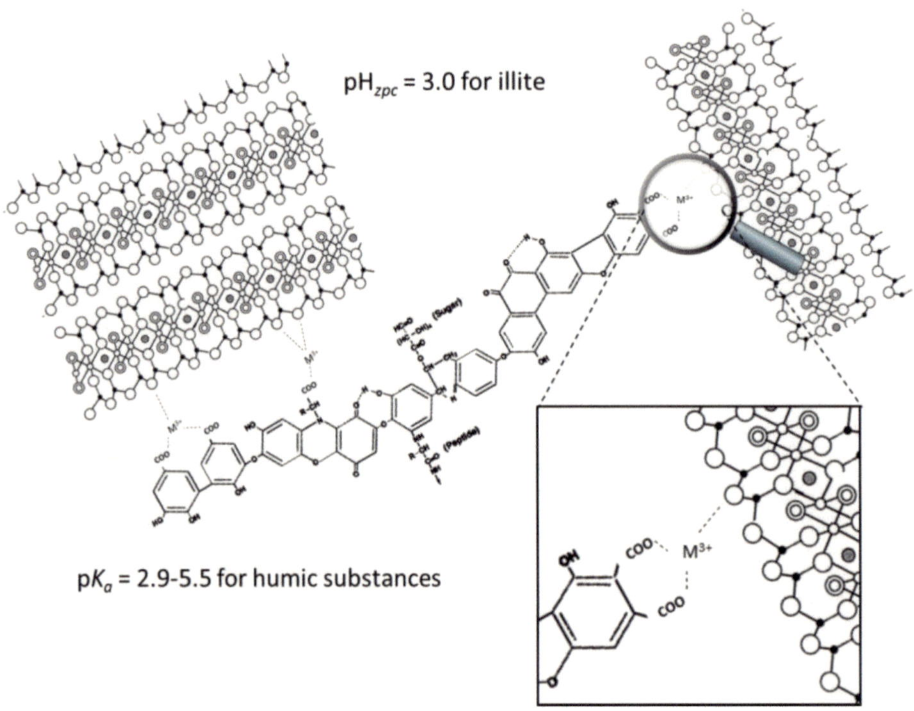

Figure 9.
Schematic illustration of agglomerated particles in DAC after enzyme treatment, which are rich in C, Al and Si. After enzyme treatment, humic-like substances are formed to interact with clay minerals through metallic ions like Fe^{3+}.

was identified by 3D fluorescence spectroscopy as humic-like substances interfering during the gold recovery process, which lead to the observed changes in the Raman spectra and increase in gold recovery to 92% (**Figure 7b**) [43]. Thus, it was proposed that sediment-like C-Si-Al phase could have been produced by an electrostatic aggregation of clay minerals like illite and the humic-like substances. Under the condition of pH 4.0, where the enzyme treatment is performed, the surface of illite is negatively charged (isoelectric point 3.0) (**Figure 9**). While the humic-like substances have a large, irregular structure, and thus, the acid dissociation constant does not have a uniform value and has a width of 2.9–5.5. At pH 4.0, if the humic-like substances have a positive charge, then it might direct form an agglomerate with the negatively charged illite. However, if charged negative, then some of the cations in the solution like Fe^{3+} might be involved in the agglomeration of C-Si-Al.

4. Conclusions

Based on the above, a sequence for DRGO bio-processing was proposed in **Figure 10**. After crushing and flotation, the biooxidation of the sulfide mineral should be performed under strongly acidic conditions to liberate the gold. Afterwards, the spent medium collected from a separate culturing bioreactor for P. chrysosporium will be applied to the residue from the sulfide biooxidation tank. This is to decompose the carbonaceous matter in the sample. The residue will then be sent for alkaline washing to remove the humic-like substances and break apart the C-Si-Al agglomerates before cyanidation to recover gold.

These steps need to be reviewed according to the content and localization of gold in the gold ore, the aromaticity of carbon, and the concentration of soluble iron components in the system. From the Raman spectroscopy analysis of the solid

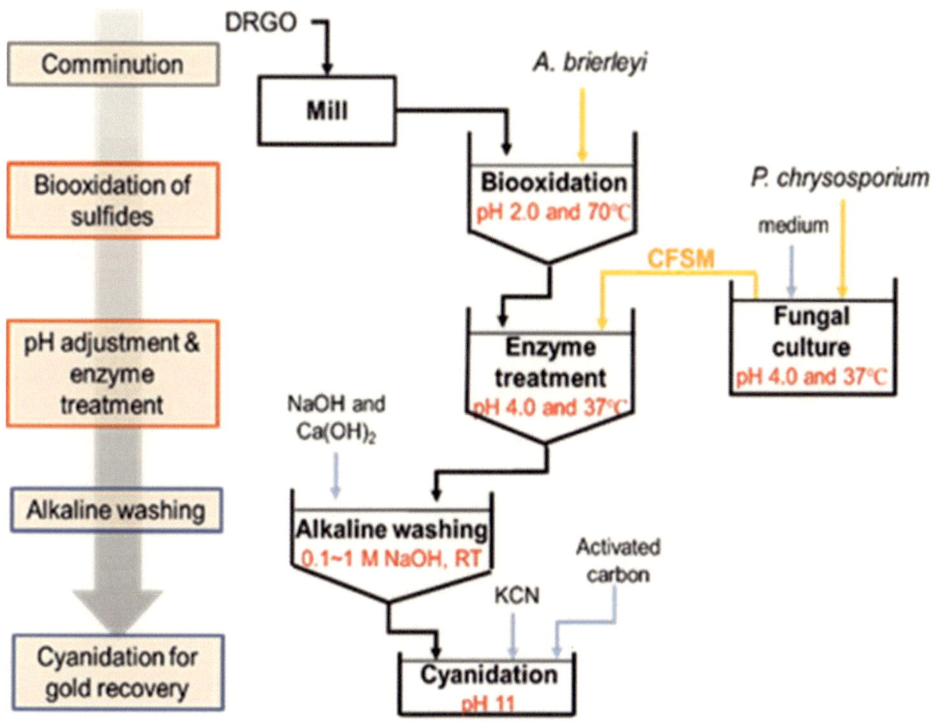

Figure 10.
Proposal of sequential biotreatment process of DRGO.

residue, after the spent medium treatment and alkaline washing, the lignin-degrading enzymes are more effective at decomposing defective graphite, and alkaline washing is more effective in converting graphitic carbon to defective form. Now it is known that it is advantageous for DRGO, which has a high aromatic attribute, to be first alkali-treated before biotreatment.

In the future, the treatment process for various types of carbonaceous gold ore will be organized and targeted for development by utilizing the characteristics of bio-treatment and chemical treatment and will contribute to the production of gold. Additional investigations will be conducted into utilizing less harmful gold lixiviants like amino acids to compared their efficacy against the cyanide system for DRGO processing (**Figure 1**) [44, 45]. Such research, in addition to using lignin-degrading enzymes for biooxidation, will greatly contribute to decreasing the environmental impact of processing DRGO.

Acknowledgements

The authors acknowledge Dr. Kwadwo Osseo-Asare and Dr. Grace Ofori-Sarpong for inspiring this work and providing the fungus, *Phanerochaete chrysosporium*, Dr. Takashi Kaneta for simultaneous determination of LiP and MnP released from *P. chrysosporium*, and Dr. Megan Becker, Dr. Susan Harrison for observing QEMSCAN map.

This work was funded by JSPS KAKENHI JP18K19045, JP 19KK0135, JSPS-NRF Bilateral Program JPJSBP 120196505, and Core-to-Core Program JPJSCCB 20200003.

Author details

Keiko Sasaki* and Kojo T. Konadu
Department of earth Resources Engineering, Kyushu University, Fukuoka, Japan

*Address all correspondence to: keikos@mine.kyushu-u.ac.jp

References

[1] Osseo-Asare K, Afenya PM, Abotsi GM. Carbonaceous matter in gold ores: Isolation, characterization and adsorption behavior in aurocyanide solutions. Proc. Met.: Min. Extrac. Process. 1984:125-144

[2] Cook NJ, Chryssoulis SL. Concentrations of invisible gold in the common sulfides. The Canadian Mineralogist. 1990;**28**:1-16

[3] Goodall WR, Scales PJ, Butcher AR. The use of QEMSCAN and diagnostic leaching in the characterization of visible gold in complex ores. Minerals Engineering. 2005;**18**:877-886. DOI: https://doi.org/10.1016/j.mineng.2005.01.018

[4] Afenya PM. Treatment of carbonaceous refractory gold ores. Minerals Engineering. 1991;**4**:1043-1055. DOI: https://doi.org/10.1016/0892-6875(91)90082-7

[5] Amankwah RK, Yen WT, Ramsay JA. A two-stage bacterial pretreatment process for double refractory gold ores. Minerals Engineering. 2005;**18**:103-108. DOI: https://doi.org/10.1016/j.mineng.2004.05.009

[6] Miller JD, Wan RY, Diaz X. Preg-robbing gold ores. Gold Ore Processing. 2016:885-907. DOI: https://doi.org/10.1016/B978-0-444-63658-4.00049-9

[7] Marsden J, House I. The chemistry of gold extraction. SME. 2006:147-224

[8] Yang HY, Qian LI, Song XL, Dong JK. Research status of carbonaceous matter in carbonaceous gold ores and bio-oxidation pretreatment. Transactions of the Nonferrous Metals Society of China. 2013;**23**:3405-3411. DOI: https://doi.org/10.1016/S1003-6326(13)62881-2

[9] Konadu KT, Sasaki K, Kaneta T, Ofori-Sarpong G, Osseo-Asare K. Bio-modification of carbonaceous matter in gold ores: Model experiments using powdered activated carbon and cell-free spent medium of *Phanerochaete chrysosporium*. Hydrometallurgy. 2017;**168**:76-83. DOI: https://doi.org/10.1016/j.hydromet.2016.08.003

[10] Yen WT, Amankwah RK, Choi Y. Inventors; microbial pre-treatment of double refractory gold ores. United States patent application US. 2009;**12**(/277):046

[11] Ofori-Sarpong G, Osseo-Asare K, Tien M. Mycohydrometallurgy: Biotransformation of double refractory gold ores by the fungus, *Phanerochaete chrysosporium*. Hydrometallurgy. 2013;**137**:38-44. DOI: https://doi.org/10.1016/j.hydromet.2013.05.003

[12] Ofori-Sarpong G, Osseo-Asare K, Tien M. Pretreatment of refractory gold ores using cell-free extracts of *P. chrysosporium*: A preliminary study. Advances in Materials Research. 2013;**825**:427-430. DOI: https://doi.org/10.4028/www.scientific.net/AMR.825.427

[13] Konadu KT, Huddy RJ, Harrison ST, Osseo-Asare K, Sasaki K. Sequential pretreatment of double refractory gold ore (DRGO) with a thermophilic iron oxidizing archeaon and fungal crude enzymes. Minerals Engineering. 2019;**138**:86-94. DOI: https://doi.org/10.1016/j.mineng.2019.04.043

[14] Doucet D, Girard D, Grondin L, Matte P. Technical Report on the December 31, 2009. Mineral Resource and Mineral Reserve Estimate and the Suuri Extension Project, Kittila Mine, Finland. 2010:81-83.

[15] Konieczny A, Pawlos W, Krzeminska M, Kaleta R, Kurzydlo P.

Evaluation of organic carbon separation from copper ore by pre-flotation. Physicochem. Problems Miner. Process. 2013;**49**:189-201. DOI: https://doi.org/10.5277/ppmp130117

[16] Mbamba CK, Franzidis JP, Harrison ST. Broadhurst JL. Flotation of coal and sulphur from South African ultrafine colliery wastes. 2013;**113**:399-405 http://hdl.handle.net/11427/21436

[17] Thella J. The influence of mineralogy and surface chemistry on flotation of Cortez complex carbonaceous double refractory gold ore, PhD dissertation, uni. Queensland. 2018:133-155. DOI: https://doi.org/10.14264/uql.2018.795

[18] Arriagada FJ, Osseo-Asare KA. Gold extraction from refractory ores: Roasting behavior of pyrite and arsenopyrite. Metallurg. Soc of AIME. 1984:367-385

[19] La Brooy SR, Linge HG, Walker GS. Review of gold extraction from ores. Minerals Engineering. 1994;7:1213-1241. DOI: https://doi.org/10.1016/0892-6875(94)90114-7

[20] Hammerschmidt J, Güntner J, Kerstiens B. Roasting of gold ore in the circulating fluidized-bed technology. Developments in mineral processing. 2005; 15:433-453. https://doi.org/10.1016/S0167-4528(05)15018-2

[21] Thomas KG, Cole AP. Roasting developments–especially oxygenated roasting. Gold Ore Processing. 2016:373-392. DOI: https://doi.org/10.1016/B978-0-444-63658-4.00023-2

[22] Li H, Ma A, Srinivasakannan C, Zhang L, Li S, Yin S. Investigation on the recovery of gold and silver from cyanide tailings using chlorination roasting process. Journal of Alloys and Compounds. 2018;**763**:241-249. DOI: https://doi.org/10.1016/j.jallcom.2018.05.298

[23] Li H, Zhang L, Koppala S, Ma A, Peng J, Li S, et al. Extraction of gold and silver in the selective chlorination roasting process of cyanidation tailing. Separation Science and Technology. 2018;**53**:458-466. DOI: https://doi.org/10.1080/01496395.2017.1388258

[24] Abotsi GM, Osseo-Asare K. Surface chemistry of carbonaceous gold ores, II. Effects of organic additives on gold adsorption from cyanide solution. Int. J. Miner, Process. 1987;**21**:225-239. DOI: https://doi.org/10.1016/0301-7516(87)90056-1

[25] Crowe JD, Olsson S. Induction of laccase activity in *Rhizoctonia solani* by antagonistic *Pseudomonas fluorescens* strains and a range of chemical treatments. App. Environ. Microbiol. 2001;**67**:2088-2094. DOI: 10.1128/AEM.67.5.2088-2094.2001

[26] Kalme S, Jadhav S, Jadhav M, Govindwar S. Textile dye degrading laccase from *Pseudomonas desmolyticum* NCIM 2112. Enzyme and Microbial Technology. 2009;**44**:65-71. DOI: https://doi.org/10.1016/j.enzmictec.2008.10.005

[27] Kuddus M, Joseph B, Ramteke PW. Production of laccase from newly isolated *Pseudomonas putida* and its application in bioremediation of synthetic dyes and industrial effluents. Biocatalysis and Agricultural Biotechnology. 2013;**2**:333-338. DOI: https://doi.org/10.1016/j.bcab.2013.06.002

[28] Ten Have R, Teunissen PJ. Oxidative mechanisms involved in lignin degradation by white-rot fungi. Chemical Reviews. 2001;**101**:397-414. DOI: 10.1021/cr000115l

[29] Tien M, Kirk TK. Lignin peroxidase of *Phanerochaete chrysosporium*. Methods in Enzymology. 1988;**161**:238-249. DOI: https://doi.org/10.1016/0076-6879(88)61025-1

[30] Wariishi H, Valli K, Gold MH. In vitro depolymerization of lignin by manganese peroxidase of *Phanerochaete chrysosporium*. Biochem. Biophy. Res. Comm. 1991;**176**:269-275. DOI: https://doi.org/10.1016/0006-291X(91)90919-X

[31] Camarero S, Sarkar S, Ruiz-Dueñas FJ. Martínez MJ, Martínez ÁT. Description of a versatile peroxidase involved in the natural degradation of lignin that has both manganese peroxidase and lignin peroxidase substrate interaction sites. The Journal of Biological Chemistry. 1999;**274**:10324-10330

[32] Koduri RS, Tien M. Kinetic analysis of lignin peroxidase: Explanation for the mediation phenomenon by veratryl alcohol. The Biochemist. 1994;**33**:4225-4230. DOI: https://doi.org/10.1021/bi00180a016

[33] Morozova OV, Shumakovich GP, Shleev SV, Yaropolov YI. Laccase-mediator systems and their applications: A review. App. Biochem. Microbio. 2007;**43**:523-535. DOI: https://doi.org/10.1134/S0003683807050055

[34] Ofori-Sarpong G, Osseo-Asare K, Tien M. Fungal pretreatment of sulfides in refractory gold ores. Minerals Engineering. 2011;**24**:499-504. DOI: https://doi.org/10.1016/j.mineng.2011.02.020

[35] Yang HY, Qian LI, Chen GB, Tong LL, Auwalu AL. Bio-dissolution of pyrite by *Phanerochaete chrysosporium*. Transactions of Nonferrous Metals Society of China. 2018;**28**:766-774. DOI: https://doi.org/10.1016/S1003-6326(18)64709-0

[36] Liu Q, Yang HY, Qiao LL. Biotransformation of arsenopyrite by *Phanerochaete chrysosporium*. Advances in Materials Research. 2013;**825**:309-313. DOI: https://doi.org/10.4028/www.scientific.net/AMR.825.309

[37] Duran N, Esposito E. Potential applications of oxidative enzymes and phenoloxidase-like compounds in wastewater and soil treatment: a review. App. Catal. B: Environ. 2000:28:83-99. https://doi.org/10.1016/S0926-3373(00)00168-5

[38] Ofori-Sarpong G, Tien M, Osseo-Asare K. Myco-hydrometallurgy: Coal model for potential reduction of preg-robbing capacity of carbonaceous gold ores using the fungus, *Phanerochaete chrysosporium*. Hydrometallurgy. 2010;**102**:66-72. DOI: https://doi.org/10.1016/j.hydromet.2010.02.007

[39] Liu Q, Yang HY, Tong LL, Jin ZN, Sand W. Fungal degradation of elemental carbon in carbonaceous gold ore. Hydrometallurgy. 2016;**160**:90-97. DOI: https://doi.org/10.1016/j.hydromet.2015.12.012

[40] Ngubane WT, Baecker AA. Oxidation of gold-bearing pyrite and arsenopyrite by *Sulfolobus acidocaldarius* and *Sulfolobus BC* in airlift bioreactors. Biorecovery. 1990;**1**:255-269

[41] Konadu KT, Mendoza DM, Huddy RJ, Harrison ST, Kaneta T, Sasaki K. Biological pretreatment of carbonaceous matter in double refractory gold ores: A review and some future considerations. Hydrometallurgy 2020:105434. https://doi.org/10.1016/j.hydromet.2020.105434

[42] Ferrari AC. Raman spectroscopy of graphene and graphite: Disorder, electron–phonon coupling, doping and nonadiabatic effects. Solid State Comm. 2007;**143**:47-57. DOI: https://doi.org/10.1016/j.ssc.2007.03.052

[43] Konadu KT, Harrison ST, Osseo-Asare K, Sasaki K. Transformation of the carbonaceous matter in double refractory gold ore by crude lignin peroxidase released from the white-rot fungus. Int. Biodeter.

Biodegrad. 2019:143:104735. https://doi.org/10.1016/j.ibiod.2019.104735

[44] Oraby EA, Eksteen JJ. The leaching of gold, silver and their alloys in alkaline glycine–peroxide solutions and their adsorption on carbon. Hydrometallurgy. 2015;**152**:199-203. DOI: https://doi.org/10.1016/j.hydromet.2014.12.015

[45] Tauetsile PJ, Oraby EA, Eksteen JJ. Activated carbon adsorption of gold from cyanide-starved glycine solutions containing copper. Part 1: Isotherms. Separation and Purification Technology. 2019;**211**:594-601. DOI: https://doi.org/10.1016/j.seppur.2018.09.024

Chapter 12

Rare Earth Elements Biorecovery from Mineral Ores and Industrial Wastes

Laura Castro, M. Luisa Blázquez, Felisa González and Jesús A. Muñoz

Abstract

Rare earth elements (REEs) are critical raw materials and are attracting interest because of their applications in novel technologies and green economy. Biohydrometallurgy has been used to extract other base metals; however, bioleaching studies of REE mineral extraction from mineral ores and wastes are yet in their infancy. Mineral ores have been treated with a variety of microorganisms. Phosphate-solubilizing microorganims are particularly relevant in the bioleaching of monazite because transform insoluble phosphate into more soluble form which directly and/or indirectly contributes to their metabolism. The increase of wastes containing REEs turns them into an important alternative source. The application of bioleaching techniques to the treatment of solid wastes might contribute to the conversion towards a more sustainable and environmental friendly economy minimizing the amount of tailings or residues that exert a harmful impact on the environment.

Keywords: bioleaching, rare earth elements, recycling, wastes, minerals

1. Introduction

Rare earth elements (REEs) are strategic materials in order to facilitate the transition from current economy based on petroleum to an efficient circular economy based on clean energy. Although often needed in small quantities, these metals are essential to produce a huge number of technologically sophisticated products for electronic, optical and magnetic applications. Among other applications, these elements play a crucial role in the development of clean emerging low-carbon energy technologies [1].

In spite of the archaic term, most of rare earths are common elements and some of them are even more abundant than other metals, such as copper or lead. Though moderately abundant in the Earth's crust, rare earth elements are scarcely concentrated in mineral deposits and this fact complicates its Extractive Metallurgy which is complex and requires economic solutions. The world production of REEs has undergone an exponential growth since its discovery in the 18th century, with a notably increase overtime from 1,000 t in 1930 to 133,600 t in 2010 [2]. The rising REEs production has been the consequence of an escalating demand for REEs as well.

IntechOpen

Based on their strong affinity for oxygen, REEs resources are mostly present in oxidic form, mainly as rare earth oxides, phosphates, carbonates and silicates. According to recent estimates, 100 Mt of rare earth oxides are accessible in more than thirty countries all around the world. More than 200 REEs ores have been identified as rock-forming minerals, however, only three are considered mineral ores for economic extraction: bastnasite ($(Ce,La)(CO_3)F$), monazite ($(Ce,La,Nd,Th)PO_4$) and xenotime (YPO_4) [3]. Thus, the primary sources of REEs are mineral deposits spread out worldwide, but confined mainly in China, Australia and USA.

Furthermore, REEs are also found in industrial wastes in vast amounts and they have been investigated as potential resources for rare earth metals [4–6]. Among REEs-bearing industrial residues, phosphogypsum, generated during the wet phosphoric acid process from fertilizer production, and red mud residues from the digestion of bauxites in the Bayer are rich in valuable rear earth metals as to be economically treated.

In addition, end-of-life materials can be recycled due to their significant quantities of REE, among them: magnets (38%), lamp phosphors (32%) and metal alloys (13%), retain more than 80% of REE market. Modern fluorescent lamps typically contain more than 20% (w/w) REE (Ce, Eu, La, Tb and Y) [7].

After ore and/or industrial waste concentration processing, rare earth metals are dissolved selectively from raw materials. Actinides, such as uranium and thorium, with similar chemical properties to REEs, are often co-dissolved during hydrometallurgical processes and this could pose a problem. REE primary ores are leached using acid (H_2SO_4, HCl, HNO_3, H_3PO_4) or alkaline (Na_2CO_3, $NaHCO_3$) reagents and $NaCl$ or $(NH_4)_2SO_4$ for REE-ion adsorbed clays [8–10]. Nonetheless, the hydrometallurgical treatment is ore-dependent and has been well established in the case of some REE ores, especially monazite, but is less evident for other key minerals with a very complex mineralogy.

Biohydrometallurgy and more specifically its application to the extraction of metals through bioleaching processes have been successfully practiced at industrial level for the recovery of uranium, copper and gold [11, 12]. Biohydrometallurgical technologies could play a fundamental role for the treatment of REE-bearing materials since they offer an alternative to physico-chemically based methods. Bioleaching is connected to the development of more cost-effective, less energy demanding and less polluting metal extraction processes than pyro- and hydrometallurgical processes and is able to treat low-grade mineral ores and a great variety of residues. These biotechnological processes involve interactions between microorganisms and metal-bearing ores that render valuable metals in solution. REE mobilization from solid matrices has been done with a wide range of microorganisms, both autotrophic and heterotrophic, and using both pure and mixed microbial cultures [13–15].

This chapter provide an insight into the global situation of REEs and the potential application of microorganisms in the extraction of REEs from both REEs-bearing minerals and industrial residues.

2. Global situation of REEs: market and technology

The demand of REEs have increased in the past decade because of their extensive use in several fields related to electronics, in renewable energy capture technologies, biomedical devices, and other industrial products. In 2018, a list of 35 critical minerals, including rare earth elements, was published by the U.S. Department of the Interior and other executive branch agencies [16]. Likewise, the European

Commission developed a critical assessment on non-energy and non-agricultural raw materials in 2017 including heavy rare earth elements, light rare earth elements and platinum group metals [17].

Global mine production was estimated to be 210,000 tons of rare earth oxide (REO) equivalent, which means an 11% increase in comparison with 2018 (**Figure 1a**). China dominates the global production of rare earth minerals, separated compounds and metals. China exports REEs to United States (31%), Japan (27%), the Republic of Korea (11%), the Netherlands (9%) and Germany (6%).

Other countries are making efforts to increase their domestic production of mineral concentrates. For example, United States enhanced the production, all of which was exported, a 44% in 2019 compared with 2018 [18].

Rare earths are relatively abundant in the Earth's crust; however, REEs resources with minable concentrations are less common. Nowadays, about 850 REE deposits have been identified worldwide, which are mainly located in China, Vietnam, Brazil, Russia, India and Australia (**Figure 1b**) [18, 19].

Prices for most rare earth products are increasing compared with those in 2016 reversing the falling trend that began after prices spikes in 2011. Gadolinium, praseodymium and neodymium experienced the greatest increase in the price, while the yttrium and dysprosium prices decreased. The estimated unit value of rare-earth compounds was $11.60 per kg in 2017 based on information from the U.S. Census Bureau on imports [20].

The estimation of global consumption of rare earth varies significantly due to the limited data transparency and it generally ranges about 140,000 and 170,000 tons of REO equivalent. Furthermore, the global consumption of scandium was estimated in 10–20 tons per year [21].

The amount of specific REEs used strongly depends on the market sector and application. Lanthanum and cerium, and lower amounts of neodymium, are consumed in the catalysts sector. There are different types of permanent magnets but neodymium-iron-boron magnets are fabricated with neodymium and praseodymium, and samarium-cobalt magnets uses samarium and gadolinium. Batteries contain mainly lanthanum, and ceramics, yttrium. Europium, yttrium and terbium are commonly associated with the phosphors sector.

The global growth rate of REEs consumption is expected to be 5–7% per year through 2022. The magnet materials sector would grow more than other sector such as catalysts, ceramic or phosphors. The increasing global demand of REEs as well as the enforcement of environmental and production legislation beyond 2022 lead to higher prices and, consequently, this situation may drive the mining and processing development outside China.

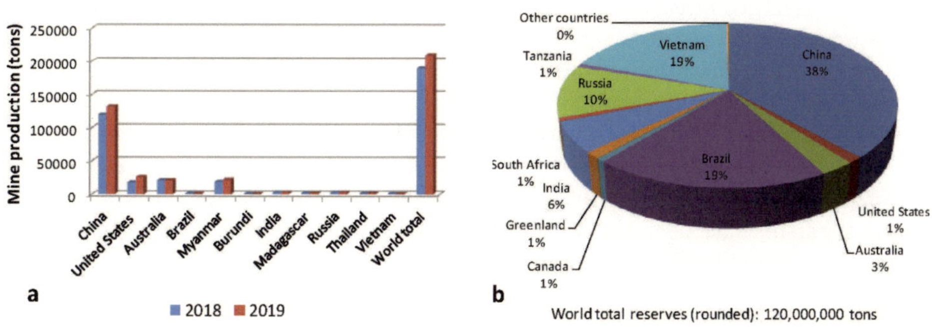

Figure 1.
World mine production (a) and reserves (b).

3. Bioleaching of REEs mineral ores

The application of biohydrometallurgy for extracting REEs offer a green alternative to the conventional methods, which are complex and energy intensive. The main economic REEs-bearing minerals are bastnaesite, loparite, monazite, xenotime, and the lateric ion-adsorption clays. Both autotrophic and heterotrophic microorganisms are capable of solubilize REEs and the selection of these microbes for bioleaching processes depends on the type of mineral. Autotrophic bacteria have been employed for the extraction of scandium from ore minerals containing metal sulfides, whereas heterotrophic bacteria are mostly employed for REEs carbonates and phosphates [22].

Bastnaesite is a rare earth fluoro-carbonate ore ($REE(CO_3)F$) containing commonly cerium, lanthanum or yttrium. REE fluorides from bastnaesite are removed by sulfuric acid roasting (400–500°C) emitting CO_2 and HF gases or alkaline pre-treatment (96°C). An alternative method consist on a thermal activation step at 400°C followed by HCl leaching that reduce the release of fluorine but not carbonate [23]. Until now only one attempt has been made to leach biologically bastnaesite minerals. Four actinobacterial strains were able to leach REEs from a bastnaesite-bearing rock and bastnaesite reaching. a total REEs concentration that ranged from 56 to 342 μg/l when grew in a nutrient-rich medium. Only the strain Streptomyces sp. grew in an oligotrophic medium in the presence of the bastnaesite-bearing rock and bastnaesite, leaching 548 μg/l of REEs. The leaching efficiency of total REEs was very low (0.008–0.08%); however, this amount should be due to the REEs precipitation and/or sorption onto residual rock and cell surfaces. Several molecules, such as organic acids, complexing ligands, and siderophores secreted by the actinobacteria, are involved in the mechanism of REEs bioleaching from bastnaesite-bearing rock [13].

Xenotime is a phosphate ore and its content of yttrium is high and the presence of other light rare earths is low. Furthermore, the composition of xenotime is very complicated containing large amount of tungsten and other impurities such as iron and silicon. Generally, xenotime is more difficult to decompose than monazite and the mineral has been treated using concentrated sulfuric acid, alkaline solution at elevated pressure, fusing with molten caustic soda, and mixing with sodium carbonate and roasting [24]. These technological difficulties are probably linked to the fact that there are no biological approaches for REEs leaching from xenotime yet.

Monazite is also a phosphate mineral and is the major commercial source of cerium. Alkali treatment is also used for monazite dissolution and the phosphate is recovered as a marketable by-product, trisodium phosphate, by using caustic soda at high temperature and high pressure.

The bioleaching of REEs has been developed especially on monazite using microorganisms able to dissolve phosphorous from inorganic rocks, named phosphate solubilizing microorganisms (**Figure 2**). Numerous organisms including bacteria, fungi, actinomycetes and algae mobilize insoluble phosphorous. Bacterial species belonging to the genera *Bacillus*, *Pseudomonas*, *Burkholderia*, *Enterobacter* or *Azotobacter*, and fungi species of the genus *Aspergillus*, *Penicillium*, *Trichoderma* or *Rhizoctonia* are reported as phosphate solubilizers [25]. Phosphate solubilizing microorganisms have been previously used as biofertilizers in agriculture to promote crop production [26]; nevertheless, there are few works related to the recovery of valuable metals from phosphate minerals [27, 28].

Bacteria such as *Enterobacter aerogenes*, *Pantoea agglomerans* and *Pseudomonas putida* are able to grow in presence of natural rare earth phosphate minerals dissolving phosphorous, iron, thorium and REEs generating numerous organic acids [3].

Figure 2.
SEM image of phosphate solubilizing bacteria grown on monazite.

Species such as *Azospirillum brasilense*, *Azospirillum lipoferum*, *Pseudomonas rhizosphaerae* and *Mesorhizobium cicero*, but especially *Acetobacter aceti*, solubilize cerium and lanthanum from monazite, although the efficiency of the process was low [28].

Several fungal strains able to solubilize phosphate minerals have been also used to leach monazite releasing rare earth elements to the aqueous phase, such as *Aspergillus niger* ATCC 1015, *Aspergillus terreus* strain ML3-1 and a *Paecilomyces* spp. strain WE3-F [14]. *Aspergillus niger* has been extensively studied because it generates organic acids such as gluconic, citric and oxalic acids [29, 30]. Nevertheless, the formation of REE oxalates limits the long-term dissolution of these elements [31].

The fungus *Penicillum* sp. reached a total concentration of 12.32 mg l^{-1} REEs in aqueous phase after 8 days. Although monazite also contains radioactive thorium, bioleaching by these fungi preferentially solubilized rare earth elements over thorium that remained in the solid residual [3]. Fungus from the genus *Paecilomyces* produces several metabolites potentially associated with bioleaching, and citric and citramalic acids present a significant contribution to REEs dissolution, 6.4 and 15.0 mg l^{-1} total REEs solubilized respectively. Furthermore, citramalic acid exhibits more interesting properties to those of citric acid due to the lower radioactive thorium release [32].

Not only heterotrophic but also autotrophic bacteria has been used for the treatment of monazite. The bioleaching of REEs using *Enterobacter aerogenes* and *Acidithiobacillus ferrooxidans* from various grades of monazite and the effect of glycine have been examined. Nevertheless, a decreased REEs bioleaching was observed in comparison to abiotic leaching or bioleaching in the absence of glycine [33].

Previous works have determined remarkable alterations in the natural microbial populations during bioleaching of monazite ores. The existence of native *Firmicutes* on the monazite seems to have significantly contributed to the increase REEs leaching observed when using non-sterilized monazite. A syntrophic effect between the native consortia and introduced microorganisms leads to a greater amount of leached REEs reaching 23.7 mg l^{-1} [34, 35].

There are a number of mechanisms explaining the inorganic phosphate solubilization. The main mechanism is the production of biological dissolving compounds such as organic acids, siderophores, extracellular polymeric substances, protons, hydroxyl ions and CO_2.

The production of organic acids leads to the chelation of cations with their carboxyl and hydroxyl groups and/or to the reduction of the pH releasing phosphorous. The decrease of the pH results in the release of phosporous by substitution of H^+ for cations in the rocks [36]. Nevertheless, there is no correlation between pH and the concentration of solubilized phosphate and the cation assimilation is also required in the process [37] or the action of the H^+ traslocation ATPase [38].

Gluconic acid is the most common organic acid involved in the mineral phosphate solubilization. This acid is generated by bacteria by direct oxidation of glucose and chelates the cations bound to phosphates [3]. Microorganisms not only produce organic acids but also inorganic acids and siderophores; however, these mechanisms are less effective in the release of phosphorous.

Extracellular polymeric substances have an indirect impact on phosphate solubilization due to their ability to bind metals influencing solubility of metal phosphates in soil. However, further investigation is needed to elucidate the role of high-molecular-weight polysaccharides in phosphate mobilization [39, 40].

Microbial phosphate solubilization can also take place through the liberation of enzymes like phosphatase or phosphohydrolase, phytase, phosphonatase, and C–P lyase. For example, phosphatase enzymes transform high-molecular-weight organic phosphate into low-molecular-weight products by the hydrolysis of ester phosphate bonds, releasing phosphate ions [41].

Despite of the recent work performed in this field, the selection of novel strains and a deeper study of bioleaching mechanisms is required to optimize the extraction of REEs. Biotechnology advances could have a key role in the development of cleaner strategies for the recovery of REEs.

4. Biorecovery of REEs from industrial and electronic wastes

Base metals as well as precious metals recycling have achieved high rates; however, the recycling rates of REEs are still very low (<1%). The low recycling of REEs can be explained by different factors, such as technological difficulties, low toxicities of the REEs, and, until few years ago, low prices and lack of incentives. The technological issues of the recycling of rare earths are due to the low concentrations of these elements in consumer goods.

The growing generation of industrial and electronic wastes and its significant content in critical metals has become these materials in an alternative economic source for the recovery of REE. Recently, secondary sources of REE including industrial wastes, mine wastes, and electronic wastes are being treated using bioprocess technology for the metal recovery. Nevertheless, bioleaching studies of REEs extraction from wastes are yet in their infancy.

Among REE-bearing industrial residues, phosphogypsum is worthy to be mentioned. REEs are often associated with phosphate deposits and phosphogypsum wastes are generated during the wet phosphoric acid process from fertilizer production in large amounts (100–280 Mt per year) with an estimate of 21 Mt of REEs locked into the total of phosphogypsum wastes accumulated to date [2]. A biolixiviant produced by the growth of the bacterium *Gluconobacter oxydans* on glucose containing organic acids was used for the leaching of synthetic phosphogypsum doped with six rare earth elements (yttrium, cerium, neodymium, samarium, europium and ytterbium). The pH of the lixiviant generated by the bacteria was 2.1 and the main organic acid present in the bioreagent was gluconic acid (220 mM). REEs leaching yield from phosphogypsum using the biolixiviant was compared with the REEs leaching results using sulfuric acid, phosphoric acid, and commercial gluconic acid. The lixiviant produced by *Gluconobacter oxydans* was more efficient at REEs disolution than gluconic acid and phosphoric acid but less efficient than sulfuric acid [42].

Other important waste material as REE resource is the red mud from the digestion of bauxites in the Bayer process. According to estimates, about 2700 Mt of red mud residues have been accumulated in Bayer plants all over the world and its generation increases at a rate of 120 Mt per year [43]. These residues are harmful due to its alkalinity, but also are rich in valuable rare earth metals as to be economically treated. Scandium represents about 95% of the economic value of the REEs present in red mud containing between 130 and 390 ppm [44].

Penicillium tricolor RM-10, a filamentous acid-producing fungi isolated from red mud, was used in bioleaching experiments. The optimum leaching ratios of the REEs were obtained under one-step bioleaching process at 2% pulp density due to the production of citric and oxalic acids. Nevertheless, the highest extraction yields were achieved under two-step process at 10% (w/v) pulp density [45]. The chemo-heterotrophic bacterium *Acetobacter* sp. leached 53% of Lu, 61% of Y, and 52% of Sc under one-step process at 2% pulp density. In addition, this bacterium was suitable for red mud bioleaching at high pulp densities, because the production of organic acids excreted by the bacteria increased with the waste concentration [46].

Spent cracking catalysts are solid wastes generated in large amounts in oil refining and biocombustible industries. It is estimated that 700,000–900,000 tons of spent fluid catalytic cracking catalyst per year are generated worldwide and the management of this waste and the recycling of rare earth metals have become a challenge [47]. Cerium and especially lanthanum are the main REEs present in cracking catalyst. Cell-free culture supernatants of *Gluconobacter oxydans* containing gluconic acid were used for REEs leaching from spent cracking catalysts and dissolved 49% of the total REE with preferential recovery of lanthanum over cerium [4]. Furthermore, several studies have optimized this biolixiviant reaching a yield up to 56% and a continuous bioreactor system was developed achieving leaching efficiencies of 51% [48]. In addition, techno-economic analysis indicated that the use of agricultural wastes as substrate for bacterial growth instead of glucose provides a cost-effective process REEs recovery [49].

Bioleaching of spent cracking catalysts using the fungus *Aspergillus niger* at 1, 3 and 5% pulp densities achieved 63%, 52%, and 33% of lanthanum recovery, respectively; while the cell-free supernatant at 1% pulp density resulted in 30.8% leaching recovery. Furthermore, chemical leaching by strong inorganic (hydrochloric, sulfuric and nitric) and organic (citric, gluconic and oxalic) acids at 100 mM was evaluated. In chemical leaching, sulfuric, nitric and oxalic acids showed lower efficiency than bioleaching. Hydrochloric, citric and gluconic acids dissolved 68%, 65%, and 64% of REEs, respectively, which is nearly the same results than bioleaching with the fungus [50].

The recovery of REEs from fluid catalytic cracking spent catalyst by biochemical processes using *Yarrowia lipolytica* while also examining a route for the valorization of biodiesel-derived glycerin, which is the main carbon source, has been investigated. Remarkable bioleaching yields were obtained, 53% of La, and 99% of Ce and Nd, using *Y. lipolytica* IM-UFRJ 50678 at 50°C [51].

Compact fluorescent lamps contain on average glass (88 wt.%), metals (5 wt.%), plastic (4 wt.%), lamp phosphor powder (3 wt.%) and mercury (0.005 wt.%). The lamp phosphor fraction contains about 10% of rare-earth phosphors bound in the triband dyes [52]. Consequently, owing to the presence of REEs along with mercury, the disposal of lamp phosphors not only would lead to loss of resources but also to environmental hazards. Some countries collect great amounts of fluorescent phosphors as a distinct fraction from the recycling of fluorescent lamps (175 tons per year in Germany) and thus it is suitable as a secondary resource of REEs.

The most common rare-earth phosphors in these lamps are: $Y_2O_3:Eu^{3+}$, $LaPO_4:Ce^{3+}$, $(Gd,Mg)B_5O_{12}:Ce^{3+},Tb^{3+}$, $(Ce,Tb)MgAl_{11}O_{19}$ and $BaMgAl_{10}O_{17}:Eu^{3+}$ [52] and it is possible to dissolve these REE-compounds through microbial processes (**Figure 3**).

A symbiotic mixed culture from tea Kombucha, consisting of yeasts and acetic acid bacteria, was used to leach REE from fluorescent powder. The highest leaching yields were achieved using the entire Kombucha-consortium or its supernatant as leaching agent compared to experiments using the isolates *Zygosaccharomyces lentus* and *Komagataeibacter hansenii*. The pH decreased during the microbial growth as a result of organic acid production, acetic and gluconic acid [53].

Other study tested a broad spectrum of different microorganisms to evaluate their potential to dissolve REE from the lamps residues. Larger amounts of REE were leached with the strains *Komagataeibacter xylinus*, *Lactobacillus casei*, *and Yarrowia lipolytica* comprising a total release of REEs of 12.6%, 10.6%, and 6.1%, respectively. Yttrium and europium were dissolved selectively during the experiments [54]. The supernatant generated by the bacterium *Gluconobacter oxydans* was also tested for the recovery of REEs from phosphor powder; however, only about 2% of the total REEs was leached [4].

The phosphor powder solubilization is probably linked to the carboxyl-functionality or a proton excess. Among the different REE components preferably the red dye $Y_2O_3:Eu^{3+}$ was shown to be preferentially solubilized in accordance with the higher solubility of REE-oxides compared to REE-phosphates and –aluminates.

Electronic wastes are discarded devices that are at the end of their economic use and cannot be utilized by consumers anymore. The total global e-waste generation in 2021 is expected to achieve 52.2 Mt. The biggest economic interest is focused on gold with 50% of the possible revenue, but e-wastes contain other metals in significant amounts and still worth to be recovered. Investigations on bioleaching associated to printed circuit boards (PCB) recycling has mostly centered on copper and gold recovery. Ferric iron generated by iron-oxidizing bacteria is involved in

Figure 3.
SEM image of the fungus Aspergillus niger *grown on fluorescent lamp powder.*

copper extraction. Cyanogenic bacteria or fungi have been tested to recover gold from e-wastes [55]. Nevertheless, these processes need to be upscaled and optimized. Recently, a two-step reactor has been developed to separate the production of biogenic ferric iron from the valuable metals leaching reaction achieving a 96% recovery of Cu [56]. Many e-wastes contain magnets with an important amount of REEs, 20–30%. The content of Nd, Dy, and Pr in NdFeB magnets is 259.5, 42.1 and 3.4 ppm, respectively [57]. *Acidithiobacillus ferrooxidans* and *Leptospirillum ferrooxidans* showed high leaching efficiencies when grew in presence of magnets but also in abiotic controls. Consequently, leaching mainly took place by chemical processes due to the presence of H_2SO_4 [58]. Furthermore, biodismantling is a new application of bioleaching in the recycling process of electronic waste to enhance the concentration of critical and precious materials imbedded in the electronic components. Considering sufficient sorting of the components after separation, some rare earth elements may become economically available by reaching a grade similar to commercial ores. A concentration of 9000 μg/g of dysprosium has been detected in one of these separated fractions [59].

The growing demand and applications of REEs in many fields lead to an increasing generation of industrial and electronic wastes. Thus, these materials become an alternative economic for REE recovery due to mineral scarcity and the environmental degradation, and the developments in bioprocess technology have a key role in sustainable mining for the green economy.

5. Conclusions

Rare earth elements possess unique properties that make them useful in a wide variety of applications, such as catalysts, magnets, batteries, phosphors and polishing compounds. Consequently, the demand of these elements is growing and the REEs extraction become an important issue. Biohydrometallurgy is a potential technological solution to conventional chemical processes that employ corrosive reagents with harmful effects on the environment. Bioleaching for the extraction of rare earth metals from mineral ores and industrial wastes can be performed by autotrophic and heterotrophic microorganisms, although the phosphate solubilizing microorganisms have been especially investigated. Several mechanisms are involved in the mobilization of REEs: organic acids, enzymes, bacterial attachment, phosphate regulation, siderophores... Moreover, the application of biotechnological strategies to the treatment of solid wastes might contribute to maximize the amount of resources minimizing the amount of tailings or residues that exert a harmful impact on the environment. Bioleaching of REEs is in its infancy, but the development of global market and the environmental policies as well as the appearance of new drivers such as synthetic biology and digital revolution could influence the evolution of biohydrometallurgy.

Acknowledgements

This work was supported by the Complutense University of Madrid (project PR75/18-21576).

Conflict of interest

The authors declare no conflict of interest.

Author details

Laura Castro*, M. Luisa Blázquez, Felisa González and Jesús A. Muñoz
Department of Chemical and Materials Engineering, Complutense University of Madrid, 28040 Madrid, Spain

*Address all correspondence to: lcastror@ucm.es

References

[1] Balaram V. Rare earth elements: A review of applications, occurrence, exploration, analysis, recycling, and environmental impact. Geoscience Frontiers. 2019;**10**(4):1285-1303

[2] Habashi F. Extractive metallurgy of rare earths. Canadian Metallurgical Quarterly. 2013;**52**(3):224-233

[3] Corbett MK, Eksteen JJ, Niu X-Z, Croue J-P, Watkin ELJ. Interactions of phosphate solubilising microorganisms with natural rare-earth phosphate minerals: a study utilizing Western Australian monazite. Bioprocess and Biosystems Engineering. 2017;**40**(6):929-942

[4] Reed DW, Fujita Y, Daubaras DL, Jiao Y, Thompson VS. Bioleaching of rare earth elements from waste phosphors and cracking catalysts. Hydrometallurgy. 2016;**166**:34-40

[5] Borra CR, Pontikes Y, Binnemans K, Van Gerven T. Leaching of rare earths from bauxite residue (red mud). Minerals Engineering. 2015;**76**:20-27

[6] Zhuang W-Q, Fitts JP, Ajo-Franklin CM, Maes S, Alvarez-Cohen L, Hennebel T. Recovery of critical metals using biometallurgy. Current Opinion in Biotechnology. 2015;**33**:327-335

[7] Sethurajan M, van Hullebusch ED, Fontana D, Akcil A, Deveci H, Batinic B, et al. Recent advances on hydrometallurgical recovery of critical and precious elements from end of life electronic wastes - a review. Critical Reviews in Environmental Science and Technology. 2019;**49**(3):212-275

[8] Whitty-Léveillé L, Reynier N, Larivière D. Rapid and selective leaching of actinides and rare earth elements from rare earth-bearing minerals and ores. Hydrometallurgy. 2018;**177**:187-196

[9] Wu S, Wang L, Zhao L, Zhang P, El-Shall H, Moudgil B, et al. Recovery of rare earth elements from phosphate rock by hydrometallurgical processes – A critical review. Chemical Engineering Journal. 2018;**335**:774-800

[10] Jha MK, Kumari A, Panda R, Rajesh Kumar J, Yoo K, Lee JY. Review on hydrometallurgical recovery of rare earth metals. Hydrometallurgy. 2016;**165**:2-26

[11] Brierley CL, Brierley JA. Progress in bioleaching: part B: applications of microbial processes by the minerals industries. Appl Microbiol Biotechnol. 2013;**97**(17):7543-7552

[12] Norris PR, Burton NP, Clark DA. Mineral sulfide concentrate leaching in high temperature bioreactors. Minerals Engineering. 2013;**48**:10-19

[13] Zhang L, Dong H, Liu Y, Bian L, Wang X, Zhou Z, et al. Bioleaching of rare earth elements from bastnaesite-bearing rock by actinobacteria. Chemical Geology. 2018;**483**:544-557

[14] Brisson VL, Zhuang WQ, Alvarez-Cohen L. Bioleaching of rare earth elements from monazite sand. Biotechnology and Bioengineering. 2016;**113**(2):339-348

[15] Beolchini F, Fonti V, Dell'Anno A, Rocchetti L, Vegliò F. Assessment of biotechnological strategies for the valorization of metal bearing wastes. Waste Management. 2012;**32**(5):949-956

[16] Critical mineral resources of the United States—Economic and environmental geology and prospects for future supply. Report. Reston, VA; 2017. Report No.: 1802.

[17] European Commission. Communication from the Commission to the European Parliament, the

Council, the European Economic and Social Committee and the Committee of the Regions on the 2017 list of Critical Raw Materials for the EU. 2017.

[18] Survey USG. Rare Earths, Mineral Commodity Summaries. 2020:132-133

[19] Zhou B, Li Z, Chen C. Global Potential of rare earth resources and rare earth demand from clean technologies. Minerals. 2017;7(11):203

[20] U.S. Census Bureau. Annual Survey of Manufactures (ASM). 2017.

[21] Gambogi J. Rare Earths. In: U.S. Department of Interior, U.S. Geological Survey, 2017 Minerals Yearbook. 2017

[22] Rasoulnia P, Barthen R, Lakaniemi A-M. A critical review of bioleaching of rare earth elements: The mechanisms and effect of process parameters. Critical Reviews in Environmental Science and Technology. 2020:1-50

[23] Peelman S, Sun ZHI, Sietsma J, Yang Y. Chapter 21 - Leaching of rare earth elements: review of past and present technologies A2 - Lima, Ismar Borges De. In: Filho WL, editor. Rare earths industry. Boston: Elsevier; 2016. pp. 319-334

[24] Qi D. Chapter 1 - Extraction of rare earths from RE concentrates. In: Qi D, editor. Hydrometallurgy of Rare Earths. Elsevier; 2018. pp. 1-185

[25] Prabhu N, Borkar S, Garg S. Chapter 11 - Phosphate solubilization by microorganisms: Overview, mechanisms, applications and advances. In: Meena SN, Naik MM, editors. Advances in biological science research. Academic Press; 2019. pp. 161-176

[26] Alori ET, Glick BR, Babalola OO. Microbial phosphorus solubilization and its potential for use in sustainable agriculture. Frontiers in Microbiology. 2017;8(971).

[27] Delvasto P, Valverde A, Ballester A, Muñoz JA, González F, Blázquez ML, et al. Diversity and activity of phosphate bioleaching bacteria from a high-phosphorus iron ore. Hydrometallurgy. 2008;**92**(3):124-129

[28] Shin D, Kim J, Kim BS, Jeong J, Lee JC. Use of phosphate solubilizing bacteria to leach rare earth elements from monazite-bearing ore. Minerals. 2015;**5**(2):189-202

[29] Keekan KK, Jalondhara JC, Abhilash. Extraction of Ce and Th from monazite using REE tolerant *Aspergillus niger*. Mineral Processing and Extractive Metallurgy Review. 2017;38(5):312-320.

[30] Castro L, Blázquez ML, González F, Muñoz JA. Bioleaching of phosphate minerals using *Aspergillus niger*: Recovery of copper and rare earth elements. Metals. 2020;**10**(7):978

[31] Kang X, Csetenyi L, Gadd GM. Monazite transformation into Ce- and La-containing oxalates by *Aspergillus niger*. Environmental Microbiology. 2020;**22**(4):1635-1648

[32] Brisson VL, Zhuang W-Q, Alvarez-Cohen L. Metabolomic analysis reveals contributions of citric and citramalic acids to rare earth bioleaching by a *Paecilomyces* fungus. Frontiers in Microbiology. 2020;**10**(3008)

[33] Fathollahzadeh H, Khaleque HN, Eksteen J, Kaksonen AH, Watkin ELJ. Effect of glycine on bioleaching of rare earth elements from Western Australian monazite by heterotrophic and autotrophic microorganisms. Hydrometallurgy. 2019;**189**:105137

[34] Corbett MK, Eksteen JJ, Niu X-Z, Watkin ELJ. Syntrophic effect of indigenous and inoculated microorganisms in the leaching of rare earth elements from Western Australian monazite. Research in Microbiology. 2018;**169**(10):558-568

[35] Corbett MK, Watkin EL. Microbial cooperation improves bioleaching recovery rates. Microbiology Australia. 2018;**39**(1):50-52

[36] Goldstein AH. Recent progress in understanding the molecular genetics and biochemistry of calcium phosphate solubilization by gram negative bacteria. Biological Agriculture & Horticulture. 1995;**12**(2):185-193

[37] Illmer P, Schinner F. Solubilization of inorganic calcium phosphates—Solubilization mechanisms. Soil Biology and Biochemistry. 1995;**27**(3):257-263

[38] Rodríguez H, Fraga R. Phosphate solubilizing bacteria and their role in plant growth promotion. Biotechnology Advances. 1999;**17**(4):319-339

[39] Yi Y, Huang W, Ge Y. Exopolysaccharide: a novel important factor in the microbial dissolution of tricalcium phosphate. World Journal of Microbiology and Biotechnology. 2008;**24**(7):1059-1065

[40] Singh AK, Cameotra SS. Rhamnolipids production by multi-metal-resistant and plant-growth-promoting Rhizobacteria. Appl Biochem Biotechnol. 2013;**170**(5):1038-1056

[41] Eivazi F, Tabatabai MA. Phosphatases in soils. Soil Biology and Biochemistry. 1977;**9**(3):167-172

[42] Antonick PJ, Hu Z, Fujita Y, Reed DW, Das G, Wu L, et al. Bio- and mineral acid leaching of rare earth elements from synthetic phosphogypsum. The Journal of Chemical Thermodynamics. 2019;**132**:491-496

[43] Klauber C, Gräfe M, Power G. Bauxite residue issues: II. options for residue utilization. Hydrometallurgy. 2011;**108**(1):11-32

[44] Akcil A, Akhmadiyeva N, Abdulvaliyev R, Abhilash MP. Overview on extraction and separation of rare earth elements from red mud: Focus on scandium. Mineral Processing and Extractive Metallurgy Review. 2018;**39**(3):145-151

[45] Qu Y, Lian B. Bioleaching of rare earth and radioactive elements from red mud using Penicillium tricolor RM-10. Bioresource Technology. 2013;**136**:16-23

[46] Qu Y, Li H, Wang X, Tian W, Shi B, Yao M, et al. Bioleaching of major, rare earth, and radioactive elements from red mud by using indigenous chemoheterotrophic bacterium *Acetobacter* sp. Minerals. 2019;**9**(2):67

[47] Ferella F, Innocenzi V, Maggiore F. Oil refining spent catalysts: A review of possible recycling technologies. Resources, Conservation and Recycling. 2016;**108**:10-20

[48] Thompson VS, Gupta M, Jin H, Vahidi E, Yim M, Jindra MA, et al. Techno-economic and life cycle analysis for bioleaching rare-earth elements from waste materials. ACS Sustainable Chemistry & Engineering. 2018;**6**(2):1602-1609

[49] Jin H, Reed DW, Thompson VS, Fujita Y, Jiao Y, Crain-Zamora M, et al. Sustainable bioleaching of rare earth elements from industrial waste materials using agricultural wastes. ACS Sustainable Chemistry & Engineering. 2019;**7**(18):15311-15319

[50] Mouna HM, Baral SS. A bio-hydrometallurgical approach towards leaching of lanthanum from the spent fluid catalytic cracking catalyst using Aspergillus niger. Hydrometallurgy. 2019;**184**:175-182

[51] Azevedo DMF, Silva JAS, Servulo EFC, Frescura VLA, Dognini J, Oliveira FJS. Recovery of lanthanides from hydrocarbon cracking spent catalyst through chemical and biotechnological strategies. Journal of

Environmental Science and Health, Part A. 2019;**54**(7):686-693

[52] Binnemans K, Jones PT. Perspectives for the recovery of rare earths from end-of-life fluorescent lamps. Journal of Rare Earths. 2014;**32**(3):195-200

[53] Hopfe S, Flemming K, Lehmann F, Möckel R, Kutschke S, Pollmann K. Leaching of rare earth elements from fluorescent powder using the tea fungus Kombucha. Waste Management. 2017;**62**:211-221

[54] Hopfe S, Konsulke S, Barthen R, Lehmann F, Kutschke S, Pollmann K. Screening and selection of technologically applicable microorganisms for recovery of rare earth elements from fluorescent powder. Waste Management. 2018;**79**:554-563

[55] Kaliyaraj D, Rajendran M, Angamuthu V, Antony AR, Kaari M, Thangavel S, et al. Bioleaching of heavy metals from printed circuit board (PCB) by *Streptomyces albidoflavus* TN10 isolated from insect nest. Bioresources and Bioprocessing. 2019;**6**(1):47

[56] Hubau A, Minier M, Chagnes A, Joulian C, Silvente C, Guezennec A-G. Recovery of metals in a double-stage continuous bioreactor for acidic bioleaching of printed circuit boards (PCBs). Separation and Purification Technology. 2020;**238**:116481

[57] Dev S, Sachan A, Dehghani F, Ghosh T, Briggs BR, Aggarwal S. Mechanisms of biological recovery of rare-earth elements from industrial and electronic wastes: A review. Chemical Engineering Journal. 2020;**397**:124596

[58] Auerbach R, Bokelmann K, Stauber R, Gutfleisch O, Schnell S, Ratering S. Critical raw materials – Advanced recycling technologies and processes: Recycling of rare earth metals out of end of life magnets by bioleaching with various bacteria as an example of an intelligent recycling strategy. Minerals Engineering. 2019;**134**:104-117

[59] Monneron-Enaud B, Wiche O, Schlömann M. Biodismantling, a novel application of bioleaching in recycling of electronic wastes. Recycling. 2020;**5**(3):22

Chapter 13

Electrochemical and Optical Methods for the Quantification of Lead and Other Heavy Metal Ions in Liquid Samples

Samrat Devaramani, Banuprakash G., Doreswamy B.H. and Jayadev

Abstract

Minerals and elementary compounds of heavy metals are part of the ecosystem. Because of their high density and property to accumulate in stable forms, they are considered to be highly toxic to animals, plants and humans. Continuous mining activities and industrial effluents are the major sources which are adding toxic heavy metal ions into ecosystem and biota. Hence it is of utmost importance to quantify the levels of heavy metal ions in environmental and biological samples. On the other hand, it is equally important to remove the heavy metal ions and their compounds from the environmental and biological samples. That facilitates the environmental samples to be fit for using, consumption. In this regard, promising quantification methods such as electrochemical, spectrophotometric, naked eye sensing, test strips for spot analysis of heavy metal ions are considered for discussion. The main objective of this chapter is to give the overview of the most practiced quantification approaches available in the literature. Please note that reader cannot find the pin to pin publications regarding the same and that is not the aim of this book chapter.

Keywords: heavy metal ions, quantification, electrochemical and optical

1. Introduction

Metals possess higher density, atomic numbers are considered as heavy metals. They are part of the ecosystem present in the form of minerals and also in the elemental form. Heavy metal ions [HMIs] are proved to be toxic to humans, animals and even to plants as they accumulate to form stable compounds. Over a period of exposure or accumulation to HMIs lead to serious health issues with respect to [w.r.t.] skin, neurological system, kidney etc. [1]. HMIs toxicity and its adverse effects are magnified due to the human activities such as mining, rigorous industrial activities adding HMIs to the various ecological system. Once it happens then animals and humans are prone to HMIs that is really disaster. Considering these facts seriously world health organization and environmental protection agency set the permissible levels for the HMIs in various samples [2]. Above that level that particular sample is not fit for usage or consumption. In this scenario there is a pressing need for the analytical methods through which the exact quantity of the HMI[s] can be measured. There is a scope for the analysis of HMIs in the wide variety of samples such as water, air, food, and biological samples etc.

It is well known fact that, there exist efficient methods either individual or coupled with detection techniques for the quantification of HMIs. Atomic absorption spectroscopy, inductively coupled mass spectroscopy, inductively coupled atomic emission spectroscopy, Ion selective high-performance liquid chromatography etc. All the mentioned methods possess good linear range, sensitive down to ppb level concentration and accurate. At the same time, they are expensive, needs to be operated by skilled person, cannot be carried to field, and may require greater volume of sample. Hence, even today researchers are putting their efforts to develop analytical methods to overcome the mentioned shortcomings. There exist few techniques such as electrochemical, optical, methods based on electrical conductivity, refractive index etc. Compared to the earlier mentioned methods latter are less expensive, easy to handle and smaller in size. But there are few challenges with the latter mentioned methods such as sensitivity and selectivity. Thrust to achieve these goals is still alive, hence, one can witness lot of publications on the same topic. But, it does not mean that electrochemical, spectroscopic methods are failure. Researchers are striving to match the sensitivity and selectivity of the electrochemical and spectroscopic methods with that of earlier mentioned methods. Another improved aspect is the size. True size of both i.e. instrument and sample [volume] required for the analysis are significantly minimized. Naked eye sensing of heavy metal ions is made possible. Just addition of sample containing HMI to the reagent system in a culture tube results the color [change] that can be recognized with naked eyes. Moving a step ahead electrochemical and spectroscopic methods are miniaturized down to small paper strip. Paper based electrochemical and colorimetric methods are trending now. Important Analytical parameters for all the above discussed methods have been summarized with broader perspective in the **Table 1** for the comparison purpose [6, 7].

	Flame AAS	Graphite Furnace AAS	ICP-MS	ICP-OES	EC* Methods	SP* Methods	SF* Methods
Sample volume required	Large	Small	Medium	Medium	Small	Medium to Small	Medium to Small
Detection Range	nM	nM	pM to nM	pM to nM	nM to μM	μM	pM to μM
Detection limits	nM	nM	pM	pM	nM to μM	nM	pM to nM
Interference	few	few	few	More	few	few	few
Spot analysis	Not possible	Not possible	Not possible	Not possible	Possible**	Possible**	Not possible
Cost	Expensive	Expensive	Expensive	Expensive	Affordable	Affordable	Affordable
Accuracy	Good	High	Very High	Very High	Good	High	High
Precision	Good	High	Very High	Very High	Good	High	High
Recovery	Good	High	Very High	Very High	Good	High	High

**EC-Electrochemical, SP -Spectrophotometric, SF-Spectrofluorimetric.*
***Classical Potentiostat and Spectrophotometers are miniaturized to portable and handy devises with customized facilities especially for the spot analysis [3–5].*

Table 1.
Generalized comparison of the parameters of analytical significance of widely followed methods for HMIs quantification.

Because of this solid background HMIs quantification by electrochemical and optical methods will be a topic of interest for large group of readers. Keeping the beginners in mind discussion is focused on the fundamentals of electrochemical approaches, types of materials and strategies used for the quantification of HMIs by both electrochemical and optical methods. It must be noted that exhaustive literature citation is avoided because of the page limits and also to avoid the monotony.

2. Electrochemical sensing of heavy metal ions

2.1 Different electrochemical techniques

2.1.1 Voltammetry

Generally, in this case resultant current due to faradaic reaction [s] of the analyte is recorded by sweeping the potential between the two chosen potentials. As a result, the graph of current vs. voltage will be obtained that is referred to as voltammogram. Below are the various Voltammetric techniques commonly followed to measure the heavy metal ion[s] concentration.

2.1.1.1 Cyclic voltammetry [CV]

Current is recorded in the forward and backward directions by sweeping the potential in the fixed potential window. Faradaic reaction of the metal ion will result the oxidation and [or] reduction peaks at a particular potential where it underwent redox reaction. By referring to cyclic voltammogram

- potential required to oxidize/reduce the metal ion can be found. That information is helpful in performing amperometric measurement of metal ion samples, to decide the potential window in the other voltammetric techniques such as differential pulse voltammetry, square wave voltammetry etc.

- reversibility of the faradaic reaction can be understood.

In this three-electrode system platinum wire is used as counter electrode, calomel or Ag/AgCl electrode is used as reference electrode and glassy carbon electrode (GCE)/gold or platinum electrode/carbon paste electrode etc. is used as working electrode. Devi et al. used gold nanoparticles modified GCE as the working electrode for the quantification of Hg^{2+} ions using CV [8]. Authors exploited the well-known interaction of thiol and gold for the functionalization of gold on the GCE. Micro molar concentration of Hg^{2+} can be quantified using this CV method.

2.1.1.2 Pulse voltammetry

In this technique series of super imposing pulse of voltage are generated to result the potential sweep. Because of the applied voltage, HMI will undergo redox reaction to result the faradaic current and that is measured. Differential pulse voltammetry (DPV) is more opted out of various pulse voltammetric techniques such as normal pulse voltammetry and reverse pulse voltammetry. Xia et al. proposed DPV method for the simultaneous determination of Pb^{2+}, Cd^{2+} and Cu^{2+}. Wherein they used carbon paste electrode modified with hexagonal mesoporous silica and quercetin [9].

2.1.1.3 Square wave voltammetry [SWV]

SWV is another voltammetric technique which is also very often used to quantify the HMIs with some advantages like fast scan rate, less adsorption on the working electrode and reduced measurement time over DPV.

In the case of CV, DPV and SWV more than one HMI can be detected simultaneously.

Generally, HMIs quantification by voltammetric techniques is performed in conjunction with electrochemical deposition followed by stripping. Hence, few examples for the above mentioned voltammetric techniques are discussed under stripping voltammetry section.

2.1.1.4 Stripping voltammetry

It is a two-step procedure. First step involves the electrodeposition of the HMI[s] from the electrolyte solution onto the working electrode surface. Second, by applying any of the voltammetric techniques [discussed above], HMI[S] on the electrode will be stripped off into electrolyte solution. Based on the applied scan i.e. anodic or cathodic or during which stripping take place technique is named as anodic stripping or cathodic stripping voltammetry [10]. Following cases can be considered as an example for how stripping step is combined with various voltammetric techniques. Yao et al. proposed square wave anodic stripping voltammetry [SWASV] for the quantification of Zn^{2+}, Cd^{2+}, Pb^{2+}, $Cu^{2+}$and Hg^{2+}. Fe_3O_4 nanocrystals of two different shapes are used as a modifier to obtain a sensitive and selective signal for the HMI [11]. **Figure 1** depicts the nature of SWASV. Serious interference of Cu^{2+} ions in the electrochemical detection of Cd^{2+} ions was effectively overcome by introducing Bi film on the GCE. As a result, stripping peaks were found to be intact even in the presence of Cu^{2+} ions for SWASV [12]. Raghu et al. developed DPASV method to achieve the quantification of Hg^{2+} ions down to picomolar concentration. Multiwalled carbon nanotubes [MWCNT] were functionalized with Fast Violet B salt through diazotization. Functionalized MWCNT then drop casted onto GCE to sense Hg^{2+} ions by DPASV in drinking water and industrial effluents [13]. Pandurangappa Malingappa and coworkers have published few exemplary works in which stripping voltammetric analysis has been systematically utilized for the analysis of HMIs from the various samples [14–16].

2.1.2 Amperometry

It is a potentiostatic technique. Electrochemical measurements are carried out at a fixed potential to measure the resultant current due to the redox reaction at the

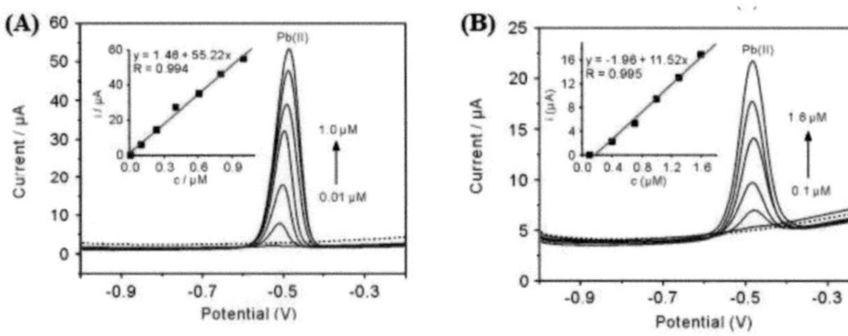

Figure 1.
SWASV recorded for the (A) octahedral and (B) cubic Fe_3O_4 modified electrodes in the presence of varied concentration of Pb^{2+}ion (reprinted with permission from [11] copyright 2014 American Chemical Society).

electrode electrolyte interface. Based on the analyte, here a particular metal ion, to be detected value of the potential needs to be applied will be decided. Hence, the measured current will be exclusively due to faradaic reaction of that particular analyte. To quote the recent example, Sannegowda and his coworkers developed iminephthalocyanine based amperometric sensor for the quantification of Pb^{2+} ions. That exhibited the linear range and detection limit in the nanomolar Pb^{2+} ion concentration [17]. Amperometric biosensor based on the urease was developed for the detection of Pb^{2+} and Hg^{2+} ions in river water samples also exhibited the analytical figures of merits closer to nanomolar levels [18]. Simultaneous quantification of more than one HMI is not possible i.e. the drawback of this method.

Another similar work can be quoted here. This case acridono-crown ether played a role of ionophore and. Poly(vinyl chloride) membrane again acted as a host. The potentiometric sensor works in a range of pH 4-7 but suffered a much-required sensitivity [19].

2.1.3 Potentiometry

Developed potential or electromotive force (EMF) is measured without applying external current. Experimental setup required for the potentiometric measurements is inexpensive. But, sensitivity of this technique is not appreciable when the routine electrodes are used. Efforts are in progress to improve the sensitivity by making use of electrodes constructed out of advanced materials such as graphene, CNT, and nanomaterials or reducing the size of the electrode itself i.e. nanoelectrodes [20, 21]. Ionophore is a corner stone of the potentiometric experimental setup that decides the selectivity and sensitivity of the procedure. Xin-Gui Li et al. developed a ionophore based on conducting copolymer microparticles. Poly[vinyl chloride] membrane acted as a platform to host the ionophore. Presence of functional molecules such as –NH–, –N, $-NH_2$, and $-SO_3H$ in the microparticles resulted high selectivity towards Pb^{2+} ions. It's worth mentioning that potentiometric sensor exhibited sub micromolar detection limit towards Pb^{2+} ions [22]. Another similar work can be quoted here. This case acridono-crown ether played a role of ionophore and poly[vinyl chloride] membrane again acted as a host. The potentiometric sensor works in a range of pH 4-7 but suffered a much-required sensitivity [23].

2.2 Electrochemical preconcentration

Generally, concentration of HMIs is very low in drinking water, food and biological samples. In addition to that, sample matrices will be complex in nature and many other ions and molecules will be present. In this regard it is very important to separate the HMI[S] from the sample matrix by enriching the same on to the working electrode surface. There by the interference from various electrochemically active species can be overcome and sensitivity can be significantly enhanced. Below two important electrochemical preconcentration methods are discussed in brief.

2.2.1 Electrochemical deposition

Electrochemical deposition is done by taking sample solution containing the HMI[s]in a three-electrode electrochemical cell. Then, suitable potential is applied to working electrode (most of the times modified working electrode) w.r.t. reference electrode. As a result of the applied potential metal ion will get reduced to metal atom and simultaneously deposited onto the working electrode. Prior to electrochemical deposition cyclic voltammetric experimental data will be helpful in deciding the deposition potential. Suitable buffer solution and pH are necessary to fine tune the selective deposition of the particular HMI(S) [24].

2.2.2 Electrochemical adsorption

As the HMs exist as ions in the solution same nature can be exploited to achieve the electrochemical adsorption. It works on the electrostatic force of attraction principle. Basically, the working electrode will be modified with a material which has got sufficient opposite charges w.r.t. HMIs or its compounds. In addition to this material which also got greater surface area, hence, more charges on it will be preferred. It is obvious that nanomaterials are the competent candidates for this purpose. First comes the carbon nano substrates such as graphene, CNT, graphite flakes etc. Reason is twofold; Inertness of the carbon substrate hence it can just act as a platform for the HMIs adsorption and for the same reason it can be used for the construction of electrode as well. Ease with which the charge bearing functional groups can be covalently bonded on to it. There exist well-established procedures to introduce various functional groups such as carboxylic acid, amine etc. onto the carbon substrates. Now the materials, functionalized graphene or CNT, having greater surface area and also charge on it are suitable for the electrodeposition of the HMIs. Apart from these qualities, defects introduced during the functionalization process and their inherent good electrical conductivity have the added advantage for this application [25–27]. Likewise, nanomaterials other than carbon substrate can also be used. Positive or negative potential is applied to the electrode to enhance the rate of adsorption.

2.3 Different materials based electrochemical sensors

2.3.1 Carbon substrates

Graphitic carbon, glassy carbon, can be considered as the bulk form, itself has got good conductivity, stability have been used as an electrode material for a long time. Nanomaterials (NMs) derived out of graphite's single or few layers resulted in graphene, CNT, fullerene, carbon nanoflakes, etc. possessed extraordinary conductivity, excellent electrocatalytic property. Each carbon NM has been extensively studied for the electrochemical sensing of HMIs either in their pristine or modified form or as composite. Considering few examples out of large number of articles each carbon NM is discussed below.

2.3.1.1 Graphene

Two-dimensional single atomic thick monolayer is metallic in nature which inherited large surface area and very high conductivity [28]. In addition, hydroxyl and carboxylic groups originally present to a more or less extent depending on how the material is synthesized. If not present, those functional groups can be introduced on its surface with ease by following well established methods. Mentioned functional groups facilitate the interaction between the graphene and HMIs through either coordination or electrostatic or both. On the other hand, these inherent functional groups are exploited to introduce organic molecules, NMs etc. to enhance the sensitivity and selectivity. Li et al. reviewed the synthesis protocols and analytical applications of the graphene with exhaustive literature [29]. GCE modified with fluorinated graphene oxide was used to detect four HMIs such as Cu^{2+}, Pb^{2+}, Cd^{2+}, and Hg^{2+}simultaneously [30]. Nafion is used in many of the reported works to assemble the graphene onto the electrode i.e. it acts as a binder. At the same time, it also plays a role of ion exchange membrane hence, interference can be considerably overcome. Combination of the duo lead to many methods for the simultaneous determination of HMIs [31–33]. Simultaneous analysis of Zn^{2+}, Cd^{2+}, Pb^{2+}, and Cu^{2+} was exhibited by a composite nafion, graphene and in situ prepared mercury film. Distinct SWASV peaks were observed for the mentioned HMIs [34]. It is worth mentioning that despite the mentioned synergetic advantages of the nafion & graphene there is problem of restacking of the graphene layers due to van der Waals force of attraction and also irreversible adsorption of HMIs onto the film was observed [35]. Gong et al. developed a strategy to overcome the restacking a graphene layers by introducing Au nanoparticles in between them. In addition, Au nanoparticles improved the analytical figures of merit for the sensing of Hg^{2+}ions [36]. In a similar approach SnO_2 and reduced graphene oxide composite lead to simultaneous determination of Cd^{2+}, Pb^{2+}, Cu^{2+}, and Hg^{2+} as shown in **Figure 2** [37]. Composite of reduced graphene oxide with cysteic acid lead to highly sensitive DPASV method for Ag^{+} ion with detection limit of 1 nM [38]. Jingbo Chang et al. exclusively reviewed the sensing strategies of HMIs using graphene and its composites their more examples on this topic can be found [35].

2.3.1.2 CNTs

Properties, modification strategies discussed w.r.t. graphene holds good for CNTs. Since CNTs can be considered as rolled up structure of graphene and carbon atom is sp^3 hybridized in both the cases. CNTs are supposed to be equally competent for the analysis of HMIs compared to graphene but there are minor differences [39]. But, hybrid of graphene and CNTs yielded much better results compared to CNTs alone. Three-dimensional structure of graphene and CNT was able to simultaneously analyze pb^{2+} and Cd^{2+}ions [40]. CNTs are used in combination with bismuth film for the detection of HMIs [41]. CNTs and Pt nanoparticle together resulted in a highly sensitive electrochemical method for the ppb to ppt level As^{3+} determination [42].

2.3.2 Nanoparticles

As is well known, large surface area, more exposed catalytic sites, enhanced conductivity, greater electron & mass transport and faster electrode kinetics attracts

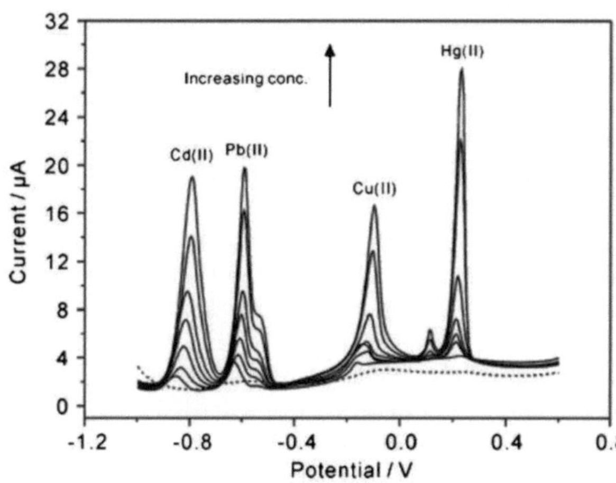

Figure 2.
SWASV peaks recorded in the presence of HMIs for SnO2-reduced graphene oxide modified electrodes (reprinted with permission from [37] copyright 2002 American Chemical Society).

the scientists to modify the working electrode with nanomaterials (NMs) for the better electrochemical sensing of HMIs. Though wide verity of NMs have been reported in the literature but metal and metal oxide nanoparticles are considered for discussion as their contribution is major among other NMs.

2.3.2.1 Metal nanoparticles

Gold nanoparticles and its derivatives have been extensively investigated as a transducer material for sensing HMIs because of its excellent conductivity, catalytic properties and inertness. There is a one more property i.e. affinity of gold towards thiol. This one property lead to plenty of works which all played around the triangular interaction between gold-thiol molecules-HMIs. As it can be observed gold component is common in below discussed cases but based on the material with which it is composed, capping agent, functionalized molecule and other experimental conditions sensitivity and selectivity towards HMIs will be totally different. Bin Zhang et al. modified carbon nanofibers with gold nanoparticles' size around 15 nm. That material was used on the GCE to simultaneously detect Cd^{2+}, Pb^{2+} and Cu^{2+} by SWASV [43]. Whereas the same Au nanoparticles capped with tannic acid modified GCE lead to highly sensitive and selective electrochemical sensor for Hg^{2+}down to 100 fM in the presence of Zn^{2+}, Al^{3+}etc. [44]. Handful number of articles can be found in which Au nanoparticles are used for the estimation of HMIs [45–47]. Inspired by the excellent electrochemical results two noble metals are combined to obtain the bimetallic composite. Same has shown promising results for the estimation of Hg^{2+} ions in the ppb range with the limit of detection down to ppt [48].

Earlier mercury electrodes were used to detect the HMIs because of the amalgamation reaction between the two. Thanks to the multicomponent alloy formation property of the bismuth and antimony with the HMIs. Because, highly toxic mercury electrodes were successfully replaced by the bismuth and antimony film modified electrodes. Nanoparticles of bismuth and antimony are proved to be an environmentally friendly and efficient platform for the quantification of HMIs [49–52]. Even the experiments were carried out to understand the effect of different morphology of the bismuth nanoparticles on the HMIs detection [53].

2.3.2.2 Metal oxide nanomaterials (MONMs)

Come into focus as a result of finding an inexpensive alternative for the noble metal nanoparticles in spite of their excellent electrochemical results as the latter are highly expensive. Other technical reasons are being as same as that any nanomaterials such as greater surface area and hence greater adsorption of HMIs, enhanced electrocatalytic activity etc. Oxides of iron, nickel, magnesium, manganese, zirconium etc. have been extensively studied for the quantification of HMIs. Majority of the reported methods are focused on experimenting with the morphology of the MONMs. Generally wide variety of morphologies can be achieved through hydrothermal synthesis. Then the MONM powder will be drop casted onto GCE for the electrochemical sensing of HMIs. Problem with this approach is modified electrode will not be sufficiently robust. Abdul Waheed et al. discussed the same with detailed literature in their review [54]. To overcome this, Lie et al. electrochemically deposited Co_3O_4 nanoparticles onto the indium tin oxide electrode followed by annealing. As prepared modified electrode is further used for the electrochemical sensing of Pb^{2+} ions in the presence of various divalent ions [55].

2.3.3 Mesoporous materials

Ordered mesoporous silicas (OMSs) can be considered as a representative candidate for mesoporous materials and their application in HMIs sensing. Because OMS exhibit uniform pore size, highly ordered pore, high volume and surface area which was explored by Mobil oil corporation [56, 57]. OMS provide a better platform for the physisorption of HMIs. As the OMS can be synthesized through various approaches different functional groups are inherently present on their surface. Out of which silanol is often present and most useful because of its reactive nature. Wide variety of organic molecules can be introduced through silanol group to result hybrid OMS. Because of the newly introduced organic molecules chemisorption of HMIs is achieved in addition to physisorption onto the hybrid OMS [58]. In majority of the reported methods particular OMS is combined with graphitic powder and mineral oil to obtain the modified carbon paste electrode to sense the HMIs. Thiol self-assembled monolayers on mesoporous supports [SAMMS]are used to prepare the carbon paste electrode for simultaneous analysis of Pb^{2+} and Hg^{2+} ion in 0.2 M HNO_3 [59]. Glycinylurea-SAMMS lead to SWASV method for the detection of Pb^{2+} with a detection limit down to 1 μg/L. It must be noted that detection can be performed over a wide range of pH 4.5 to 6.5 without using any specific buffer solution [60]. Yantasee et al. developed advanced remote accessible automated DPASV coupled with flow injection analysis of Pb^{2+} ions using Phosphonic-SAMMS [61]. Detailed literature on synthetic approaches and analytical applications of mesoporous materials can be found elsewhere [62].

3. Optical sensing of heavy metal ions

3.1 Spectrophotometric/ Fluorometric methods

Wide range of materials including organic molecules, nanomaterials, hybrid of earlier two, and quantum dots are used to sense HMIs by generating either color or fluorescence signal. Simultaneous signal generation for multiple HMIs is more often observed in electrochemical sensing. Whereas in case of optical methods, generally, signal will be generated exclusively due to target HMI. Hence, below are the sections categorized w.r.t. type of the material used for sensing. Lead i.e. Pb is considered as an example to explain the commonly followed strategies for the optical sensing of HMIs.

3.1.1 Cyclodextrins

In the year 1996 Czarnik et al. proposed a bench mark work for the fluorescence sensing of Pb^{2+} ions using a small molecule, heteroatom containing ligands. On complexation with Pb^{2+} ions proposed ligand exhibited enhanced fluorescence signal for about 15-fold [63]. Cyclodextrin molecule in binary solvent system exhibited 20-fold enhancement in the fluorescence signal after the addition of Pb^{2+} ions [64]. Similar cyclodextrin molecules are proposed by Chen et al. and Hayashita et al. for fluorometric sensing of the HMI as shown in **Figure 3** [65, 66].

3.1.2 Organic molecules

Color resulting dibromo-p-methyl-methylsulfonazo molecule was explored by Li et al. for spectrophotometric detection of the Pb^{2+}ions [67]. Inspired by this strategy Meng et al. worked out similar organic molecule for the colorimetric sensing of HMI [68].

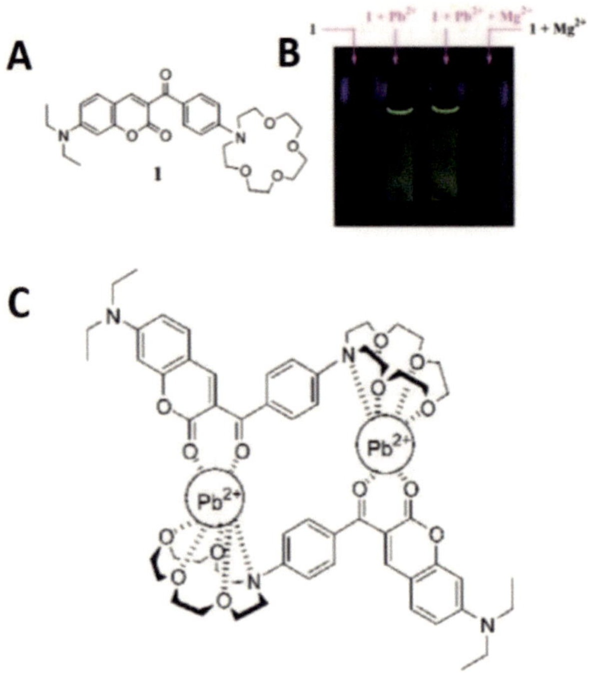

Figure 3.
(A) Representation of the structure of the ligand. (B) Image of the fluorescence response of the proposed ligand. (C) Binding mechanism of the Pb^{2+} ions with the proposed ligand (reprinted with permission from [65] copyright 2002 American Chemical Society).

3.1.3 Rhodamines

Rhodamine family molecules and its derivatives have been utilized extensively for the optical sensing of not only Pb^{2+} ions but most of the HMIs. Majority of such works result the optical signal through spirolactum ring opening or forming mechanism after interaction between the rhodamine and HMI. Yoon et al. reported rhodamine B derivative for both fluorometric and colorimetric sensing of Pb^{2+} ion in methyl cyanide medium [69].

3.1.4 Calixarenes

Calixarene family molecules are another category contributed significantly for the HMIs sensing. Calixarene structures generally from a dative bond with HMI through the functional groups to result an optical signal. Calix [4]arene derivatives are reported for the Pb^{2+} ion sensing can be observed as a regular method [70]. Switchable i.e. on–off–on fluorescent sensor observed due to the binding capacity of the calixarene molecule with two different cations i.e. Pb^{2+} and K^{+} as shown in **Figure 4**. Observed phenomenon is due to the interaction of the K^{+} ions with the molecular orbitals whereas that is absent in the Pb^{2+} ions case [71].

Apart from the above-mentioned categories there exist another variety of organic and bioorganic molecule extensively studied for the HMI. To name a few polymers based, peptide based, DNAzyme based sensors.

3.1.5 Carbon quantum dots

Carbon quantum dots (CQDs): Again, inherent functional groups on its surface and their deceptive role in anchoring desired molecules, nanoparticles etc.

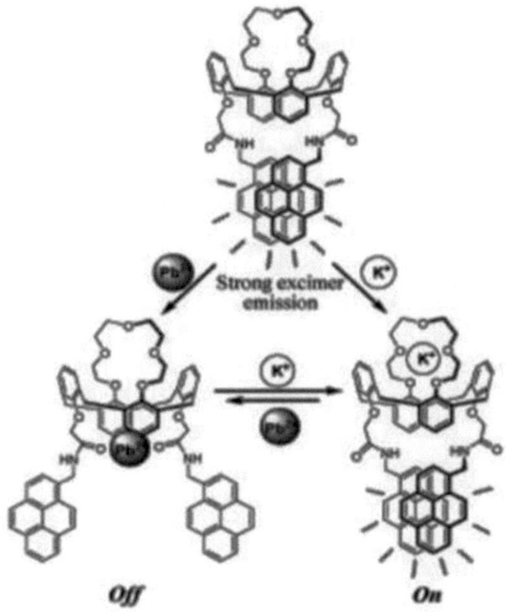

Figure 4.
Representation of the switchable fluorescence of the calix [4] arene derivative and its binding mechanism with Pb^{2+} and K^+ (reprinted with permission from [71] copyright 2004 American Chemical Society).

made CQDs as emerging optosensors for HMIs. It is prerequisite to understand the mechanism of sensing of HMIs using CQDs before considering the individual articles on the same. Interaction between the light (electromagnetic radiation) of suitable wavelength and CQDs leads to the generation of charge carriers i.e. due to the excitation of electron from valence band to the conduction band. Generated charge carriers are utilized in the sensing of HMIs depending on the interaction of HMI with the ligand on the CQDs surface. Photo induced charge transfer, fluorescence enhancing/quenching, inner filter effects, phosphorescence etc. [72]. Sodium citrate and polyacrylamide sodium citrate resulted blue fluorescent CQDs which was quenched upon the addition of Pb^{2+} ion. By adding pyrophosphoric acid to the quenched solution, fluorescence was regenerated. This fluorescence off–on method was able to detect the Pb^{2+} of the order of 4.6 nM [73]. CQDs synthesized from the green approach using *Lantana camara* berries were exhibited sensitive and wide linear range up to 200 nM Pb^{2+} ion concentration [74]. CQDs synthesized from chocolate source also exhibited nanomolar HMI detection [75]. Household sugar was used to prepare the CQDs and found to be sensitive and selective for the naked eye sensing of Pb^{2+} ions in water [76].

3.1.6 Nanoparticles

Nanoparticles forms an interesting domain for the HMIs sensing. Though there exist a wide range of metal nanoparticles majority of the work has been done on Ag and noble metals Pt, Au nanoparticles in spite of their costly affair. Surface plasmon resonance is the corner stone of the optical sensing of HMIs using Ag, Au, and Pt. Hupp et al. used 11-mercaptoundecanoic acid capped Au nanoparticles as color generating agent for the detection of Pb^{2+}, Cd^{2+}, and Hg^{2+} ions [77]. Thomas et al. proposed quite a different approach wherein they started with Au^{3+}/Ag^{+} ions and gallic acid solution. After the addition of Pb^{2+} Au/Ag nanoparticles are formed to result pink or red color respectively [78]. Non-aggregation-based sensor is developed by Huang et al. Originally Gold nanoparticles surrounded by thiosulphate and

2-mercaptoethanol in a solution. After the addition of Pb^{2+} due to the formation of Pb-Au alloy gold nanoparticles dissolved into solution and hence Surface plasmon resonance is decreased [79]. Opposite to that, i.e. aggregation based, sensing method was developed by Su et al. using glutathione modified Au nanoparticles. Upon the addition of HMI Au nanoparticles aggregate to result a color change from red to blue [80]. Theme of this work is summarized in the **Figure 5**.

Above discussed category of organic molecules, bioorganic molecules, nanoparticles have also been extensively used for the sensing of remaining HMIs. Functional moieties in the sensing molecules are different for different HMIs. But, conceptual wise it will be a repetition if the discussion is extended for the other HMIs. Ha Na Kim et al. has reviewed all the above discussed type of materials except CQDs for HMIs with an exhaustive literature survey and detailed discussion [81]. One can get more details and literature on CQDs for HMIs sensing in the review article presented by Pooja Devi et al. [72].

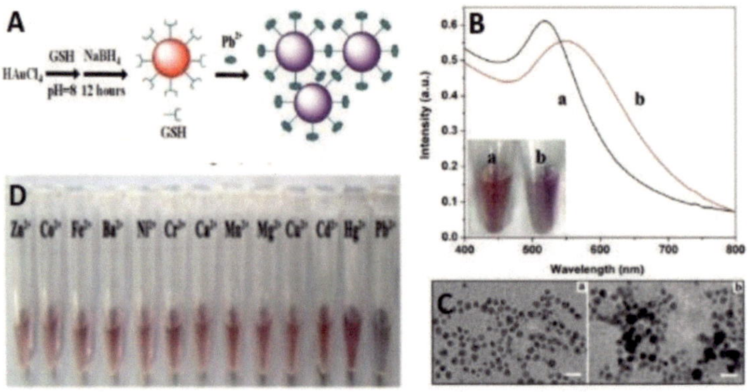

Figure 5.
(A) Scheme represents the preparation of glutathione capped Au nanoparticles. (B) Uv–Vis spectra, inset shows the images, (C) TEM images of Au nanoparticles in the absence (a) and presence of Pb^{2+} ions (b). (D) Images of the Au nanoparticle in the presence of various metal cations and Pb^{2+} ions (reprinted with permission from [80] copyright 2010 American Chemical Society).

4. Paper based optical and electrochemical sensing of HMIs

It is striking from the above discussions that both electrochemical and optical methods are good enough to quantify HMIs from various sample matrices. Selectivity, sensitivity and reproducibility of the most of the methods are sufficient enough i.e. can be used to quantify the HMIs within the permissible limits fixed by the world health organization. Apart from this there are some shortcomings for both the methodologies. It must be noted that both the instruments i.e. classical potentiostat and spectrophotometer cannot be carried away to the place at which the samples have to be analyzed. Hence, spot analysis of the HMIs is quite not possible using these methods. Also, both the instruments are expensive. Keeping these shortcomings in mind researchers are working out nonconventional methods. As it is already mentioned in **Table 1** miniaturized potentiostat and spectrophotometers are developed for spot analysis. Affordable and mobile phone installable softwares are developed to readout the color intensity on the spot [82]. On the other hand, sample holders i.e. electrochemical cells and cuvettes are replaced by paper strips. Below is the glimpse of such non-trivial methods.

4.1 Paper/strip

Making use of paper for the spot analysis of copper can be observed way back to 1945 [83]. Lateral flow through capillary action is the basis for the development of test strips not only for HMIs but various analytes. Pioneering work was published by Whiteside's group in the year 2007. In which photolithography was used to pattern the paper for the bioanalysis on the spot [84]. Paper based analytical device (PAD) is modified with resorufin thionocarbonate-based dye and it can detect μM concentration Hg^{2+} ions by using buffer solution of pH 8 [85]. 3,3′,5,5′-tetramethylbenzidine (TMB) and gold nanoparticles were used to develop a PAD for semiquantitative analysis of Hg^{2+} ions. Based on the enzyme like action of the gold and mercury nanoparticles TMB will give blue color [86]. In a similar work, platinum nanoparticles and TMB combination is used to detect Hg^{2+} ion using PAD. In this case decrease in the color intensity indicates the concentration of HMI. Enzymatic activity of the platinum nanoparticles to turn the TMB to blue is inhibited by the Hg^{2+} ions in this case [87]. Many such PAD based naked eye detection of HMIs can be found in literature [88–90]. Paper based electrochemical sensors (PESs) are coined as they inherit the simplicity and advantages of the PADs with better sensitivity. Whiteside's group developed PES by adopting the commercial electrochemical readers [91]. In case of PES electrodes are printed on a paper with carbon inks (working and counter electrodes), silver-silver chloride inks (reference electrode). These electrodes connected to a respective terminal of the electrochemical reader to measure a signal. Mariana Medina-Sánchez et al. developed PES for the quantification of Pb^{2+} and Cd^{2+} ion in a range 10 to 100 ppb. Developed sensor was reagent free and ecofriendly [92]. Two substrates i.e. plastic film and paper were used to coat graphene and polyaniline composite for the simultaneous detection of Zn^{2+}, Cd^{2+}, and Pb^{2+} ions [93]. Going one step ahead Poomrat Rattanarat et al. came up with a multi-layer-based device capable of producing both optical and electrochemical signal for iron, nickel, copper, chromium, lead and cadmium ions [94]. Iron, nickel, copper, chromium can be detected using optical signal whereas lead and cadmium ions are detected by electrochemical output as shown in **Figure 6**.

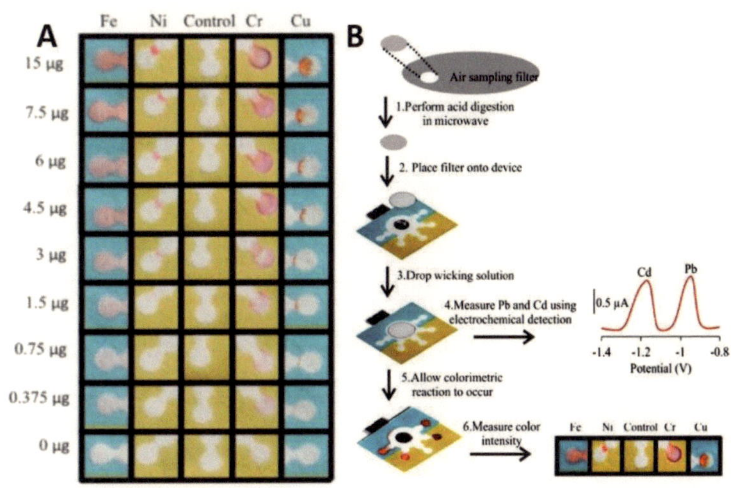

Figure 6.
(A) Image of the multilayer paper-based device's color changing response for the given HMIs. (B) Flow diagram details the preparation, electrochemical, and colorimetric response of the multilayer device (reprinted with permission from [94] copyright 2014 American Chemical Society).

5. Conclusions

Importance of HMIs quantification is stated followed by efficient methods and their drawbacks. No doubt that electrochemical and optical methods are promising alternatives. Because of the advantages such as user friendly, economical and competency, electrochemical and optical methods are further considered for the discussion. Different electrochemical techniques followed for the HMIs quantification are described with the fundamentals and reasoning in brief. Among them stripping voltammetric techniques are found to be efficient. Many electrochemical methods are reported to be capable of sensing HMIs down to ppb level and few can detect down to pico molar concentration. Another advantage of the electrochemical methods is the simultaneous analysis of multiple HMIs. Electrochemical methods are yet to be adopted for the field [spot] analysis. On the other hand, optical methods in particular colorimetric methods are metamorphosed to naked eye sensing of HMIs which will help in field analysis. PADs are coined to further simplify and miniaturize both electrochemical and optical methods. Alternatives for the often-used expensive color inducing materials such as gold, platinum and silver are still need to be explored in case of PADs for optical sensing of HMIs. It can be concluded that PESs are highly sensitive and selective but cannot be carried to field. Whereas PADs used for optical sensing are sensitive and can be used for field analysis are semiquantitative.

Acknowledgements

Dr. Samrat D thank TEQIP CELL, Visvesvaraya Technological University (VTU) for receiving fund under Competitive Research Grant with Ref. No. VTU/TEQIP 3/2019/321.

Dr. Samrat D thank Dr. Prasad B E, Dr. Rudresh Kumar K J, RV Institute of Technology and Management and Dr. Ramakrishnappa Thippeswamy, Dr. Suresh Kumar K from BMS institute of Technology and Management for the useful discussion and inputs in fine tuning the book chapter.

Dr. Banuprakash G thank VGST Grant (GRD No. 532) Under CISEE Scheme (2016-17).

Author details

Samrat Devaramani[1*], Banuprakash G.[2], Doreswamy B.H.[3] and Jayadev[2]

1 Department of Chemistry, RV Institute of Technology and Management, Bengaluru, India

2 Department of Chemistry, SJB Institute of Technology, Bengaluru, India

3 Department of Physics, SJB Institute of Technology, Bengaluru, India

*Address all correspondence to: samchem83@gmail.com

References

[1] U.S. Environmental Protection Agency Framework for metals risk assessment. Office of the Science Advisor, Risk Assessment Forum. EPA 120/R-07/001. Washington, D.C. 2007

[2] U.S. Environmental Protection Agency. Risk Assessment, Management and Communication of Drinking Water Contamination. US EPA 625/4-89/024, EPA: Washington, DC, 1989

[3] Ying X, Yan D, Chao L, Haijing Z, Miao G, Yong Y, PC software-based portable cyclic voltammetry system with PB-MCNT-GNPs-modified electrodes for *E. coli* detection. Review of Scientific Instruments. 2020:91:1-9. DOI:10.1063/1.5113655

[4] Johann FO, Crhistian CS, Miniaturization of cyclic voltammetry electronic systems for remote biosensing. Journal of Biosensors and Bioelectronics. 2017:3:297-299. DOI: 10.15406/ijbsbe.2017.03.00068

[5] Katrina L, Aleksejs Z, Mercedes V, Kirsty Mc D, Janis L, Hadar BY, Varis K, Krisjanis S, Portable low-cost open-source wireless spectrophotometer for fast and reliable measurements. HardwareX. 2020:7:e00108. DOI:10.1016/j.ohx.2020.e00108

[6] Mansour M, Jafar E N D, Mohammadali T, Aziz H R, Nanomaterials based surface plasmon resonance signal enhancement for detection of environmental pollutions. Biosensors and Bioelectronics. 2019:127:72-84. DOI: 10.1016/j.bios.2018.12.023

[7] ABM Helaluddin, Reem SK, Mohamed A, Syed AA, Main Analytical Techniques Used for Elemental Analysis in Various Matrices. Tropical Journal of Pharmaceutical Research February. 2016:15:427-434. DOI:10.4314/tjpr.v15i2.29

[8] Ramila D, Manickam S, Ashok KS, Gold nanoparticles-thiol-functionalized reduced graphene oxide coated electrochemical sensor system for selective detection of mercury ion. Journal of Electrochemical Society. 2018:165:3046-3053. DOI: 10.1149/2.0081808jes

[9] Xia F, Zhang X, Zhou C, Sun D, Dong Y, Liu Z, Simultaneous determination of copper, lead, and cadmium at hexagonal mesoporous silica immobilized quercetin modified carbon paste electrode. J Autom Methods Manag Chem. 2010: DOI:10.1155/2010/82419 7

[10] Jothimuthu P, Wilson RA, Herren J, Haynes EN, Heineman WR, Papautsky I, Lab-on-a-chip sensor for detection of highly electronegative heavy metals by anodic stripping voltammetry. Biomedical Microdevices. 2011:13:695-703. DOI: 10.1007/s10544-011-9539-1

[11] Yao X, Guo Z, Yuan Q, Liu Z, Liu J, Huang X, Exploiting differential electrochemical stripping behaviors of Fe_3O_4 nanocrystals toward heavy metal ions by crystal cutting. ACS Appl Mater Interfaces. 2014:6:12203-12213. DOI:10.1021/am501617a

[12] Guo Zhao, Hui Wang, Gang Liu, Direct Quantification of Cd2+ in the Presence of Cu2+ by a Combination of Anodic Stripping Voltammetry Using a Bi-Film-Modified Glassy Carbon Electrode and an Artificial Neural Network, Sensors [Basel]. 2017: 17: 1558. DOI:10.3390/s17071558

[13] Raghu GK, Pandurangappa M, Covalently modified multiwalled carbon nanotubes as a new voltammetric interface for the determination of mercury at picomolar level. Electrochemistry Communications. 2012: 25:83-86. DOI:10.1016/j.elecom.012.09.005

[14] Raghu GK, Pandurangappa M, Diazonium functionalized exfoliated graphitic carbon as a binderless and covalently modified electrochemical interface for mercury sensing. Sensors and Actuators B. 2013:186:478-485. DOI: 10.1016/j.snb.2013.06.028.

[15] Raghu GK, Pandurangappa M, Covalent modification of glassy carbon spheres through ball milling under solvent free conditions: A novel electrochemical interface for mercury[II] quantification. Talanta.2014:126:54-60. DOI: 10.1016/j.talanta.2014.02.043.

[16] Raghu GK, Pandurangappa M, A binderless, covalently bulk modified electrochemical sensor: Application to simultaneous determination of lead and cadmium at trace level. Analytica Chimica Acta. 2012:728:9-17. DOI:10.1016/j.aca.2012.03.053.

[17] Veeresh AS, Shambhulinga A, Manjunatha N, Manjunatha P, C.P. Keshavananda P, Lokesh KS, Nanomolar detection of lead using electrochemical methods based on a novel phthalocyanine. Inorganica Chimica Acta. 2020:506:119564, DOI:10.1016/j.ica.2020.119564

[18] Manju BG, Uma MK, John BBR, Design and development of amperometric biosensor for the detection of lead and mercury ions in water matrix—a permeability approach. Analytical and Bioanalytical Chemistry. 2017:409:4257-4266. DOI:10.1007/s00216-017-0376-9

[19] Ádám G, Viola H, Péter H, Tünde T, Fast Potentiometric Analysis of Lead in Aqueous Medium under Competitive Conditions Using an Acridono-Crown Ether Neutral Ionophore. Sensors (Basel). 2018:18:1407. DOI:10.3390/s18051407

[20] Ali D, Gustavo A. Z G, Gastón A C, Santiago M, Jordi R F. Xavier R, Nanostructured materials in potentiometry. Analytical and Bioanalytical Chemistry. 2011:399:171-181. DOI: 10.1007/s00216-010-3974-3

[21] Eric B, Ernö P, Nanoscale potentiometry. Trends in Analytical Chemistry. 2008; 27: 612-618. DOI: 10.1016/j.trac.2008.04.007

[22] Mei-Rong H, Yong-Bo D, Xin-Gui L, Lead-ion potentiometric sensor based on electrically conducting microparticles of sulfonic phenylenediamine copolymer. Analyst. 2013:138:3820-3829. DOI:10.1039/C3AN00346A

[23] Ádám G, Viola H, Péter H, Tünde T, Fast Potentiometric Analysis of Lead in Aqueous Medium under Competitive Conditions Using an Acridono-Crown Ether Neutral Ionophore. Sensors (Basel). 2018:18:1407. DOI:10.3390/s18051407

[24] Joseph W, Stripping Analysis, Encyclopedia of Electrochemistry: Online, 125-133. DOI:10.1002/9783527610426.bard030203.

[25] E.Marguí BR, Skorek T, Theato I, Queralt M, Hidalgo RS. Analytical possibilities of different X-ray fluorescence systems for determination of trace elements in aqueous samples pre-concentrated with carbon nanotubes. Spectrochimica Acta Part B: Atomic Spectroscopy.2013:88:192-197. DOI: 10.1016/j.sab.2013.07.004

[26] Rafal S, Paulina J, Beata Z, Ewa T, Eva M, Ignasi Q, Green Approach for Ultratrace Determination of Divalent Metal Ions and Arsenic Species Using Total-Reflection X-ray Fluorescence Spectrometry and Mercapto-Modified Graphene Oxide Nanosheets as a Novel Adsorbent. Analytical Chemistry. 2015:87:3535-3542. DOI: 10.1021/acs.analchem.5b00283,

[27] Marek T, Analytical applications of carbon nanotubes: a review. Trends in

Analytical Chemistry.2006:25:480-489. DOI:10.1016/j.trac.2005.11.008

[28] Matthew J. Allen, Vincent C. Tung, Richard B. Kaner, Honeycomb Carbon: A Review of Graphene. Chemical Reviews. 2010:110:132-145.DOI:10.1021/cr900070d

[29] Li ZJ, Xia QF, Recent advances on synthesis and application of graphene as novel sensing materials in analytical chemistry, Reviews in Analytical Chemistry. 2012:31 :57-81. DOI:10.1515/revac-2011-0039

[30] Antony RT, Boopathi S, Werden K, Aicheng C, Facile one-pot synthesis of fluorinated graphene oxide for electrochemical sensing of heavy metal ions, ElectrochemistryCommunications. 2017:76:42-46. DOI: 10.1016/j.elecom.2017.01.015

[31] Li J, Guo SJ, Zhai YM, Wang EK, High-sensitivity determination of lead and cadmium based on the Nafion-graphene composite film. Analytica Chimica Acta. 2009: 649:196-201. DOI: 10.1016/j.aca.2009.07.030

[32] Yasunori I, Akihiro H, Masao T, Toshihiko Y, Radiation deterioration in mechanical properties and ion exchange capacity of Nafion N117 swelling in water. Journal of Membrane Science. 2008:322:249-255. DOI: 10.1016/j.memsci.2008.05.045

[33] Ta-Yung C, Johna L, Ion exchange capacity of nafion and nafion composites. Langmuir. 2000:16:2866-2871. DOI: 10.1021/la991094x

[34] Chandre MW, Khotso T, Nazeem J, Priscilla GB, Emmanuel II, Metallo-graphene nanocomposite electrocatalytic platform for the determination of toxic metal ions. Sensors. 2011:1:3970-3987. DOI: 10.3390/s110403970

[35] Jingbo C,Guihua Z, Erik RC, Robert H, Junhong C, Graphene-based sensors for detection of heavy metals in water: a review. Analytical and Bioanalytical Chemistry.2014:406: 3957-3975. DOI: 10.1007/s00216-014-7804-x

[36] Gong JM, Zhou T, Song DD, Zhang LZ, Monodispersed Au nanoparticles decorated graphene as an enhanced sensing platform for ultrasensitive stripping voltammetric detection of mercury[II]. Sensors Actuators B Chemical. 2010:150:491-497. DOI: 10.1016/j.snb.2010.09.014

[37] Wei Y, Gao C, Meng FL, Li HH, Wang L, Liu JH, Huang XJ, SnO2/reduced graphene oxide nanocomposite for the simultaneous electrochemical detection of cadmium[II], lead[II], copper[II], and mercury[II]: an interesting favorable mutual interference. Journal of Physical Chemistry C. 2012:116:1034-1041. DOI: 10.1021/jp209805c

[38] Lin L, Chengyin W, Guoxiu W, Novel cysteic acid/reduced graphene oxide composite film modified electrode for the selective detection of trace silver ions in natural waters. Analytical Methods. 2013:5:5812-5822. DOI: 10.1039/C3AY40888D

[39] Kenneth KSL, José B, Kenneth BKT, Manish C, Gehan AJA, William IM, Gareth H M, Karen KG, Superhydrophobic carbon nanotube forests, Nano Letters. 3 2003:3:1701–1705. DOI: 10.1021/nl034704t

[40] Hui H, Ting C, Xiuyu L, Houyi M, Ultrasensitive and simultaneous detection of heavy metal ions based on three-dimensional graphene-carbon nanotubes hybrid electrode materials. Analytica Chimica Acta. 2014:852:45-54. DOI: 10.1016/j.aca.2014.09.010

[41] Gil HH, Won KH, Joon SP, Sung GK, Determination of trace metals by anodic stripping voltammetry using a bismuth-modified carbon nanotube electrode. Talanta. 2008:76:301-308. DOI: 10.1016/j.talanta.2008.02.039

[42] He X, Liping Z, Sujie X, Yuezhong X, Litong J, Microwave-irradiated synthesized platinum nanoparticles/carbon nanotubes for oxidative determination of trace arsenic[III].Electrochemistry Communications. 2008:10:551. DOI: 10.1016/j.elecom.2008.01.028

[43] Bin Z, Jia DC, Han Z, Ting TY, Mei LZ, Ming Z, Ming LD, Facile and green fabrication of size-controlled AuNPs/CNFs hybrids for the highly sensitive simultaneous detection of heavy metal ions. Electrochimica Acta. 2016:196:422-430. DOI: 10.1016/j.electacta.2016.02.163

[44] Alex LS, Sabine K, Eden ELT, Neil PY, Richard GC, Electrochemical Hg^{2+} detection at tannic acid-gold nanoparticle modified electrodes by square wave voltammetry. Analyst. 2018:143:2035-2041. DOI: 10.1039/C8AN00508G.

[45] Bikash KJ, C. Retna R, Gold Nanoelectrode Ensembles for the Simultaneous Electrochemical Detection of Ultratrace Arsenic, Mercury, and Copper. Analytical Chemistry. 2008:80:4836-4844. DOI: 10.1021/ac071064w.

[46] Jahir O, César FS, Cecilia JJ, Underpotential Deposition–Anodic Stripping Voltammetric Detection of Copper at Gold Nanoparticle-Modified Ultramicroelectrode Arrays. Environmental Science and Technology. 2008:42: 4877-482. DOI: 10.1021/es8005964

[47] Benzhi L, Liyuan L, Min W, Yanqin Z, A study of nanostructured gold modified glassy carbon electrode for the determination of trace Cr[VI]. Journal of Chemical Sciences. 2008:120: 493-498. DOI: 10.1007/s12039-008-0077-1

[48] Jingming G, Ting Z, Dandan S, Lizhi Z, Xianluo H, Stripping Voltammetric Detection of Mercury[II] Based on a Bimetallic Au–Pt Inorganic–Organic Hybrid Nanocomposite Modified Glassy Carbon Electrode. Analytical Chemistry. 2010:82: 567-573. DOI: 10.1021/ac901846a

[49] M Ángeles GR, Mara OM, Eduardo PG,Modification of carbon screen-printed electrodes by adsorption of chemically synthesized Bi nanoparticles for the voltammetric stripping detection of Zn[II], Cd[II] and Pb[II]. Talanta. 2009:80:631. DOI: 10.1016/j.talanta.2009.07.039

[50] Gyoung-JaL, Hi-MinL, Chang-KyuR, Bismuth nano-powder electrode for trace analysis of heavy metals using anodic stripping voltammetry. Electrochemistry Communications. 2007:9:2514-2518. DOI: 10.1016/j.elecom.2007.07.030

[51] Natalya AM, Natalya YS, Khiena ZB, Novel approach to bismuth modifying procedure for voltammetric thick film carbon containing electrodes. Electrochemistry Communications. 2007:9:221-227. DOI: 10.1016/j.elecom.2006.09.003

[52] Kathryn E. T, Lei X, Gregory GW, Electroanalytical determination of cadmium[II] and lead[II] using an antimony nanoparticle modified boron-doped diamond electrode. Richard G. Compton. Electroanalysis 2009, 21, 1113. DOI: 10.1002/elan.200904547/abstract

[53] Zhi Z, Ke Y, Dan B, Ziqiang Z. Synthesis and Electrochemical Sensing Toward Heavy Metals of Bunch-like Bismuth Nanostructures. Nanoscale Research Letters. 2010:5:398-402. DOI: 10.1007/s11671-009-9495-3.

[54] Abdul W, Muhammad M, Nisar U, Nanomaterials-based electrochemical detection of heavy metals in water: Current status, challenges and future direction.2018:105:37-51. DOI:10.1016/j.trac.2018.04.012

[55] Lei Y, PingZ, HongxiuD, LingC, Houyi M, Meng L, Dazhong S, An electrochemical sensor based on Co_3O_4 nanosheets for lead ions determination. RSC Advances. 2017:7:39611-39616. DOI: 10.1039/C7RA06269A

[56] Kresge CT, Leonowicz ME, Roth WJ, VartuliJC, Beck JS, Ordered mesoporous molecular sieves synthesized by a liquid-crystal template mechanism. Nature. 1992: 359:710-712. DOI: 10.1038/359710a0

[57] J. S. Beck JS, Vartuli JC, Roth WJ, Leonowicz ME, Kresge CT, Schmitt KD, Chu CT, Olson DH, Sheooard EW, A new family of mesoporous molecular sieves prepared with liquid crystal templates. Journal of American Chemical Society. 1992: 114:10834-10843. DOI: 10.1021/ja00053a020

[58] Frank H, Michael F, Vitalising porous inorganic silica networks with organic functions—PMOs and related hybrid materials. Chemical Society Reviews. 2011:40:608-620. DOI:10.1039/C0CS00076K

[59] Wassana Y, Yuehe L, Thomas S Z, Glen EF, Voltammetric detection of lead[ii] and mercury[ii] using a carbon paste electrode modified with thiol self-assembled monolayer on mesoporous silica [SAMMS]. Analyst. 2003:128:467-472. DOI: 10.1039/B300467H

[60] Yantasee W, Fryxell GE, Conner MM, Lin Y, Nanostructured Electrochemical Sensors Based on Functionalized Nanoporous Silica for Voltammetric Analysis of Lead, Mercury, and Copper. Journal of Nanoscience and Nanotechnology. 2005:5: 1537-1540. DOI: 10.1166/jnn.2005.324

[61] Wassana Y, Charles T, Glen EF, Brian PD, Yuehe L, Automated portable analyzer for lead[II] based on sequential flow injection and nanostructured electrochemical sensors. Talanta. 2005:68:256-261. DOI: 10.1016/j.talanta.2005.07.013

[62] Isabel S, Damia′ PQ, Heavy metal complexation on hybrid mesoporous silicas: an approach to analytical applications. Chemical Society Reviews. 2013:42:3792-3807. DOI:10.1039/c2cs35221d

[63] Mi-Young C, Juyoung Y, Anthony WC, Chelation-enhanced fluorescence chemosensing of Pb[II], an inherently quenching metal ion. Journal of Molecular Recognition. 1996:9:297-303. DOI:10.1002/[SICI]1099-1352[199607]9:4<297:AID-JMR335>3.0.CO;2-6

[64] Wen-Sheng X, Russell HS, Chao-Jun L, Joel TM, Chu-Ping L, Dirk MG, Chemosensors for Lead [II] and Alkali Metal Ions Based on Self-Assembling Fluorescence Enhancement [SAFE]. J. Phys. Chem. B, 2002:106:833-843. DOI: 10.1021/jp013274x

[65] Chao-Tsen C, Wan-Pei H, A Highly Selective Fluorescent Chemosensor for Lead Ions. Journal of Americal Chemical Society. 2002:124:6246-6247. DOI: 10.1021/ja025710e

[66] Takashi H, Dai Q, Masakazu M, Jong CL, ChangHK, Norio T, Highly selective recognition of lead ion in water by a podandfluoroionophore/γ-cyclodextrin complex sensor. Chemical Communication. 2003:2160-2161. DOI: 10.1039/B305758E

[67] Zaijun L, Zhenzhong Z, Yepu C, Chung-Gin H, Jiaomai P, Spectrophotometric determination of lead in biological samples with dibromo-p-methyl-methylsulfonazo. Talanta, 1999: 48:511-516. DOI: 10.1016/S0039-9140[98]00247-1

[68] Guozhen F, Yongwen L, Shuangming M, Yong G, Spectrophotometric determination of lead in vegetables with

dibromo-p-methyl-carboxysulfonazo. Talanta, 2002:57:1155-1160. DOI: 10.1016/S0039-9140[02]00195-9

[69] Ji YK, Yun JJ, Yoon JL, Kwan MK, Mi SS, Wonwoo N, Juyoung Y, A Highly Selective Fluorescent Chemosensor for Pb^{2+}. Journal of American Chemical Society. 2005:127:10107-10111. DOI: 10.1021/ja051075b

[70] Rémi M, Isabelle L, Bernard V,A highly sensitive and selective fluorescent molecular sensor for Pb[ii] based on a calix[4]arene bearing four dansyl groups. Chemical Communications. 2003:996-997. DOI: 10.1039/B301323E

[71] Sung KK, Seoung HL, Ji YL, Jin YL, Richard AB, Jong SK, An Excimer-Based, Binuclear, On–Off Switchable Calix[4]crown Chemosensor. Journal of American Chemical Society. 2004:126:16499-16506. DOI: 10.1021/ja045689c

[72] Pooja D, Prachi R, Anupma T, Ki-Hyun K, Praveen K, Recent advances in carbon quantum dot-based sensing of heavy metals in water.TrAC Trends in Analytical Chemistry.2019:114:171-195. DOI: 10.1016/j.trac.2019.03.003.

[73] Yongli L, Qingxiang Z, Yongyong Y, Yalin W, Hydrothermal synthesis of fluorescent carbon dots from sodium citrate and polyacrylamide and their highly selective detection of lead and pyrophosphate. Carbon.2017:115:550-560. DOI: 10.1016/j.carbon.2017.01.035

[74] Rajkumar B, Ramakrishn, Dadigalaa B, Reddy G, Veerabhadram G, Green synthesis of highly fluorescent nitrogen – Doped carbon dots from *Lantana camara* berries for effective detection of lead[II] and bioimaging. Journal of Photochemistry and Photobiology B: Biology. 2018:178:330-338. DOI: 10.1016/j.jphotobiol.2017.11.010

[75] Yongli L, Qingxiang, ZhouJing L, Man L, Xiuyi Y, Selective and sensitive chemosensor for lead ions using fluorescent carbon dots prepared from chocolate by one-step hydrothermal method. Sensors and Actuators B: Chemical.2016:237:597-604. DOI: 10.1016/j.snb.2016.06.092

[76] Ansi VA, Renuka NK, Table sugar derived Carbon dot – a naked eye sensor for toxic Pb^{2+} ions. Sensors and Actuators B: Chemical.2018:264:67-75> DOI: 10.1016/j.snb.2018.02.167.

[77] Youngjin K, Robert CJ, Joseph TH,Gold Nanoparticle-Based Sensing of "Spectroscopically Silent" Heavy Metal Ions. Nano Letters. 2001:1:165-167. DOI: 10.1021/nl0100116

[78] Karuvath Y, Binil II, Cherumuttathu HS, K. George T, In Situ Synthesis of Metal Nanoparticles and Selective Naked-Eye Detection of Lead Ions from Aqueous Media. Journal of Physical Chemistry C. 2007:111:12839-12847. DOI: 10.1021/jp073923q

[79] Yi-You C, Huan-Tsung C, Yen-Chun S, Yu-Lun H, Cheng-Kang C, Chih-Ching H, Colorimetric Assay for Lead Ions Based on the Leaching of Gold Nanoparticles. Analytical Chemistry. 2009:81:9433-9439. DOI: 10.1021/ac9018268

[80] Chai F, Wang C, Wang T, Li L, Su Z, Colorimetric Detection of Pb^{2+} Using Glutathione Functionalized Gold Nanoparticles, ACS Applied Materials and Interfaces. 2010:2:1466-1470. DOI:10.1021/am100107k

[81] Ha Na K,Wen XR,Jong SK,Juyoung Y, Fluorescent and colorimetric sensors for detection of lead, cadmium, and mercury ions. Chemical Society Reviews. 2012:41:3210-3244, DOI:10.1039/C1CS15245A.

[82] Jolyon T, Martin S, Image calibration and analysis toolbox – a

free software suite for objectively measuring reflectance, colour and pattern. Methods in Ecology and Evolution. 2015:6:1320-1331. DOI:10.1111/2041-210X.12439

[83] West PW, Selective spot test for copper.Industrial and EngineerinfChemistry:AnalyticalEdition. 1945:17:740-41. DOI: 10.1021/i560147a024

[84] Martinez AW, Phillips ST, Butte M J, Whitesides GM, Patterned paper as a platform for inexpensive, low-volume, portable bioassays. Angewandte Chemie International Edition. 2007:46:1318-1320. DOI: 10.1002/anie.200603817

[85] Myung G C, So Y P, Ka Y P, Suk-Kyu C, Novel Hg^{2+}-selective signaling Probe Based on Resorufin thionocarbonate and its μpAD Application. Scientific Reports. 2019: 9:3348-3355, DOI:10.1038/s41598-019-40169-6

[86] Kwi Nam H, Jong-Soon C, Joseph K, Gold nanozyme-based paper chip for colorimetric detection of mercury ions. Scientific Reports.2017:7:2806-2812, DOI:10.1038/s41598-017-02948-x

[87] Weiwei C, Xueen F, Hua L, Hongmei C, Jilie K, A Simple Paper-Based Colorimetric Device for Rapid Mercury [II] Assay. Scientific Reports.2016:6:31948-31954, DOI:10.1038/srep31948

[88] Yan Z, MengL, HaiyunL, Shenguang G,Jinghua Y, Label-free colorimetric logic gates based on free gold nanoparticles and the coordination strategy between cytosine and silver ions. New Journal of Chemistry. 2016:40:5516-5522. DOI: 10.1039/C5NJ03471J,

[89] Peuli N, Ravi Kumar A, Nripen C, Smart gold nanosensor for easy sensing of lead and copper ions in solution and using paper strips. RSC Advances. 2015:5:69024-69031. DOI:10.1039/C5RA14886C,

[90] Jian-feng G, Dan-qun H, Mei Y, Chang-jun H, Jun-jie L, Huan-bao F, Hui-bo L, Ping Y, Colorimetric detection of Cr [VI] based on the leaching of gold nanoparticles using a paper-based sensor. Talanta. 2016: 161: 819-825. DOI: 10.1016/j.talanta.2016.09.032

[91] Niez D F, Liux A, Whitesides G M, Integration of paper-based microfluidic devices with commercial electrochemical readers. Lab on a Chip. 2010:10:3163-3169

[92] Mariana MS, Miquel C, Josep R, Arben M, Eco-friendly electrochemical lab-on-paper for heavy metal detection. AnalyticalBioanal Chemistry. 2015:407:8445-8449.DOI:10.1007/s00216-015-9022-6

[93] Ruecha N, Rodthongkum N, Cate DM, Volckens J, Chailapakul O, Henry CS, Sensitive electrochemical sensor using a graphene-polyaniline nanocomposite for simultaneous detection of Zn[II], Cd[II], and Pb[II]. Analytica Chimica Acta. 2015: 874:40-48. DOI: 10.1016/j.aca.2015.02.064

[94] Poomrat R, Wijitar D, David C, John V, Orawon C, Charles S. H, Multilayer Paper-Based Device for Colorimetric and Electrochemical Quantification of Metal. Analytical Chemistry. 2014:86:3555–3562. DOI:10.1021/ac5000224

Chapter 14

Bio Hydrometallurgical Technology, Application and Process Enhancement

Mulugeta Sisay Cheru

Abstract

The review is in general try to see some of the major microorganism involved in bioleaching process and studied by different scholars, identify the mechanics and techniques employed to bioleach minerals and factor that enhance or to inhibit the leaching process of microorganism with major reaction taking while bioleaching. Here the methodology and different leaching technique presented with their respected pros and cons, which are commonly employed and reasons behind with justifiable evidences were presented. The values and bioleaching sulfide mineral (copper), precious metal (gold) and radioactive element (uranium) were discussed with some the known producers in the world and finally some highlight given on industrial application of bioleaching.

Keywords: bioleaching, leaching techniques, biooxidation, radioactive, pretreatment

1. Introduction

Bioleaching is the extraction of metals from ores using the principal components water, air and microorganism [1]. It is the extraction or mobilization of valuable (target) metal from the ore, can also be defined as a process of recovering metals from low grade ore [2, 3], with regard to solubility, bioleaching can be defined as a process of recovering soluble one from insoluble impurities after dissolving sulfide metal as soluble salt in a solution [4] that results toxics and heavy metals removed. It is isolation of metals from their ores, concentrates and mineral wastes under the influence of microorganisms leading to dissolution of metal solutions of leach liquor containing metals [5], followed by solvent extraction, stripping, ion exchange, electro wining to get pure metal.

Both bioleaching and biooxidation leads to recovery target mineral; but there is technical difference between the two technologies. Bioleaching refers to the use of bacteria, the common Thiobacillus Ferrooxidans and other bacterial as a leachant to leach sulfide minerals where the target elements remains in the solution during oxidation process, after the metal recovery the solid left behind regarded as residue and in the contrary biooxodation discard the solution after having metal values in solid phase [6, 7] microorganism also engaged in removal of radionuclides and leaching of metal that are regarded as toxic in some cases and good for bioremediation of soil, the process stops radio nucleation that result the removal of stability of target elements [7].

Bioleaching has been used for a long period of time without regarded as microbial leaching process; it has been used as early as 1000 BC when a man from metal laden recovered copper from a water, passes through copper ore deposit [8]. It was in 1556 at the mine located in Spain at Rio Tinto (Rd River) mine, slurry containing very high concentration of ferric ions leached due to the action of microorganisms [4]. Copper was precipitated from the solution obtained from this river, the very first bio mineralization process was copper dissolution, then the process continued to be developed in countries like Norway, Germany and English at different era of time, in the year 1947 heap and dump leaching was practiced that leads to the development of bacterial bioleaching process [9].

The gram-negative chemolithotroph bacterial, Thiobacillus Ferooxidans was first cultured and isolated from mine water by Colmer and Hinkel [9]. Thiobacillus Ferrooxidans is rod shaped ranging in diameter from 0.3 to 0.8 micrometers (μm), in length from 0.9 to 2 μm, 0.5 μm in width in which its movement is due to a single polar flagellum [10]. Since now this bacteria is the most studied. These bacteria were able to oxidize sulfur to sulfuric acid and ferrous to ferric in acidic environment where pH value is less than 5 [7, 10, 11]. From this point onwards the technology of bioleaching has shown tremendous growth, especially industrial coppers production, which makes annualized world copper production reach up 10% from 0.2%. It was in Chile the first industrial scale copper bioleaching plant was established in 1980 using Thiobacillus bacteria [12] large-scale production begins and bioleaching taken as main manufacturing process as any convection techniques in Chile 1984 [13]. Among the many microorganism involved, bacteria (autotrophic and heterotrophic), fungi and yeasts can be mentioned. The bacterium has these calcification based on their species as Thiobacillus Ferrooxidans, Leptospirillum Ferrooxidans, Thiobacillus Thiooxidans, Sulfolobus, but there are many classifications based on different characteristics reveled by the organisms.

Acidophilic Thiobacillus species are used to leach refractory elements like gold, they generally characterized as aerobic, acidophilic, and autotrophic used to leach sulfide minerals (copper, nickel, zinc and soon). Most common bacteria involved in bioleaching are Acidithiobacillus Ferrooxidans (Thiobacillus Ferrooxidans), Acidithiobacillus Thiooxidans, Leptospirillum Ferrooxidans, Sulpholobus Spp, Sulpholobus Thermosulphidoxidans and Sulpholobus Brierleyi. Acidithiobacillus Ferrooxidans is most vital one, which was named and characterized in 1951. Most common fungi are Aspergillus Niger and Penicillium Simplicissimum. The efficiency of bioleaching depends up on physiological requirement and capability of Thiobacillus to oxidize ferrous ion (Fe^{2+}) and sulfur (S). There are five main species of Thiobacillus, these are Thiobacillus Thioparus, Thiobacillus Dentrificans, Thiobacillus Thiooxidans, Thiobacillus Intermedius, and Thiobacillus Ferrooxidans. On the bases of pH values for growth genus Thiobacillus can be divided into two groups, those that can grow only in neutral pH values are T. Thioparus and T. Dentrificans. The second Thiobacillus species those grow at lower pH value are T. Thiooxidans, *T. intermedius*, and T. Ferrooxidans.

Study of different scholars at the inceptions shows the capability of bacteria (genus Thiobacillus) to oxidize sulfur compounds to sulfuric acid; it can oxidize also range of sulfur compounds (S^{2-}, S^0, S_2O_4, $S_2O_4^{2-}$, SO_4^{2-}) [11], followed by separation process of the iron and the bacteria Acidithiobacillus Ferrooxidans (Thiobacillus Ferrooxidans) from the solution. A. Ferrooxidans is found in drainage waters and it is commonly identified as pyrite oxidizer [14]. The bacterial (acidophile) obtain energy from inorganic sources, it grows in acidic medium that fixes carbon to the bacteria itself. Most economically important metals like iron, copper, gold, and uranium can be easily extracted by using acidophilic and chemo-litho-autotrophic microorganism. Acidithiobacillus Ferrooxidans is chemoautotrophic microorganism or acidophilic.

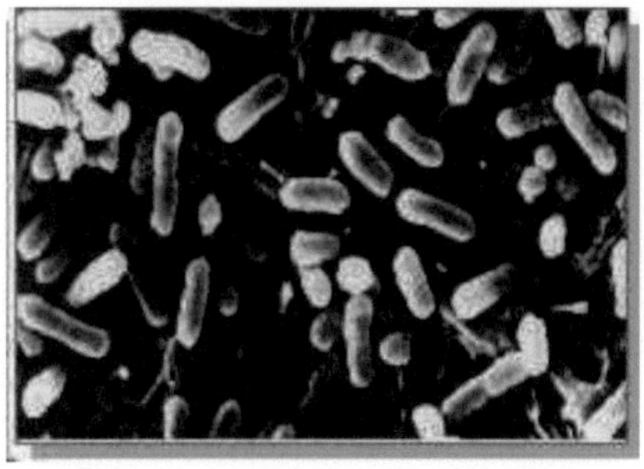

Figure 1.
Image of bioleaching bacterial [4].

Let see the ecology, physiology, availability and genetics of microorganism involved in bioleaching. There are three basic principles for microorganism to leach and mobilize target metals from ore concentrate – redox reaction, formation of organic and inorganic acid and finally the excretion of complexing agent (**Figure 1**) [4].

Here is a generalized reaction used to express biological oxidation of sulfide mineral.

MS + $2O_2 \rightarrow MSO_4$, where M is bivalent metal and reaction below show a metal sulfide directly oxidize by Acidithiobacillus Ferrooxidans to soluble metal sulfate according to the reaction.

MS + $2O_2 \rightarrow M^{+2}$ + SO_4^{+2} [15].

2. Mechanisms of bioleaching

The two majorly known mechanism in bacterial leaching are direct mechanism (involves physical contact of the organism with the insoluble sulphide) or hypothesized enzymatic reaction taking place between an attached cell and the underlying mineral surface which is independent of indirect mechanisms and it is where reduced sulfur dissolution takes place [16], it is only the direct attack of the bacteria can lead to leaching. Check the following reactions.

$$2FeS_2 + 7O_2 + 2H_2O \rightarrow 2FeSO_4 + 2H_2SO_4. \quad (1)$$

Indirect (involves the ferric-ferrous cycle) or it is a mechanism of sulfide oxidation involves non-specific oxidation of surfaces by Fe^{3+} that is generated by microorganisms that oxidize iron or oxidation of mineral by ferric ions [16]. The attached cells of bacterial oxidize the surface using either of the two mechanisms [9, 11, 14]. The reaction below shows oxidation of iron.

$$FeS_2 + Fe_2(SO_4)_3 \rightarrow 3FeSO_4 + 2S^0. \quad (2)$$

Minerals are broken due to the attack to their constituents, that results energy production for the microorganism. This energy production or oxidation passes through intermediates reaction processes. Two mechanisms have been proposed for the oxidation, viz. thiosulphate mechanism and polysulfide mechanism. Thiosulfate mechanism includes acid-insoluble metal sulfides like pyrite (FeS_2) and molybdenite (MoS_2) and polysulfide mechanism includes acid-soluble metal sulfides like chalcopyrite ($CuFeS_2$) or galena (PbS) [15]. In thiosulfate mechanism, the attack of ferric ion on acid insoluble metal sulfides brings about solubilization via thiosulfate as an intermediate and sulfates as end product. The breaking reaction shown below.

$$FeS_2 + 6Fe^{3+} + 3H_2O \rightarrow S_2O_3^{\ 2-} + 7Fe^{2+} + 6H^+. \quad (3)$$

$$S_2O_3^{\ 2-} + 8Fe^{3+} + 5H_2O \rightarrow 2SO_4^{\ 2-} + 8Fe^{2+} + 10H^+. \quad (4)$$

In polysulfide mechanism, a combined attack of ferric ion and protons on acid-soluble metal sulfides causes the solubilization with sulfur as an intermediate in its elemental form which can be oxidized to sulfate by sulfur-oxidizing microbes that the reaction is shown below [17].

$$MS + Fe^{3+} + H^+ \rightarrow M^{2+} + 0.5\,H_2Sn + Fe^{2+}\,(n \geq 2). \quad (5)$$

$$0.5\,H_2Sn + Fe^{3+} \rightarrow 0.125S_8 + Fe^{2+} + H^+. \quad (6)$$

$0.125\ S_8 + 1.5O_2 + H_2O \rightarrow SO_4^{\ 2-} + 2H^+$ the reaction show the production of sulfuric acid results hydrogen (proton) for attacking mineral.

Fe (II) is re-oxidized to Fe (III) by iron oxidizing organisms (chemotrophic bacteria), the role of microorganisms in solubilization.

$2Fe^{2+} + 0.5O_2 + 2H^+ \rightarrow 2Fe^{3+} + H_2O$ this reaction keep iron in ferric state that oxidize mineral.

The process of chemical attack takes place on a substrate or the mineral surface where the bacteria forms a composite and attach itself as firm as possible in order to increases maturity that finally detached and dispersed into the solution.

An important reaction mediated by Acidithiobacillus Ferrooxidans is:

$$4FeSO_4 + O_2 + 2H_2SO_4 \rightarrow 2Fe_2\,(SO_4)_3 + 2H_2O. \quad (7)$$

Strong oxidizing agent, ferric sulfate that basically used to dissolve metal sulfide minerals, and leaching brought about by ferric sulphate is termed indirect leaching due to the absence of both oxygen and viable bacterial. Check the following leaching mechanism of reaction on several minerals.

$$CuFeS_2\,(\text{chalcopyrite}) + 2Fe(SO_4)_3 \rightarrow CuSO_4 + 5FeSO_4 + 2S. \quad (8)$$

$$FeS_2\,(\text{Pyrite}) + Fe_2\,(SO_4)_3 \rightarrow 3FeSO_4 + 2S. \quad (9)$$

$$UO_2 + Fe_2\,(SO_4)_3 + 2H_2SO_4 \rightarrow UO_2\,(SO_4)_{4-3} + 2FeSO_4 + 4H^+. \quad (10)$$

$$4FeS_2 + 15O_2 + H_2O^{\text{Acidithiobacillus ferrooxidans}}\ 2Fe_2\,(SO_4)_3 + 2H_2SO_4. \quad (11)$$

$$4CuFeS_2 + 17O_2 + 2H_2SO_4{}^{\text{Acidithiobacillus ferrooxidans}}\ 4CuSO4 + 2Fe_2\,(SO_4)_3 + 2H_2O. \quad (12)$$

$$2Cu_2S + 5O_2 + 2H_2SO_4{}^{\text{Acidithiobacillus ferrooxidans}}\ 4CuSO_4 + 2H_2O. \quad (13)$$

$$CuS + 2O_2{}^{\text{Acidithiobacillus ferrooxidans}}\ CuSO_4. \quad (14)$$

Acidithiobacillus Ferrooxidans can convert elemental sulfur generated by indirect leaching to sulfuric acid –.

$$2SO + 3O_2 + 2H_2O^{\text{Acidithiobacillus ferrooxidans}} 2H_2SO_4. \tag{15}$$

This sulfuric acid maintains the pH value at levels, which is favorable to the growth of bacteria and also helps for effective leaching of oxide minerals:

$$CuO(\text{Tenorite}) + 2H_2SO_4 \rightarrow CuSO_4 + H_2O. \tag{16}$$

$$UO_3 + 3H_2SO_4 \rightarrow UO_2(SO_4)_3 + H_2O + 4H^+. \tag{17}$$

Chemolithotrophic (uses carbon for the synthesis of new cell material) bacteria can be categorized based on response to temperature as mesophiles, moderate thermoacidophiles and extreme thermoacidophiles.

Mesophiles-grows at a temperature values ranges (28°C -37°C) where Thiobacillus Ferrooxidans is able use the inorganic substrate to draw energy by oxidizing Fe (II) to Fe (III) and sulfur to sulfide and sulfate. The other mesophiles is Leptospirilium Ferrooxidans that use Thiobacillus Ferrooxidans to effect the oxidization of sulfur to sulfate. Moderate thermoacidophie-temperature values ranges (40–50°C), Sulfobacillus Thermosulfidooxidans is common one, which oxidize both sulfur and iron from energy production. This category includes Archaea and Eubacteria, and most of gram-positive microorganisms are included here. Extreme thermoacidophiles-temperature ranges 60–80°C, genera Sulfolobus, Acidanus, Metallosphaera and Sulfurococcus are in this category, [11, 18, 19]. Thermal value some time extends above the limitation values, it is due to exothermal reaction which is above the maximum growth temperature of microorganism, some microorganism genus like Archaea withstand thermal values up to 90° [19, 20].

This category is formed by closely related species that can act together with a common name given Sulfolobus. Sulfolobusa Acidocaldarius, Sulfolobus Sofataricus, Sulfolobus Brierley, and Sulfolobus Ambioalous that can oxidize Fe (II) to Fe (III) and sulfur to sulfate. Aspergillus Niger and Penicillium Simplicissimum are both used to leach sulfide minerals like copper with mobilization rate of 65% and aluminum, nickel, lead and zinc by more than 95% mobilization rate. Thiobacillus and Leptospirillum are characterized by the oxidation of sulphide minerals in acidic environment and temperature values less than 35°C, with regard to area of application these two are mostly used in dump and tank leaching of metal from sulphide based mixed ores [20, 21]. The other group of genus Sulphobacillus used under the same areas of application but relatively higher temperature up to 60°C, the temperature reaches up to 90°C in case of genera Sulpholobus and Acidianus, Organotrophic microorganisms like yeast, fungi and algae which destruct sulphide mineral and aluminum silicate, facilitate bio sorption of metals that solubilize gold, these microorganism uses carbonate and silicate ore for the extraction of metals and selective gold extraction from ore floatation and metal solution.

2.1 Autotrophic and heterotrophic leaching

The two bacterial leaching namely autotrophic and heterotrophic leaching has their distinct characteristics while bioleaching process takes place, in case of autotrophic bioleaching (effective on sulfide minerals) there are two proposed mechanism of Acidithiobacillus Ferrooxidans action on sulfide minerals, first the mechanism, that the bacterial oxidize ferrous ion to ferric ion in which the bulk solution where the mineral is leached counted as indirect, this mechanism which is indirect oxidation of ferrous ions to ferric ions is exopolymeric process, both takes place on the layer where the mineral is leached. The second proposed mechanism, does not require ferrous or ferric ions, the bacteria directly oxidize the minerals by biological means having direct contact mechanism of reaction. Autotrophic leaching uses both Thiobacillus Ferooxidans and Thiobacillus Thiooxidans to leach sulfide mineral and studies shows combining the two bacterial results an increase in selectivity and rate of leaching efficiency while leaching of nickel sulfide. From the heterotrophic genus of bacteria Thiobacillus and Pseudomonas are those used to leach non-sulfide minerals and from the genus of fungi Penicillium and Aspergillus (heterotrophic fungi) are those used in leaching process, a study shows 55–60% leaching rate for nickel and cobalt, some other studies indicates that 95% and 92% leaching rate achieved while using pretreated Aspergillus Niger by ultrasound for 14 and 20 days respectively which increase its stability [4, 11, 20, 22] (**Table 1**).

Heterotrophic fungi Aspergillus and Penicillium species combined to leach low-grade nickel-cobalt oxide ores, low-grade laterite ores and spudumene (aluminosilicate), these aluminosilicate (spudumene) also leached by heterotrophic yeasts (*Rhodotorula rubra*), Aspergillus Niger used to leach zinc and nickel silicate [11]. Bacterial leaching can be generalized in three mechanism redoxolysis, acidolysis, complexolysis, and in case fungal leaching bioaccumulation is important mechanism. To solubilize rock phosphorous, Aspergillus Niger has been used in many occasions due to the production of organic acids with low molecular weight and phosphorous is basically essential micronutrients for the growth of bio organism, these microorganism convert insoluble phosphate to soluble, the two filamentous fungi used in phosphate leaching are Aspergillus Niger and some Penicillium, the metabolic fungal reaction produces organic acid that result the formation of acidolysis, complex and chelate [22].

The second group of bacterial genus is Leptospirillum, which is categorized in moderate thermophilic bacteria that can only oxidize ferrous ions; it is dominate iron oxidizer, which is referred as Leptospirillum Ferrooxidans (L. Ferrooxidans). Oxidation process takes place under strong acidity and temperature up to 30°C, L. Ferrooxidans has high affinity to Fe^{2+} and low affinity to Fe^{3+} which results a working condition of high Fe^{3+}/Fe^{2+} ratio, when redox potential is low, L. Ferrooxidans has low growth rate at the initial stage of a mixed batch culture, a native strain of Leptospirillum Ferrooxidans used to leach zinc from low grade sulfide complex from La Silvita and La Resbalosa (Patagonia Argentina) [23]. The leach liquor itself has been a place where microorganism found, higher amount of

Microorganism/ both autotrophic and heterotrophic	Ore sample
Aspergillus Niger, Hyphomicrobium	Flourapatite (phosphorus ore)
Pseudomonas Oryzihabitans	Magnesite, Dolomite (magnesium ore)
Bacillus Licheniformis	Silica
Thiobacillus Ferrooxidans, Acidianus Brierleyi, Sulfobacillus, Thermosul Fidooxidans, Sulfolobus Rivotincti	Chalcopyrite (Low and high grade), Pyrite, Covellite
Penicillium Simplicissimum, Penicillium Verruculosum, Aspergillus Niger, Acidithiobacillus Ferrooxidans	Iron ore, Hematite, Zinc and nickel Silicates
Thiobacillus Thiooxidans	Pyrrhotite
Thiobacillus Caldus	Arsenopyrite
Metallosphaera Sedula, Sulfolobus Metallicus (BC),	Pyrite
Paenibacillus polymyxa	Bauxite (low grade)

Table 1.
Some of microorganism and leachable ore [4, 11].

Leptospirillum Ferriphilum were in a leach liquor, in a study conducted to leach the effect of pH on the bioleaching of a low-grade, black schist ore from Finland using Acidithiobacillus Ferrooxidans and Leptospirillum Ferrooxidans as extractant [24]. The bacteria can relatively resist high concentration of uranium, molybdenum, and silver, this is due to its affinity towards to ferrous ions or resistivity to refractory elements, but it cannot oxidize sulfur or any sulfur related compounds. By combing it with other sulfur- oxidizing acidophiles, sulfur-oxidizing process can be achieved; these are T. Caldus, T. Ferrooxidans, or T. Thiooxidans, to oxidize sulphidic gold concentrate a mixed culture of Thiobacillus and Leptospirillum has been used [11].

The third group thermophic bacteria mainly characterized by oxidation of iron to assure growth chemolithorophically, some are facultative autotrophs that require synergetic effect of other microorganism to like yeasts extract, cysteine, or glutathione. Among the microorganism in this group Sulfolobus species is the major one, these organism categorized as moderate thermophilic at an average values of temperature 40°C -60°C and the second group is extreme thermophilc at an average values of temperature 65–85°C. One of the moderate thermophilic gram positive bacteria, Sulfobacillus Thermosulfidooxidans is facultative autotrophs in which its growth stimulated by yeasts extract, where the presence of CO_2, weight and volume ratio (w/v) are factor to facilitate and inhibit growth. From of extreme theremophilic Sulfolobus Acidocaldarius and Acidianus Brierleyi are those in genera Archaebacteria, among the other four genera Sulfolobus, Acidanus, Metallosphaera, and Sulfurococcus act aerobically and categorized in extreme thermophilic acidophilic bacterial which oxidizes ferrous and elemental sulfur and sulphide based minerals. These bacteria grows under all conditions (auto, mixo, heterotrphic) depending on the yeast extract ratio (w/v), found in facultative chemolithotrophic species act in acidic medium and temperature value can be up to 90°C [11].

All the major concepts of bioleaching have been discussed, so what are factors affecting rate of bioleaching and leaching efficiency, the major factors can be summarized as microbiological, mineralogical and physiochemistry factors. A physiochemistry factor includes temperature, pH, redox potential, oxygen content, carbon dioxide content which facilitate mineral oxidation required for cell growth, mass transfer, light, surface tension which mean that the topography of mineral surface that indicate the rate adsorption and crystal structure which has direct relation on the rate of reaction. Microbiological factors includes microbial diversity that is the distinct nature of micro organisms with regard to range of unicellular organisms, variety of microorganism found in an environment suitable for bioleaching, these includes bacteria, fungi, algae, flagellates, and those found in microbial biocenosis, the other microbiological factors are population diversity, metal tolerance, spatial distribution microorganism and adaptation ability of microorganism. The third major factor is the nature of mineral processed, characteristics like grain size which affect rate of dissolution, porosity related to rate of chemical attack and digestion rate, hydrophobicity is another physiochemistry factor to determine the rate leaching, hydrophobicity is differentiating whether the elements are water hating or loving while floatation takes place. Process is the other major factor affecting leaching efficiency, techniques where bioleaching process takes place (heap, dump, in situ) which we will be discussed below, pulp density is the variable which results variation on dissolution rate, a study shows that dissolution metal increases while pulp density increases but it is based on (w/v) ratio that is between 5 and 20%, the other factor is concentration of target mineral, this can inhibit the growth of microorganism, that cause a limitation of pulp density usage [25]. Stirring speed is also another factor affecting rate of dissolution and geometry of the heap during heap leaching process, the other major factor is the presence of fluoride released from the ore sample, which inhibits the process of bioleaching, and when the release decreases the rate of inhibition eventually reduced.

Besides leaching process microorganisms are used for bioremediation of mining sites, treatment of gangue, tailing, and mineral wastes from the industry, contamination of sediments due heavy metals and soil from toxicities, sewage sludge can cured by microorganism in which the process is called bioremediation [26].

3. Bacterial leaching techniques

The successful bioleaching process is characterize by the intimacy of microorganisms to a mineral surface, strong attachment result high rate of oxidation and dissolution on a substrate (mineral surface), this is achieved by the rate success of bio film formation. In general leaching techniques are two – Percolation leaching – a solution infiltrate through a fixed mineral location, and agitation leaching - mineral bearing ore stirred by a solution but while working in large scale, percolation leaching is usually chosen [7]. The principal commercial methods are aerated stirred-tanks, in situ, dump, heap, vat, bench scale, tank, column, reactor leaching are among the many. It was dump bioleaching process taken as the first commercial bioleaching in 1950 used to leach copper from sulfide minerals, since then bioleaching bloomed by copper oxide heap leaching, industrial microbial leaching process applied for sulfidic gold and bioheap commercial leaching of copper ore (chalcocite and covellite). The high production of bio heap leaching of copper in 1980 established at Lo Aguirre mine in Chile processing 16,000 tones ore/per year at the inception [27], these wipe the way that led to Chile's industrial bio copper production in large scale especially from the year 1984 [28].

3.1 Stirred-tank biooxidation

Aerated, stirred-tank bioreactors, used in mineral concentrate feeds, involve a series of stages that can have lots of tanks connected in parallel depending on the retention of the concentrate [7] a study conducted to check Na-chloride can possibly enhances the chemical and bacterial leaching of chalcopyrite uses three bioreactors engaged with inoculum of the bacteria [29]. Other tanks needed for value adding purposes which are usually single tanks might be connected in series, since these tanks subject to chemical attack, air, heat and sulfide mineral, they should be relatively resistant to corrosion, chemical attack, and soon, in order to have these character tanks can be lined with rubber, galvanized, or other corrosion protection method like using sacrificial anode or using high grade material like stainless steel, aluminum or copper.

Temperature maintained at optimum level by cooling coil or some time tanks are equipped with water jackets depending on the required temperature by the bacteria, these values can be conditioned based of the mineral to be leached, and sometimes the chemical used to enhance the leaching process [29]. Several tanks can be continuously arranged, named as continuous stirred tank reactor (CSTR), as per the above it can be followed by a series of small equal sized reactors [16]. Example of bioleaching of sulfidic gold concentrates, that the discharge from the final stage is subjected to water washing and solid/liquid separation in thickeners. Even though there is less power consumption basically used for agitator and blower, it has linear relationship with the amount of sulfide -sulfure which is required to oxidize and recover the target metal from the parent ore, rate of recovery depends up on the metal grade also.

The main advantages of these tank over other conventional methods like pressure autoclave, roasting, smelting, calcination and soon are; it has low capital and operational cost, relatively less construction period, less complicated requiring less skilled man power and most importantly it is environmental friendly. In general

Australia, Chile, USA, Brazil and South Africa are among those countries involved in bio oxidation by stirred tank [7, 16].

3.2 Dump leaching

Dump (run of mine) [18] leaching involves uncrushed waste rock and low grade ore is piled up or changing a pit to dump by blasting it. Conventional methods would be very expensive to process these type of ores samples, except dump leaching, dump can be very huge, containing in excess of 10 million tons of waste rock, up to 60 m deep [7, 30].

In order to digest some of unwanted minerals like silicate and to promote the growth of acidophilic microorganism, acid water solution is spread on the top surface, the acid water solution percolate through the dump, the more acidic the environment the more growth of microorganisms that oxidize minerals to be recovered. The pregnant leach liquor or acid run-off is collected at the bottom of the dump, from where it is pumped to a recovery station. After collection the process followed by solvent extraction, electro wining for the metal production but dump aeration is vital for the microorganism to growth, tailing from solvent extraction recycled on the top of dump. Escondida mine found Chile is the biggest bio dump leach in world [26, 30].

3.3 Heap leaching

Heap leaching (crushed and agglomerated) [18] is composed of air, acid and microorganism where commutation takes place on rock samples to turn it to smaller particles which increases the surface area for acid digestions and conditioning it to microorganisms, particle should not be very fine and should be piled allowing a simplifies aeration pipe placed to facilitate air flow. To improve drainage of the mineral containing solution from the bottom of the ore, conditioned ore is spread on specially engineered pads (lined with high-density polyethylene (HDPE)), which consist of perforated plastic drain lines and air also supplied to optimize the growth of microorganism [7]. Heap can be large up to kilometer long, but commonly less than 500 m wide and 10 m long, the size and height of a heap depends up on air (for bacterial to grow) water, acid, heat generated due to the process and its dissipation [31]. Heap surface should be permeable enough for the sulfuric acid to infiltrate and dissolves iron to ferric solution producing ferric ion that react with copper sulfide results ferrous ion and copper solutions. Acidithiobacillus Ferrooxidans oxidize iron where the bacterial can be inoculated and works by attaching itself to ore, with having free movement. After collecting PLS (pregnant leach solution), then solvent extraction is followed where the target mineral recovered and formed into cathodes.

This aerobic bacteria works only in the presence of oxygen in the heap, those bacteria consume it from the solution where oxygen is in liquid phase. This process enhances the conversion of ferrous to ferric ions as per the reaction below.

$$Fe^{2+} + 0.25O_2 + H^+ \rightarrow Fe^{3+} + 0.5H_2O. \tag{18}$$

Heap some time can be crushed 19 mm with rotating drum with acidified water [29] aeration can be conducted using low pressure fans those directing air through piping on the pad [26]. It is clear that heap leaching requires the preparation of the ore, primarily size reduction, so as to maximize mineral-lixiviant interaction and lay of an impermeable base to prevent lixiviant loss and pollution of water bodies. Heap leaching basically used to leach low-grade ore of copper and zinc, even in the case of copper grade level can be (0.2–2%). To have an effective heap leaching process a mathematical model has been developed by taking heat, mass transfer, liquid, gas flow and chemical process in to consideration [31]. Heap also employed to bioleach silicate mineral, in a study where two microorganism were tested 'Ferroplasma acidarmanus or the common Acidithiobacillus ferrooxidans against the amenity of silicate minerals. Beside oxidation process energy was generated from flat plate solar energy collectors where heap is designed by HeapSim, heap bioleach simulation tool was used to simulate the heap and process occurring in the heap, even calculating the copper output [32].

3.4 In situ bioleaching

In situ leaching requires making the ore permeable for a solution and air to be circulated through the ore body. It does not require metal containing material to be removed from the ground [18]. It employs a method of recovering target minerals from the leach solution. The acid solution percolates until it reaches to impermeable layer. In situ includes recovery of minerals from the intact ore. The resulting metal-enriched solutions are recovered through wells drilled below the ore body. In case of in situ leaching the main concern is pollution of ground water, with this regard there are three types of ore bodies generally considered for in situ leaching: surface deposits above the water table, surface deposits below the water table and deep deposits below the water table. It is burden materials, establish permeability allowing air to pass in which metal bearing solution collected in the sumps [7]. It is combined with mineral recovery operation time and again to pull out the minerals from recovered fluid or pregnant solution or leachant. Acidified leach solutions, applied to the top surface of the entire ore zone, infiltrate through the fragmented ore due to the blast. The leaching bacteria become established and facilitate metal extraction. Metal-rich solutions or large volume of solution is circulated with the aid of gravity flow and pumped and recovered in sumps then again pumped to the surface for metal recovery, the returning fluids to the extraction operation are known as "barren solution". Metal recovery depends on two major things first the bacteria used (Acidithiobacillus Ferrooxidans) and permeability of the ore-body, which can be increased by fragmenting of ores in place, called "rubblizing". Due to the ground water pollution this leaching process becomes less used and less popular [18] on the contrary it has been said that it is a best substitute for open pit and shaft mining operation, basically when in situ leaching is applied, no gangue or tailing is byproduct, it also called green mining or mine of the future [33].

4. Bioleaching of some elements

4.1 Bioleaching of uranium

Recent study shows that elements like uranium, copper, gold, zinc and other elements are commercial focus of bioleaching and biooxidation [34]. Many studies indicate microbial leaching is more important in low-grade ore, ore sample collected from Mianhuakeng uranium mine located in northern Guangdong province in China, leached by heap, by mixed microorganism of Acidithiobacillus Ferrooxidans and Leptospirillum Ferriphilum with 88.3% leaching efficiency [35]. Uranium leaching takes places by indirect mechanism, as Acidithiobacillus Ferrooxidans does not directly interact with uranium minerals. The role of Acidithiobacillus Ferrooxidans in uranium leaching is the best example of the indirect mechanism. Bacterial activity is limited to oxidation of pyrite and ferrous iron. The process involves periodic spraying or flooding of worked-out stops and tunnels of underground mines with lixiviant [4]. The pH of lixiviant was optimized during the bioleaching of uranium from low grade Indian silicate-apatite ore with 0.024% of U_3O_8. This study uses Acidithiobacillus Ferrooxidans for leaching and biochemically generated ferric ions as an oxidant, optimizing particle, pulp density and redox potential results 98% uranium bioleaching. In this indirect bioleaching of uranium, the bacteria generate ferric sulfate and pyrite is oxidized by a lixiviant, within acidic environment the oxidations of ferrous ion to ferric ions process executed by the bacteria is fasters than chemical oxidation [36]. In case of uranium bioleaching the main drawback is to oxidize uranium (IV) since it insoluble but on this bioleaching process when ferrous sulfate produced in the process, then re-oxidized to ferric sulfate which enzymatically oxidize uranium (IV) to uranium (VI) by the energy produced by this reaction. A case study in India at Jaduguda mines proofs that use of biogenic ferric sulfate produced by the strain which was then used for efficient uranium extraction and cause no harm to the environment, while extracting uranium, use of reduced MnO_2 in Bacfox process to generate biogenic ferric sulfate, results passed air saturated ferrous sulfate solution over Acidithiobacillus Ferrooxidans which is absorbed on solid surface [36]. Since the permeability of the ore surface is a factor, the above study uses a process called "rubblizing" that increase fragmenting of ore in place which can be applied in the extraction of sulfide mineral, gold and uranium. While isolating the bacteria from mine water, the isolation media and H_2SO_4 consumption during isolation, pH variation and temperature were determinate factors, the microbial cell count and the growth of (A. Ferrooxidans) determines by rate of oxidation of iron from Fe^{2+} to Fe^{3+}, so while leaching if the amount of Fe^{2+} decrease means the bacteria is using it as energy source to convert it to Fe^{3+}, uranium bioleaching depends on the synergic effects Fe^{3+} and proton produced by the bacterial [37] that process uses either of the two energy sources to growth iron or sulfur. The reaction of making insoluble uranium to soluble form is as follows [38].

$$UO_2 + Fe_2(SO_4)_3 \rightarrow UO_2SO_4 + 2FeSO_4. \tag{19}$$

Studies indicate that microbial cell count and pulp density ranges 5–30% (w/v), particle size <75 μm has brought an optimum ore leaching but it should be clear that each ore has its own distinct behavior and no size fits all, meaning results indicated here might be different for another ore sample due to ore elemental composition, crystal structure, grade, topography and surface tension.

4.2 Bioleaching of copper

The ore is loaded on a water-resistant surface or ore is piled on an impermeable surface until a dump of suitable dimension forms. After leveling the top, then spraying a leach solution onto the dump is followed [4]. These dump is a habitat of heterogeneous microorganism. Dump can have variety particles sizes, where the bacterial annexation, which is anaerobic (microaerophilic), thermophilic begins from the top.

Dump leaching used to pretreat low-grade, refractory- sulfidic gold ores and to leach copper from chalcocite ores while ore grade is low with values ranges between 0.1–0.5%. Copper can be obtained from ore rocks from the mound then washed with dilute H_2SO_4 to facilitate the oxidation process of mineral by acidophiles, which is followed by cementation process where copper is precipitated from the drainage with scrap iron since it primary iron oxidizing process [39]. Check the leaching process of copper sulfide chalcocite (Cu_2S), which occurs with pyrite (FeS_2), leaching is due to ferric ion reacts with copper sulfide mineral processes ferrous and copper ions in solution.

$$Cu_2S + 4Fe^{3+} \rightarrow 4F^{2+} + 2Cu^{2+} + S. \quad (20)$$

$$FeS_2 + 8H_2O + 14Fe^{3+} \rightarrow 15Fe^{2+} + 2SO_4^{\ 2-} + 16H^+. \quad (21)$$

In these regions indirect leaching by ferric sulphate also prevails. The exterior of the dump is at ambient temperature and undergoes changes in temperature reflecting seasonal and diurnal fluctuations. Many different microorganisms have been isolated from copper dumps, some of which have been studied in the laboratory. These include a variety of mesophilic, aerobic iron and sulfur oxidizing microorganisms; thermophilic iron and sulfur oxidizing microorganisms; and anaerobic sulphate reducing bacteria. In copper leaching the concentration of target metal by itself is an important variable, copper concentration (100–300 mM range) is values cause difficulty for the microorganism to operate, selecting the microorganism is one of the mechanisms of copper resistant, Acidithiobacillus Ferrooxidans can resist copper concentration and strong acidic environment [40]. Thiobacillus Ferrooxidans was the main product observed after a culture study, from an ore or leach solution for the identification of composition of bacterial population and incase of low ferrous ions, it was Leptospirillum Ferrooxidan was observed, the study shows that utilization of ferrous iron as energy source is dominated by the previous bacteria as the culture shows. *Pseudomonas aeruginosa*, where heterotrophic bacteria produce various organic acids in an appropriate culture medium is used in copper leaching [41]. The addition of salt in bioleaching of copper resulted process enhancement, after designing the bioreactor the bioleaching of copper was enhanced in both stirred tank or shack flask by adding sodium chloride in leach solution, increasing the dissolution of Fe^{3+} that eventually reduces precipitation [29] addition of some elements might result inhibition of bioleaching process, fluorine in solution increase the viscosity of leach liquor that result inhibition of bioleaching [42]. It is important to understand the microbiology, which is responsible or identify a means to study bulk activity of microorganism, these features are oxygen uptake in solid and liquid samples, redox potential, pH, ferrous iron concentration and temperature. Microbial leaching has also direct relation with enrichment and culture from solution of ores. Acidithiobacillus Thiooxidans, Acidithiobacillus Ferrooxidans, and Leptospirillum Ferrooxidans have been cultured where the process run at an ambient temperature and the strain of bacterial related to the microorganism mentioned here [27, 43]. Leach solutions enriched with copper exit at the base of the dump and are conveyed to a central recovery facility. In most large-scale operations the leach solution, copper-bearing solution pumped into large cementation units containing iron scrapings for cementation and then electrolysis followed [4]. It was in Chile and Australia the commercial bio heap leaching of copper started mass production. And the first bioleach heap copper extraction plant is in China [44]. The copper extracted percentage can be calculated as,

E = Copper content in the solution/copper content in the sample X 100% [41].

4.3 Bioleaching of gold

Acidophilic bacteria are able to oxidize gold containing sulphidic ore, such a process can be ameliorated by conventional process of cyanidation, these basically reduces the complexation by increasing the capability of microorganisms to reach to the target metal. Certain sulphidic ores containing encapsulated particles of elemental gold, resulting in improved accessibility of gold to complexation by leaching agents such as cyanide. Relative to other conventional process and pretreatments like roasting, smelting and pressure oxidation, bio-oxidation demands less cost and no harm to nature [7]. Though it is under study a commercial bio-oxidation and bio heap leaching of gold prior cyanide extraction. It is the bacteria, Acidithiobacillus Ferrooxidans used to oxidize the sulphide matrix for gold recovery. Prior to extraction, gold ore must be bio-oxidize by the bacteria. In this process refractory sulphidic gold ores contain mainly two types of sulphides: pyrite and arsenopyrite where silver ion was used as a catalyst in acidic environment. Since gold is usually finely disseminated in the sulphide matrix, the objective of biooxidation of refractory gold ores is to break the sulphide matrix by dissolution of pyrite and arsenopyrite and extract 95% of iron and arsenic, the residue of both filtered through a vacuum pump. The consumption of cynide is much higher while biooxidation, the study suggested that using thiourea instead of cyanide is much less toxic but since the process require high consumption of thiourea cost increase steadily, consumption of thiourea reduced by using different agents like SO_2, bisulfite, cystine, cystine with oxygen during extraction process [45].

5. Industrial application

The mesophilic tank leaching is the most common bioleaching process in the world; thermophilc tank is favored while the temperature is high, among such tanks BioCop™ well known, In order to have effective thermophilc tank the following are basic requirements, microbial catalyzed reaction which is needed to facilitate metal dissolution by microbial oxidizing of ferrous iron to ferric iron, initial solublization of ferrous ion takes place using acid solution, oxidation of mineral sulfide takes place by the combination effects of ferric iron and acid solution followed by oxidization of reduced sulfur to sulfate by microorganisms. Reactor configuration is the other factor where the six equal size continuous reactor, three arranged in parallel considered as primary reactors, and the other three arranged in series considered as secondary reactors, in this case reactors are considers as a large continues stirred tank supplied with aeration and agitation. The other factors are oxygen, carbon dioxide, pulp density and finally even though the operational cost is much less plant location, construction material, blower or compressor to supply oxygen to the microbes, high power agitator in case of oxygen plant for oxygen dispersal in the reactor. Growth of industries results the demand of metals in very high quantity and likely go further in the years to come. This brings diminution of high grade ore with effluents and solid wastes that needs to be treated to recover the important elements and protect the environment.

Regarding to environment biohydrometallurgy is vital process, the fact that bio-process is conducted without the presence of toxic chemical and relatively required low cost makes it most needed. The direct implication of microorganisms in the reduction of uranium is of considerable interest because of its potential application in bio remediating of contaminated sites, in pretreating radioactive wastes, bioleaching is becoming a promising technology.

Author details

Mulugeta Sisay Cheru
Institution- Jimma Institute of Technology, Jimma, Ethiopia

*Address all correspondence to: mulugetajimma07@gmail.com

References

[1] B.P Gilbertson, (2013) Creating Value Through Innovation Biotechnology in Mining. Mineral Processing and Extractive Metallurgy, 109(2), pp 61-67.

[2] K. Bosecker, (2001) Microbial Leaching in Environmental Clean-Up Programmes. Hydrometalurgy, 59(2-3), pp 245-248.

[3] K. Bosecker, (1997) Bioleaching: Metal Solubilization by Microorganism. FEMS Microbiology Reviews, 20(3-4), pp 591-604.

[4] J.D.S.D Mishra, (2007) Bioleaching of Lateritic Nickel Ore Using Chemolithotrophic Micro Organisms (Acidithiobacillus Ferrooxidans). A Thesis Submitted in Partial Fulfillment of the Requirements for the Degree of Bachelor of Technology. National Institute of Technology, Rourkela, India.

[5] M.A Faramarzi, M.M Manzari, H Brandle, (2020) Bioleaching of Metals from Wastes and Low Grade Sources by HCN-forming Microorganism. Hydrometallurgy, 191, pp 105-228.

[6] Y. Songrong, X Jiyuan, Q. Guanzhou, H. Yeuhua, (2002) Research and Application of Bioleaching and Biooxidation Technology in China. Minerals Engineering, 15(5), pp 361-363.

[7] D.E Rawlings, (1997) Biomining: Theory, Microbes and Industrial Processes, Springer-Verlag Berlin Heidelberg, 1st Edition, New York.

[8] H.L Ehrich, (2002) Geomicrobiology, Revised and Expanded. Marcel Dekker Publisher, 4th Edition, New York.

[9] A.D Agate, (1996) Recent Advances in Microbial Mining. World journal of Microbiology and Biotechnology, 12, pp 487-495.

[10] A.E Yüce, H. M Tarkan, M Dogan, (2006) Effect of Bacterial Conditioning and The Flotation of Copper Ore and Concentrate. African Journal of Biotechnology, 5(5), pp 448-452.

[11] W Jonglertjunya, (2003) Bioleaching of chalcopyrite, A Thesis Submitted to the University of Birmingham for the Degree of Doctor of Philosophy, United Kingdom.

[12] C.S Demergasso, P.A.P Galleguillos, L.V.G Escudero, V.J.A Zepeda, D Castillo, E.O Casamayor, (2005) Molecular Characterization of Microbial Populations in a Low-Grade Copper Ore Bioleaching Test Heap. Hydrometallurgy, 80, pp 241-253.

[13] O Pizarro, E Jedlicki, O Orellana, J Romero, R.T. Espejo, (1996) Bacterial Populations in Samples of Bioleached Copper Ore as Revealed by Analysis of DNA Obtained Before and After Cultivation. Applied and Environmental Microbiology, 62(4), pp 1323-1328.

[14] K.J Edwards, B Hu, R.J Hamers, J.F Banfeld, (2001) A New Look at Microbial Leaching Patterns on Sulphide Minerals. FEMS Microbiology Ecology, 34, pp 197-206.

[15] A. Schippers, W.Sand, (1999) Bacterial Leaching of Metal Sulfides Proceeds by Two Indirect Mechanisms via Thiosulfate or via Polysulfides and Sulfur. Appl Environ Microbiol, 65(1), pp 319-321.

[16] R Gonzalez, J.C Gentina, F Acevedo, (2004) Biooxidation of a Gold Concentrate in a Continuous Stirred Tank Reactor: Mathematical Model and Optimal Configuration. Biochemical Engineering Journal, 19, pp 33-42

[17] T Das, S Ayyappan, G.R Chaudhury, (1999) Factors Affecting Bioleaching Kinetics of Sulfide Ores Using Acidophilic Micro-Organism. Biometals, 12, pp 1-10.

[18] D.E Rawlings, B.D Johnson, (2007) Biomining. Springer-Verlag Berlin Heidelberg, 1st Edition, New York.

[19] A.K Halinen, N Rahunen, A.H Kaksonen, J.A Puhakka, (2009) Heap Bioleaching of a Complex Sulfide Ore: Part II. Effect of Temperature on Base Metal Extraction and Bacterial Compositions, Hydrometallurgy 98, pp 101-107.

[20] M Jafari, H Abdollahi, S.Z Shafaei, M. Gharabaghi, H Jafari, A Akcil, S Panda, (2018) Acidophilic Bioleaching: A Review on the Process and Effect of Organic–Inorganic Reagents and Materials on its Efficiency. Mineral Processing and Extractive Metallurgy Review, 40(2), pp 87-107.

[21] J.A Diaz, J Serrano, E Leiva, (2018) Bioleaching of Arsenic-Bearing Copper Ores. Minerals, 8(5), pp 215.

[22] B.M Goebel, E Stackebrandt, (1994) Applied and Environmental Microbiology Cultural and Phylogenetic Analysis of Mixed Microbial. Populations Found in Natural and Commercial Bioleaching Environments, 60(5), pp 1614-1621.

[23] A. Giaveno, L. Lavalle, P. Chiacchiarini, E. Donati, (2007) Bioleaching of Zinc from Low-Grade Complex Sulfide Ores in an Airlift by Isolated Leptospirillum ferrooxidans. Hydrometallurgy, 89, pp 117-126.

[24] A.K Halinen, N Rahunen, A.H Kaksonen, J.A Puhakka, (2009) Heap Bioleaching of a Complex Sulfide Ore Part I: Effect of pH on Metal Extraction and Microbial Composition in pH Controlled Columns. Hydrometallurgy, 98, pp 92-100.

[25] H.N Bhatti, F Anjum, S Saeed, K.U Rehman, (2009) Bioleaching of Rock Phosphate by Aspergillus Niger. Asian Journal of Chemistry, 21(8), pp 6029-6040.

[26] A.J Francis, (2006) Microbial Transformations of Radionuclides and Environmental Restoration Through Bioremediation, for Presentation at the Symposium on "Emerging Trends in Separation Science and Technology" SESTEC 2006 Bhabha Atomic Research Center (BARC), Trombay, Mumbai, India.

[27] J.C Gentina, F Acevedo, (2016) Copper Bioleaching in Chile. Minerals, 6(23), pp 1-9.

[28] J.C Gentina, F Acevedo, (2013) Application of Bioleaching to Copper Mining in Chile. Electronic Journal of Biotechnology, 16(3), pp 16.

[29] D Bevilaqua, H Lahti, P.H Suegama, O.J Garcia, A.V Benedetti, J.A Puhakka, O.H Tuovinen, (2013) Effect of Na-chloride on the Bioleaching of a Chalcopyrite Concentrate in Shake Flasks and Stirred Tank Bioreactors. Hydrometallurgy, 138, pp 1-13.

[30] C. L Brierley, (2008) How will Biomining be Applied in Future?. Trans.Nonferrous Met Soc China, 18, pp 1303-1310.

[31] M.J Leahy, M.R Davidson, M.P Schwarz, (2005) A Two-Dimensional CFD Model for Heap Bioleaching of Chalcocite. Anziam.J 46(E), pp C439-C457.

[32] C Murray, W Platzer, J Petersen, (2017) Potential for Solar Thermal Energy in the Heap Bioleaching of Chalcopyrite in Chilean Copper Mining, Minerals Engineering, 100, pp 75-82.

[33] M Haschke, J Ahmadian, L Zeidler, T Hubrig, (2016) In-Situ Recovery of Critical Technology Elements. Procedia Engineering, 138, pp 248-257.

[34] A.H Kaksonen, N.J Boxall, Y Gumulya, H.N Khaleque, C Morris, T Bohu, K Y Cheng, K Usher, A.M Lakaniemi, (2018) Recent Progress in Biohydrometallurgy and Microbial Characterization. Hydrom, DOI: 10.1016/j.hydromet.2018.06.018.

[35] X Wang, Y Liu, Z Sun, J Li, L Chai, X Min, Y Guo, P Li, Z Zhou, (2017) Heap Bioleaching of Uranium from Low-Grade Granite-Type Ore by Mixed Acidophilic Microbes. J Radioanal Nucl Chem , DOI: 10.1007/s10967-017-5406-8.

[36] Abhilash, S Singh, K.D Mehta, V Kumar, B.D Pandey, V.M Pandey, (2009) Dissolution of Uranium from Silicate-Apatite Ore by Acidithiobacillus Ferrooxidans. Hydrometallurgy, 95, pp 70-75.

[37] Abhilash, B.D Pandey, (2013) Microbialy Assisted Leaching of Uranium – A Review. Mineral Processing and Extractive Metallurgy Review, 34(2), pp 81-113.

[38] Abhilash, K.D Mehta, V Kumar, B.D Pandey, P.K Tamrakar, (2011) Bioleaching -An Alternate Uranium Ore Processing Technology for India. Energy Procedia, 7, pp 158-162.

[39] N Pradhan, K.C Nathsarma, K.S Rao, L.B Sukla, B.K Mishra, (2008) Heap Bioleaching of Chalcopyrite: A Review. Minerals Engineering, 21, pp 355-365.

[40] C.M Bussenius, C.A Navarro, C.A Jerez, (2017) Microbial Copper Resistance: Importance in Biohydrometallurgy. Microb Biotechnol, 10(2), pp 279-295.

[41] M.A. Shabani, M. Irannajad, A.R. Azadmehr, and M. Meshkini, (2013) Bioleaching of Copper Oxide Ore by *Pseudomonas aeruginosa*. International Journal of Minerals, Metallurgy and Materials, 20(12), pp 1130.

[42] M Dopson, L Lövgren, D Boström, (2009) Silicate Mineral Dissolution in the Presence of Acidophilic Microorganisms: Implications for Heap Bioleaching. Hydrometallurgy, 96, pp 288-293.

[43] D.E Rawlings, (2005) Characteristics and Adaptability of Iron- and Sulfur-Oxidizing Microorganisms Used for the Recovery of Metals from Minerals and Their Concentrates. Microbial Cell Factories, 4(13), pp 1-15.

[44] L Xingyu, C Bowei, W Jiankang, R Renman, (2010) Leptospirillum Forms a Minor Portion of the Population in Zijinshan Commercial Non-Aeration Copper Bioleaching Heap Identified by 16S rRNA Clone Libraries and Real-Time PCR. Hydrometallurgy, 104, pp 399-403.

[45] T Deng, M Liao, (2002) Gold Recovery Enhancement from a Refractory Flotation Concentrate by Sequential Bioleaching and Thiourea Leach. Hydrometallurgy, 63, pp 249-255.

Chapter 15

Modern Technologies for Pest Control: A Review

Meenu Agarwal and Ayushi Verma

Abstract

The major concern for farmers is important loss due to pests and diseases, which is regardless of any production system adopted. Plant pathogens, insects, and weed pests devastate over 40% of all possible sustenance creation every year. This loss happens despite utilizing approximately 3 million tons of pesticide per year in addition to the use of a variety of nonchemical controls such as biological controls and crop rotations. If some of this food could be saved from pest attack, it could be utilized to bolster an excess of 3 billion people who are malnourished in the world today. Expansive range of conventional insecticides such as carbamates, organophosphates, pyrethroids, and organochlorines were developed. They have been used to control insect pests in the course of recent decades, resulting in the reduction of the loss of agricultural yield. However, problems of resistance reaching crisis proportions, the extreme unfavorable impacts of pesticides on the environment, and public complaints led to stricter protocols and regulations directed to reduce their utilization. The pest control industry is continuously examining novel technologies and products that will improve the way to manage and prevent pests. The general objective is to likewise diminish the effects of various available pesticides on the environment and on nontarget creatures, besides the economic influence on bottom lines.

Keywords: pesticides, insects, agriculture, biological control

1. Introduction

Pests are organisms which can damage the crops and compete with them. They cause decrease in the plant density, stunted growth of plant, a lower production capacity, and lessen the yield or nature of horticultural products. A standout among the most well-known approaches to regulate or dispense pests is to kill pests using pesticides. The pesticides may cause biological or physical damage to pest organisms. Some pesticides are applied indirectly or by spraying on a plant which can be consumed by an insect.

Several techniques for crop protection were developed to prevent and minimize the loss of crops due to pests in the field (preharvest losses) and during storage (postharvest losses). Crop protection involves products, tools, and practices which can be used by farmers to protect their harvest against insects, disease, and weeds. Weeds steal sunlight, nutrients, and water from crops. The food production can be influenced by insects and disease. Farmers around the globe settle on various choices every day with respect to how best to secure their crops by using different practices like biological control, microbial pesticides, pest behavior, genetic manipulation,

IntechOpen

Methods	Actions
Biological control	Suppression of pest populations by natural enemies (predators, parasites, competitors, diseases)
Microbial pesticides	Relatively stable formulations of microorganisms that suppress pests by producing poisons, causing diseases, preventing establishment of other microorganisms, or other mechanisms
Pest behaviour-modifying chemicals	Exploitation of the chemical signals used by living organisms to evoke specific behaviours from other organisms
Genetic manipulation of pest populations	Release into the pest population of individuals genetically altered to carry genes that interfere with the pest's reproduction or impact
Plant immunization	Enhancement of plant resistance to pests by means other than breeding or genetic engineering

Table 1.
Different methods to control pests and diseases.

and plant immunization of pest population. Fortunately, a variety of solutions are available with advances in modern agriculture. **Table 1** shows different methods to control pests and diseases.

2. Literature review

It was studied that the farmer takes into account to control the steady-state mice populations in order to use no or a moderate amount of pesticide. It depends on the prices as well as the shape of the grain production function which is determined by the microparameters of grain reproduction [1].

It was found that the barn owls (*Tyto alba*) can be used for the purpose of vertebrate pest control, and farmers in some agricultural regions breed barn owls and hunt on their farms by installing artificial nest boxes. The barn owl diet and nest box occupancy in an intensive agricultural landscape in the Central Valley of California were used to measure whether agricultural land use affected the barn owl diet [2].

It has been studied that heat treatment (HT) methods can be used to control insect pests in flour mills by thermal analyses and temperature trend models. The results show that to achieve an air temperature above 45°C, the time period of HT should be increased by 9 h. In thermal bridges, the surface temperatures should be lethal to insects and be capable of dropping sensible heat flux loss with the help of insulating materials [3].

It was observed that lablab bean (*Lablab purpureus*) faces major losses due to insect pests in spite of heavy dependence on conventional insecticides. It was managed by testing biorational insecticides as substitutes for conventional insecticides to control pod borers (*Maruca vitrata*) and aphids (*Aphis craccivora*) [4].

It has been stated that the important life-sustaining actions of phytopathogenic fungi, parasitic nematode, and phytophagous insects are the preeminent constraining variables of agricultural production organizations [5].

It has been stated that integrated pest management (IPM) is broadly stimulated among the European Union (EU) member states. The effectiveness of natural enemies for pest control is low as compared with insecticides, especially under high pest damage levels. Farmers with greater income have more financial flexibility to adopt either pest control method. The environment surrounding a farm may also influence its owner's willingness to promote natural pest control [6].

It has been investigated that crop damage can be minimized by using vertebrates which could consume numerous crop pests [7].

It has been studied that to obtain healthy crop as well as high yield, the control of pest attack is a significant feature in agriculture. The growth of awareness level is supposed to be relational to density of the healthy pests in the crop field. Global sources like televisions, radio, etc. can enhance the awareness level. The basic reproduction number represents the existence and the stability criteria of the equilibria. The Hopf-bifurcation analysis was done at the endemic equilibrium by considering time delay as the bifurcation parameter. The analytical results were justified by numerical simulations [8].

This overview provides a valuable insight to various stakeholders in the food chain on how food handlers and companies perceive pest control [9].

It was found that there are several methods to control pests, which involve the use of molting hormone (20-hydroxyecdysone) analogs (e.g., tebufenozide), chitin synthesis inhibitors (e.g., diflubenzuron), and juvenile hormone analogs (e.g., methoprene). They have been collectively called as "insect growth regulators" or IGRs, as they harmfully affect the normal growth and development in one way or another like they affect insect reproduction [10].

3. New methods of pest control

Following are the new advanced methods for pest control.

3.1 Biorational products

Biologically based pesticide products or biorational products contain pheromones (a chemical substance released by an animal into the environment, particularly an insect or a mammal, affecting the physiology or behavior of its other species) or living microorganisms, which have little negative impact on the environment and are nontoxic. *Bacillus thuringiensis*-based (*Bt*-based) microbial pesticides are most effective biorational pesticides. These types of biorational pesticides can be used in the place of competitive chemical products, which are banned in export markets where the concerns about food residue are high, and in environmentally sensitive areas.

"Pheromone-assisted technique" increases the efficiency of insecticides against ants. When a combination of insecticide and pheromones is used, it can actually trap away ants from their trails and nests to the insecticide. Thus, it reduces the contact of insecticide with the environment, while increasing the exposure of ants to the insecticide for eradication.

3.2 Insect growth regulators (IGRs)

Insect growth regulators (IGRs) are synthetic insect hormones, which can be used as insecticides to regulate the populations of harmful insect pests. IGRs prevent an insect from reaching maturity by interfering with the molting process. Many IGR products are mixed with other insecticides that kill adult insects. IGRs are generally less toxic to humans.

3.3 Nontoxic heat treatments

New methods of pest control are based on low-toxic solutions that can be more sustainable and effective than harsh chemicals; one such method is to use heat to eliminate insects. Rentokil's Entotherm is a new solution that eradicates bedbugs and cockroaches. Rather than using sprays, which often do not penetrate the hard outer shell, the heat delivered by the Entotherm system kills insects from the inside through dehydration and damage to essential physiological processes. Another benefit is that the heat effectively kills all the life stages of insects—egg, larva, pupa, and adult, without needing to go over 56–60°C, which is high enough to kill the pests rapidly.

3.4 CRISPR technology to control pests

CRISPR gene editing tool is a method to alter key genes that regulate the fertility and sex determination of insects. CRISPR technology has allowed researchers to invent a new effective control technology which can be safe, self-limiting, and scalable genetic population for a specific species. It has the potential to be developed and utilized for a plethora of insect pests and disease vectors. This technology can be safely used in the field to suppress and even destroy target species.

3.5 Second-generation green products

Since last few years, the major focus surrounding green products has been from a public health perspective. The demand of green products is increasing and shifting to eco-protection. Nowadays, second-generation green products are emerging. These second-generation green products will have better ingredients, greater efficacy, and less disadvantages. For example, a first-generation insecticide containing plant essential oils may have an unpleasant smell. In contrast to this, the second-generation insecticide contains different amounts of the original ingredients or other materials, which reduces the smell, while it is still being effective.

3.6 Communication tools

The mobile data-capture devices can be introduced for pest control, which are able to capture real-time data about the location and nature of the pest problems and then store it in a central database online, which both facility managers and pest control professionals could access. This also made it simple to track multiple facilities and allowed for more timely corrective actions and more targeted pest-control treatments.

3.7 Rodent birth control

For the past few years, pest management used birth control for pest birds like pigeons as a means of managing populations. The rodent birth control may soon provide an effective way to control prolific breeders like mice and rats without negatively impacting many nontarget creatures. Be on the lookout for rodent birth control in the near future.

3.8 Fly baits

The fly bait stickers can be used near the food storage area or where the flies congregate. The small sticker is coated with insect food and an insecticide that has the ability to knockdown flies in just 1 min upon contact.

4. Conclusions

Conventional pesticides are toxic residues, pest resistance, secondary pests, and pest resurgence. Thus, they are not only dangerous to the environment but also hazardous to human health. In order to encourage organic farming, plant protection, and reduce pesticides' usage in food products, pheromone traps and other modern technologies can play a crucial role. So, it necessary to apply novel technologies that result in producing high quality of food and better income generation.

Author details

Meenu Agarwal and Ayushi Verma*
Department of Biotechnology, Noida Institute of Engineering and Technology, Greater Noida, U.P., India

*Address all correspondence to: drayu18iitr@gmail.com

References

[1] Christiaans T, Eichner T, Pethig R. Optimal pest control in agriculture. Journal of Economic Dynamics and Control. 2007;**31**:3965-3985. DOI: 10.1016/j.jedc.2007.01

[2] Kross SM, Bourbour RP, Martinico BL. Agricultural land use, barn owl diet, and vertebrate pest control implications. Agriculture, Ecosystems & Environment. 2016;**223**:167-174. DOI: 10.1016/j.agee.2016.03.002

[3] Porto SMC, Valenti F, Arcidiacono C. Improving the effectiveness of heat treatment for insect pest control in flour mills by thermal simulations. Biosystems Engineering. 2017;**164**:189-199. DOI: 10.1016/J.BIOSYSTEMSENG.2017.10.015

[4] Douglas MR, Chang J, Ramasamy S. Evaluation of biorational insecticides and DNA barcoding as tools to improve insect pest management in lablab bean (*Lablab purpureus*) in Bangladesh. Journal of Asia-Pacific Entomology. 2018;**21**:1326-1336. DOI: 10.1016/j.aspen.2018.10.007

[5] Subbanna ARNS, Pattanayak A. Pesticidal prospectives of chitinolytic bacteria in agricultural pest management. Soil Biology and Biochemistry. 2018;**116**:52-66. DOI: 10.1016/j.soilbio.2017.09.019

[6] Zhang H, Potts SG, Breeze T, Bailey A. European farmers' incentives to promote natural pest control service in arable fields. Land Use Policy. 2018;**78**:682-690. DOI: 10.1016/j.landusepol.2018.07.017

[7] Lindell C, Eaton RA, Howard PH, Roels SM, Shave ME. Enhancing agricultural landscapes to increase crop pest reduction by vertebrates. Agriculture, Ecosystems & Environment. 2018;**257**:1-11. DOI: 10.1016/j.agee.2018.01.028

[8] Basir FA, Banerjee A, Ray S. Role of farming awareness in crop pest management—A mathematical model. Journal of Theoretical Biology. 2019;**461**:59-67. DOI: 10.1016/j.jtbi.2018.10.043

[9] Djekic L, Kavallieratos NG, Rajkovic A. Pest control in Serbian and Greek food establishments—Opinions and knowledge. Food Control. 2019;**98**:281-289. DOI: 10.1016/j.foodcont.2018.11.045

[10] Smagghe G, Zotti M, Retnakaran A. Targeting female reproduction in insects with biorational insecticides for pest management: A critical review with suggestions for future research. Current Opinion in Insect Science. 2019;**31**: 65-69. DOI: 10.1016/j.cois.2018.10.009